Selected Titles in This Series

58 **Samuel J. Lomonaco, Jr., Editor,** Quantum computation: A grand mathematical challenge for the twenty-first century and the millennium (Washington, DC, January 2000)

57 **David C. Heath and Glen Swindle, Editors,** Introduction to mathematical finance (San Diego, California, January 1997)

56 **Jane Cronin and Robert E. O'Malley, Jr., Editors,** Analyzing multiscale phenomena using singular perturbation methods (Baltimore, Maryland, January 1998)

55 **Frederick Hoffman, Editor,** Mathematical aspects of artificial intelligence (Orlando, Florida, January 1996)

54 **Renato Spigler and Stephanos Venakides, Editors,** Recent advances in partial differential equations (Venice, Italy, June 1996)

53 **David A. Cox and Bernd Sturmfels, Editors,** Applications of computational algebraic geometry (San Diego, California, January 1997)

52 **V. Mandrekar and P. R. Masani, Editors,** Proceedings of the Norbert Wiener Centenary Congress, 1994 (East Lansing, Michigan, 1994)

51 **Louis H. Kauffman, Editor,** The interface of knots and physics (San Francisco, California, January 1995)

50 **Robert Calderbank, Editor,** Different aspects of coding theory (San Francisco, California, January 1995)

49 **Robert L. Devaney, Editor,** Complex dynamical systems: The mathematics behind the Mandlebrot and Julia sets (Cincinnati, Ohio, January 1994)

48 **Walter Gautschi, Editor,** Mathematics of Computation 1943–1993: A half century of computational mathematics (Vancouver, British Columbia, August 1993)

47 **Ingrid Daubechies, Editor,** Different perspectives on wavelets (San Antonio, Texas, January 1993)

46 **Stefan A. Burr, Editor,** The unreasonable effectiveness of number theory (Orono, Maine, August 1991)

45 **De Witt L. Sumners, Editor,** New scientific applications of geometry and topology (Baltimore, Maryland, January 1992)

44 **Béla Bollobás, Editor,** Probabilistic combinatorics and its applications (San Francisco, California, January 1991)

43 **Richard K. Guy, Editor,** Combinatorial games (Columbus, Ohio, August 1990)

42 **C. Pomerance, Editor,** Cryptology and computational number theory (Boulder, Colorado, August 1989)

41 **R. W. Brockett, Editor,** Robotics (Louisville, Kentucky, January 1990)

40 **Charles R. Johnson, Editor,** Matrix theory and applications (Phoenix, Arizona, January 1989)

39 **Robert L. Devaney and Linda Keen, Editors,** Chaos and fractals: The mathematics behind the computer graphics (Providence, Rhode Island, August 1988)

38 **Juris Hartmanis, Editor,** Computational complexity theory (Atlanta, Georgia, January 1988)

37 **Henry J. Landau, Editor,** Moments in mathematics (San Antonio, Texas, January 1987)

36 **Carl de Boor, Editor,** Approximation theory (New Orleans, Louisiana, January 1986)

35 **Harry H. Panjer, Editor,** Actuarial mathematics (Laramie, Wyoming, August 1985)

34 **Michael Anshel and William Gewirtz, Editors,** Mathematics of information processing (Louisville, Kentucky, January 1984)

33 **H. Peyton Young, Editor,** Fair allocation (Anaheim, California, January 1985)

32 **R. W. McKelvey, Editor,** Environmental and natural resource mathematics (Eugene, Oregon, August 1984)

31 **B. Gopinath, Editor,** Computer communications (Denver, Colorado, January 1983)

30 **Simon A. Levin, Editor,** Population biology (Albany, New York, August 1983)

(Continued in the back of this publication)

AMS SHORT COURSE LECTURE NOTES
Introductory Survey Lectures
published as a subseries of
Proceedings of Symposia in Applied Mathematics

Proceedings of Symposia in APPLIED MATHEMATICS

Volume 58

Quantum Computation:
A Grand Mathematical Challenge for the Twenty-First Century and the Millennium

American Mathematical Society
Short Course
January 17–18, 2000
Washington, DC

Samuel J. Lomonaco, Jr.
Editor

American Mathematical Society
Providence, Rhode Island

Editorial Board

Peter S. Constantin (Chair)   Eitan Tadmor

LECTURE NOTES PREPARED FOR THE
AMERICAN MATHEMATICAL SOCIETY SHORT COURSE
QUANTUM COMPUTATION
HELD IN WASHINGTON, DC
JANUARY 17–18, 2000

The AMS Short Course Series is sponsored by the Society's Program Committee for National Meetings. The series is under the direction of the Short Course Subcommittee of the Program Committee for National Meetings.

2000 *Mathematics Subject Classification.* Primary 81–01, 81–02, 81P68, 68Q05, 94A60; Secondary 22E70, 57M99, 81V80, 94A15.

Library of Congress Cataloging-in-Publication Data
American Mathematical Society. Short course (2000 : Washington, DC)
   Quantum computation : a grand mathematical challenge for the twenty-first century and the millennium : American Mathematical Society, Short Course, January 17–18, 2000, Washington, DC / Samuel J. Lomonaco, editor.
   p. cm. – (Proceedings of symposia in applied mathematics, ISSN 0160-7634; v. 58 AMS short course lecture notes)
   Includes bibliographical references and index.
   ISBN 0-8218-2084-2 (alk. paper)
   1. Quantum computers–Congresses. I. Lomonaco, Samuel J. II. American Mathematical Society. III. Title. IV. Proceedings of symposia in applied mathematics; v. 58. V. Proceedings of symposia in applied mathematics. AMS short course lecture notes.
   QA76.889.A44 2000
   004.1–dc21   2001055366

   Copying and reprinting. Material in this book may be reproduced by any means for educational and scientific purposes without fee or permission with the exception of reproduction by services that collect fees for delivery of documents and provided that the customary acknowledgment of the source is given. This consent does not extend to other kinds of copying for general distribution, for advertising or promotional purposes, or for resale. Requests for permission for commercial use of material should be addressed to the Assistant to the Publisher, American Mathematical Society, P. O. Box 6248, Providence, Rhode Island 02940-6248. Requests can also be made by e-mail to reprint-permission@ams.org.
   Excluded from these provisions is material in articles for which the author holds copyright. In such cases, requests for permission to use or reprint should be addressed directly to the author(s). (Copyright ownership is indicated in the notice in the lower right-hand corner of the first page of each article.)

   © 2002 by the American Mathematical Society. All rights reserved.
   The American Mathematical Society retains all rights
   except those granted to the United States Government.
   Printed in the United States of America.

   ∞ The paper used in this book is acid-free and falls within the guidelines
   established to ensure permanence and durability.
   Visit the AMS home page at URL: http://www.ams.org/

   10 9 8 7 6 5 4 3 2 1   07 06 05 04 03 02

Contents

Preface	ix
Acknowledgements	xiii
Original AMS Short Course Announcement	xv

Chapter I. An Invitation to Quantum Computation

A Rosetta stone for quantum mechanics with an introduction to quantum computation SAMUEL J. LOMONACO, JR.	3
Qubit devices HOWARD E. BRANDT	67

Chapter II. Quantum Algorithms and Quantum Complexity Theory

Introduction to quantum algorithms PETER W. SHOR	143
Shor's quantum factoring algorithm SAMUEL J. LOMONACO, JR.	161
Grover's quantum search algorithm SAMUEL J. LOMONACO, JR.	181
A survey of quantum complexity theory UMESH VAZIRANI	193

Chapter III. Quantum Error Correcting Codes and Quantum Cryptography

An introduction to quantum error correction DANIEL GOTTESMAN	221
A talk on quantum cryptography: Or how Alice outwits Eve SAMUEL J. LOMONACO, JR.	237

Chapter IV. More Mathematical Connections

Topological quantum codes and anyons ALEXEI KITAEV	267

Quantum topology and quantum computing
LOUIS H. KAUFFMAN 273

An entangled tale of quantum entanglement
SAMUEL J. LOMONACO, JR. 305

Index 351

Preface

This book arose from an American Mathematical Society Short Course entitled:

Quantum Computation
*A Grand Mathematical Challenge
for the
Twenty-First Century and the Millennium*

held in Washington, DC, January 17–18, 2000 in conjunction with the Annual Meeting of the American Mathematical Society in Washington, DC, January 19–22, 2000.

This AMS Short Course was created with the objective of sharing with the scientific community the many exciting mathematical challenges arising from the new and emerging field of quantum computation and quantum information science. In so doing, it was hoped that this AMS Short Course would act as a catalyst to encourage, to entice, and, yes, ... to challenge and ultimately to dare mathematicians into working on the many diverse grand mathematical challenges arising in this field.

To meet this objective, the AMS Short Course was designed to demonstrate the great breadth and depth of this mathematically rich research field. Interrelationships with existing mathematical research areas were emphasized as much as possible. Moreover, the entire AMS Short Course was designed in such a way that course participants with little, if any, background in quantum mechanics would, on completion of the course, be prepared to begin reading the research literature on quantum computation and quantum information science.

Much to my surprise, the response to this AMS Short Course exceeded my most ambitious expectations. I thank the audience for their encouragement and enthusiastic response, and for their many helpful questions and suggestions in regard to the material presented.

This book is a written version of the eight lectures given in this AMS Short Course. Based on audience feedback and questions, the written versions of these lectures have been greatly expanded, and supplementary material has been added.

Chapter I of this book begins with two papers extending an invitation to this research field. The first by Samuel J. Lomonaco, Jr., entitled "A Rosetta Stone to Quantum Mechanics with an Introduction to Quantum Computation," provides the reader with an overview of the relevant parts of quantum mechanics, and with an entrée into the world of quantum computation. The second by Howard E. Brandt, entitled, "Qubit Devices," provides an overview of many potential quantum mechanical computing devices.

Chapter II consists of four papers devoted to quantum algorithms and quantum complexity theory. The first paper by Peter W. Shor, entitled "Introduction to Quantum Algorithms," gives an in-depth overview of quantum algorithms, explaining why recent results in this area are so surprising. It also illustrates the general technique of using the Fourier transform to find periodicity and discusses the quantum algorithms of Simon, Shor, and Grover. The second and third papers by Samuel J. Lomonaco, Jr., entitled respectively "Shor's Quantum Factoring Algorithm" and "Grover's Quantum Search Algorithm," were created in response to audience feedback, and focus more closely on the mathematical inner workings and underpinnings of Shor's and Grover's algorithms. The chapter closes with a fourth paper by Umesh Vazirani, entitled "Quantum Complexity Theory," devoted to the many intriguing and challenging issues of quantum complexity theory. The paper begins with a discussion of the current-day challenge to the Church-Turing thesis and then proceeds into a discussion of quantum complexity classes.

Chapter III is comprised of two papers focusing on quantum error correcting codes and quantum cryptography. The first paper by Daniel Gottesman, entitled "An Introduction to Quantum Error Correction," gives an introduction to quantum information science's first line of defense against the ravages of quantum decoherence, i.e., quantum error correcting codes. The paper begins with a discussion of error models, moves on to quantum error correction and the stabilizer formalism, ending with a quantum error correction sonnet. The second paper by Samuel J. Lomonaco, Jr., entitled "A Talk on Quantum Cryptography, or How Alice Outwits Eve," gives an in-depth analysis of quantum cryptographic protocols, including the BB84, B92, and possible eavesdropping countermeasures. All of this is interwoven into the context of a fictional story about how Alice invents quantum cryptographic protocols to outwit the archvillainess Eve, ... or does she?

The final Chapter IV consists of three papers discussing many more diverse connections between quantum computation and mathematics and physics. The first paper by Alexei Kitaev, entitled "Topological Quantum Codes and Anyons," discusses how anyons can be applied to quantum computation. Anyons are two-dimensional particles that have been found in two-dimensional electronic liquids exhibiting the fractional quantum Hall effect. This paper discusses how Aharonov-Bohm-like interactions of anyons can be used to create topological obstructions to quantum decoherence. Such topological obstructions can be used to construct quantum error correcting codes. Both abelian and nonabelian anyons are discussed. The second paper by Louis H. Kauffman, entitled "Quantum Topology and Quantum Computing," explores some of the tantalizing relationships among knots, links, three manifold invariants, and quantum information science. Many possible applications of quantum topology to quantum computing and vice versa are discussed. The third and final paper by Samuel J. Lomonaco, Jr., entitled "An Entangled Tale of Quantum Entanglement," discusses how Lie group invariant theory can be used to quantify a physical phenomenon that many believe to be at the very core of quantum computation, namely, quantum entanglement. The paper shows how to lift the big adjoint action of the group of local unitary transformations to the corresponding infinitesimal action to produce a system of partial differential equations whose solution is a complete set of entanglement invariants. Examples are given.

It is hoped that this book will encourage its readers to embrace and pursue the grand challenge of quantum computation.

> Samuel J. Lomonaco, Jr.
> Lomonaco@umbc.edu
> http://www.csee.umbc.edu/~lomonaco
>
> August 15, 2001

Acknowledgements

This work was partially supported by Army Research Office (ARO) Grant #P-38804-PH-QC, by the National Institute of Standards and Technology (NIST), by the Defense Advanced Research Projects Agency (DARPA) and Air Force Materiel Command USAF under agreement number F30602-01-0522, and by L-O-O-P Fund No. 2000WADC. The author gratefully acknowledges the hospitality of the University of Cambridge Isaac Newton Institute for Mathematical Sciences, Cambridge, England, where some of this work was completed.

Thanks are due to the other AMS Short Course lecturers, Howard Brandt, Dan Gottesman, Lou Kauffman, Alexei Kitaev, Peter Shor, Umesh Vazirani, and the many Short Course participants for their support and helpful discussions and insights. I would also like to thank Jeffrey Bub, Gilles Brassard, Lov Grover, Lucien Hardy, Tim Havel, Richard Jozsa, David Meyer, John Myers, and Nolan Wallach for their helpful suggestions and discussions. Thanks should also be given to Paul Black, Ron Boisvert, and Carl Williams of NIST for their encouragement and support.

At UMBC, I would like to thank the UMBC CSEE Quantum Computation Faculty Seminar participants Richard Chang, Dhananjay Phatak, John Pinkston, Alan Sherman, Jon Squire, Brooke Stephens, Yaacov Yesha, and my UMBC graduate course students Bianca Benincasa, Ali Bicak, Justin Brody, Mark Colangelo, Koustuv Dasgupta, Stan Finkler, Joan Grindell, Bryan Jacobs, Lalana Kagal, Yoon-Ho Kim, Brett Kurtin, Kimball Martin, Chris McCubbin, Jim Parker, Ravindra Peravali, Scott Rose, Andrew Skinner, and Anocha Yimsiriwattana for their many helpful questions, suggestions and insights.

Many thanks to Sergei Gelfand, Wayne Drady, Gil Poulin, Shirley Hill, and Christine Thivierge of the American Mathematical Society for their editorial support in this endeavor. I am also indebted to Christopher Martin whose computer support made this paper possible.

Finally, I would like to thank my wife Bonnie for her support during this endeavor.

Original AMS Short Course Announcement

Quantum Computation
*A Grand Mathematical Challenge
for the
Twenty-First Century and the Millennium*

AMS Short Course Overview

The Nobel Laureate Richard Feynman was one of the first individuals to observe that there is an exponential slowdown when computers based on classical physics, i.e., classical computers, are used to simulate quantum systems. Richard Feynman then went on to suggest that it would be far better to use computers based on quantum mechanical principles, i.e., quantum computers, to simulate quantum systems. Such quantum computers should be exponentially faster than their classical counterparts.

Interest in quantum computation suddenly exploded when Peter Shor devised an algorithm for quantum computers that could factor integers in polynomial time. The fastest known algorithm for classical computers factors much more slowly, i.e., in superpolynomial time. Shor's algorithm meant that, if quantum computers could be built, then cryptographic systems based on integer factorization, e.g., RSA, could easily be broken. These cryptographic systems are currently extensively used in banking and in many other areas. Lov Grover then went on to create a quantum algorithm that could search databases faster than anything possible on a classical computer. These algorithms are based on physical principles not implementable on classical computers, quantum superposition and quantum entanglement.

As a result, the race to build a quantum computer is on. But the mathematical, physical, and engineering challenges to do so are formidable, and are a worthy challenge for the best scientific minds. One of the chief obstacles to creating a quantum computer is quantum decoherence. By this we mean that quantum systems want to wander from their computational paths and quantum entangle with the rest of the environment.

This short course focuses on the mathematical challenges involved in the development of quantum computers and quantum algorithms, challenges worthy of the best mathematical minds. It is hoped that, as a result of this course, many mathematicians will be enticed into working on the grand challenge of quantum computation.

The Short Course will begin with an overview of quantum computation, given in an intuitive and conceptual style. No prior knowledge of quantum mechanics will be assumed.

In particular, the Short Course will begin with an introduction to the strange world of the quantum. Such concepts as quantum superposition, Heisenberg's uncertainty principle, the "collapse" of the wave function, and quantum entanglement (i.e., EPR pairs) will be introduced. This will also be interlaced with an introduction to Dirac notation, Hilbert spaces, unitary transformations, and quantum measurement.

Some of the topics covered in the course will be:

- Quantum teleportation
- Shor's quantum factoring algorithm
- Grover's algorithm for searching a database
- Quantum error-correcting codes
- Quantum cryptography
- Quantum information theory
- Quantum complexity theory, including the quantum Turing machine
- The problems of quantum entanglement and locality
- Implementation issues from a mathematical perspective

Each topic will be explained and illustrated with simple examples.

Chapter I

An Invitation to Quantum Computation

A Rosetta Stone for Quantum Mechanics with an Introduction to Quantum Computation

Samuel J. Lomonaco, Jr.

ABSTRACT. The purpose of these lecture notes is to provide readers, who have some mathematical background but little or no exposure to quantum mechanics and quantum computation, with enough material to begin reading the research literature in quantum computation, quantum cryptography, and quantum information theory. This paper is a written version of the first of eight one hour lectures given in the American Mathematical Society (AMS) Short Course on Quantum Computation held in conjunction with the Annual Meeting of the AMS in Washington, DC, USA in January 2000.

Part 1 of the paper is a preamble introducing the reader to the concept of the qubit

Part 2 gives an introduction to quantum mechanics covering such topics as Dirac notation, quantum measurement, Heisenberg uncertainty, Schrödinger's equation, density operators, partial trace, multipartite quantum systems, the Heisenberg versus the Schrödinger picture, quantum entanglement, EPR paradox, quantum entropy.

Part 3 gives a brief introduction to quantum computation, covering such topics as elementary quantum computing devices, wiring diagrams, the no-cloning theorem, quantum teleportation.

Many examples are given to illustrate underlying principles. A table of contents as well as an index are provided for readers who wish to "pick and choose." Since this paper is intended for a diverse audience, it is written in an informal style at varying levels of difficulty and sophistication, from the very elementary to the more advanced.

2000 *Mathematics Subject Classification.* Primary 81P68, 81-02, 81-01, 81P15, 22E70; Secondary 15A90, 94A15.

Key words and phrases. Quantum mechanics, quantum computation, quantum algorithms, entanglement, quantum information.

This work was partially supported by Army Research Office (ARO) Grant #P-38804-PH-QC, by the National Institute of Standards and Technology (NIST), by the Defense Advanced Research Projects Agency (DARPA) and Air Force Materiel Command USAF under agreement number F30602-01-0522, and by the L-O-O-P Fund. The author gratefully acknowledges the hospitality of the University of Cambridge Isaac Newton Institute for Mathematical Sciences, Cambridge, England, where some of this work was completed. I would also like to thank the other AMS Short Course lecturers, Howard Brandt, Dan Gottesman, Lou Kauffman, Alexei Kitaev, Peter Shor, Umesh Vazirani and the many Short Course participants for their support.

© 2002 American Mathematical Society

Contents

Part 1. Preamble

1. Introduction

2. **The classical world**
   2.1. Introducing the Shannon bit.
   2.2. Polarized light: Part I. The classical perspective

3. **The quantum world**
   3.1. Introducing the qubit – But what is a qubit?
   3.2. Where do qubits live? – But what is a qubit?
   3.3. A qubit is ...

Part 2. An Introduction to Quantum Mechanics

4. **The beginnings of quantum mechanics**
   4.1. A Rosetta stone for Dirac notation: Part I. Bras, kets, and bra-(c)-kets
   4.2. Quantum mechanics: Part I. The state of a quantum system
   4.3. Polarized light: Part II. The quantum mechanical perspective
   4.4. A Rosetta stone for Dirac notation: Part II. Operators
   4.5. Quantum mechanics: Part II. Observables
   4.6. Quantum mechanics: Part III. Quantum measurement — General principles
   4.7. Polarized light: Part III. Three examples of quantum measurement
   4.8. A Rosetta stone for Dirac notation: Part III. Expected values
   4.9. Quantum Mechanics: Part IV. The Heisenberg uncertainty principle
   4.10. Quantum mechanics: Part V. Dynamics of closed quantum systems: Unitary transformations, the Hamiltonian, and Schrödinger's equation
   4.11. The mathematical perspective

5. **The Density Operator**
   5.1. Introducing the density operator
   5.2. Properties of density operators
   5.3. Quantum measurement in terms of density operators
   5.4. Some examples of density operators
   5.5. The partial trace of a linear operator
   5.6. Multipartite quantum systems
   5.7. Quantum dynamics in density operator formalism
   5.8. The mathematical perspective

6. **The Heisenberg model of quantum mechanics**

7. **Quantum entanglement**
   7.1. The juxtaposition of two quantum systems
   7.2. An example: An n-qubit register $\mathcal{Q}$ consisting of the juxtaposition of n qubits.
   7.3. An example of the dynamic behavior of a 2-qubit register

7.4. Definition of quantum entanglement
7.5. Einstein, Podolsky, Rosen's (EPR's) grand challenge to quantum mechanics.
7.6. Why did Einstein, Podolsky, Rosen (EPR) object?
7.7. EPR's objection
7.8. Quantum entanglement: The Lie group perspective

8. **Entropy and quantum mechanics**
8.1. Classical entropy, i.e., Shannon Entropy
8.2. Quantum entropy, i.e., Von Neumann entropy
8.3. How is quantum entropy related to classical entropy?
8.4. When a part is greater than the whole, then Ignorance = uncertainty

9. **There is much more to quantum mechanics**

Part 3. Part of a Rosetta Stone for Quantum Computation

10. **The Beginnings of Quantum Computation – Elementary Quantum Computing Devices**
10.1. Embedding classical (memoryless) computation in quantum mechanics
10.2. Classical reversible computation without memory
10.3. Embedding classical irreversible computation within classical reversible computation
10.4. The unitary representation of reversible computing devices
10.5. Some other simple quantum computing devices
10.6. Quantum computing devices that are not embeddings
10.7. The implicit frame of a wiring diagram

11. **The No-Cloning Theorem**

12. **Quantum teleportation**

13. **There is much more to quantum computation**

References

Part 1. Preamble

1. Introduction

These lecture notes were written for the American Mathematical Society (AMS) Short Course on Quantum Computation held 17-18 January 2000 in conjunction with the Annual Meeting of the AMS in Washington, DC in January 2000. They are intended for readers with some mathematical background but with little or no exposure to quantum mechanics. The purpose of these notes is to provide such readers with enough material in quantum mechanics and quantum computation to begin reading the vast literature on quantum computation, quantum cryptography, and quantum information theory.

The paper was written in an informal style. Whenever possible, each new topic was begun with the introduction of the underlying motivating intuitions, and then followed by an explanation of the accompanying mathematical finery. Hopefully, once having grasped the basic intuitions, the reader will find that the remaining material easily follows.

Since this paper is intended for a diverse audience, it was written at varying levels of difficulty and sophistication, from the very elementary to the more advanced. A large number of examples have been included. An index and table of contents are provided for those readers who prefer to "pick and choose." Hopefully, this paper will provide something of interest for everyone.

Because of space limitations, these notes are, of necessity, far from a complete overview of quantum mechanics. For example, only finite dimensional Hilbert spaces are considered, thereby avoiding the many pathologies that always arise when dealing with infinite dimensional objects. Many important experiments that are traditionally part of the standard fare in quantum mechanics texts (such as for example, the Stern-Gerlach experiment, Young's two slit experiment, the Aspect experiment) have not been mentioned in this paper. We leave it to the reader to decide if these notes have achieved their objective.

2. The classical world

2.1. Introducing the Shannon bit.

Since one of the objectives of this paper is to discuss quantum information, we begin with a brief discussion of classical information.

The Shannon bit is so well known in our age of information that it needs little, if any, introduction. As we all know, the Shannon bit is like a very decisive individual. It is either 0 or 1, but by no means both at the same time. The Shannon bit has become so much a part of our every day lives that we take many of its properties for granted. For example, we take for granted that Shannon bits can be copied.

2.2. Polarized light: Part I. The classical perspective.

Throughout this paper the quantum polarization states of light will be used to provide concrete illustrations of underlying quantum mechanical principles. So we also begin with a brief discussion of polarized light from the classical perspective.

Light waves in the vacuum are transverse electromagnetic (EM) waves with both electric and magnetic field vectors perpendicular to the direction of propagation and also to each other. (See figure 1.)

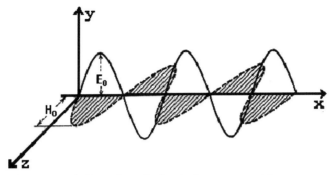

Figure 1. A linearly polarized electromagnetic wave.

If the electric field vector is always parallel to a fixed line, then the EM wave is said to be **linearly polarized**. If the electric field vector rotates about the direction of propagation forming a right-(left-)handed screw, it is said to be **right** (**left**) **elliptically polarized**. If the rotating electric field vector inscribes a circle, the EM wave is said to be right- or left-**circularly polarized**.

3. The quantum world

3.1. Introducing the qubit – But what is a qubit?

Many of us may not be as familiar with the quantum bit of information, called a **qubit**. Unlike its sibling rival, the Shannon bit, the qubit can be both 0 and 1 at the same time. Moreover, unlike the Shannon bit, the qubit can not be duplicated[1]. As we shall see, qubits are like very slippery, irascible individuals, exceedingly difficult to deal with.

One example of a qubit is a spin $\frac{1}{2}$ particle, which can be in a spin-up state $|1\rangle$ which we label as "1", in a spin-down state $|0\rangle$ which we label as "0", or in a **superposition** of these states, which we interpret as being both 0 and 1 at the same time. (The term "superposition" will be explained shortly.)

Another example of a qubit is the polarization state of a photon. A photon can be in a vertically polarized state $|\updownarrow\rangle$. We assign a label of "1" to this state. It can be in a horizontally polarized state $|\leftrightarrow\rangle$. We assign a label of "0" to this state. Or, it can be in a superposition of these states. In this case, we interpret its state as representing both 0 and 1 at the same time.

Anyone who has worn polarized sunglasses is familiar with the polarization states of light. Polarized sunglasses eliminate glare by letting through only vertically polarized light, while filtering out the horizontally polarized light. For that reason, they are often used to eliminate road glare, i.e., horizontally polarized light reflected from the road.

---

[1]This is a result of the no-cloning theorem of Dieks[**24**], Wootters and Zurek[**93**]. A proof of the no-cloning theorem is given in Section 11 of this paper.

3.2. Where do qubits live? – But what is a qubit?

But where do qubits live? They live in a Hilbert space $\mathcal{H}$. By a Hilbert space, we mean:

A **Hilbert space** $\mathcal{H}$ is a vector space over the complex numbers $\mathbb{C}$ with a complex valued inner product

$$(-,-) : \mathcal{H} \times \mathcal{H} \to \mathbb{C}$$

which is complete with respect to the norm

$$\|u\| = \sqrt{(u,u)}$$

induced by the inner product.

REMARK 1. *By a complex valued inner product, we mean a map*

$$(-,-) : \mathcal{H} \times \mathcal{H} \to \mathbb{C}$$

from $\mathcal{H} \times \mathcal{H}$ into the complex numbers $\mathbb{C}$ such that:
1) $(u,u) = 0$ *if and only if* $u = 0$
2) $(u,v) = (v,u)^*$
3) $(u, v + w) = (u,v) + (u,w)$
4) $(u, \lambda v) = \lambda (u,v)$

where '*' denotes the complex conjugate.

REMARK 2. *Please note that* $(\lambda u, v) = \lambda^*(u,v)$.

3.3. A qubit is ... [2]

> A **qubit** is a quantum system $\mathcal{Q}$ whose state lies in a two dimensional Hilbert space $\mathcal{H}$.

Part 2. An Introduction to Quantum Mechanics

---

[2]Barenco et al in [**3**] define a qubit as a quantum system with a two dimensional Hilbert space, capable of existing in a superposition of Boolean states, and also capable of being entangled with the states of other qubits. Their more functional definition will take on more meaning as the reader progresses through this paper.

4. The beginnings of quantum mechanics

4.1. A Rosetta stone for Dirac notation: Part I. Bras, kets, and bra-(c)-kets.

The elements of a Hilbert space $\mathcal{H}$ will be called **ket vectors**, **state kets**, or simply **kets**. They will be denoted as:
$$|\,label\,\rangle$$
where '*label*' denotes some label, i.e., some chosen string of symbols that designate a state[3].

Let $\mathcal{H}^*$ denote the Hilbert space of all Hilbert space morphisms of $\mathcal{H}$ into the Hilbert space of all complex numbers $\mathbb{C}$, i.e.,
$$\mathcal{H}^* = Hom_\mathbb{C}\left(\mathcal{H}, \mathbb{C}\right).$$
The elements of $\mathcal{H}^*$ will be called **bra vectors**, **state bras**, or simply **bras**. They will be denoted as:
$$\langle\,label\,|$$
where once again '*label*' denotes some label.

Also please note that the complex number
$$\langle\,label_1\,|\,(|\,label_2\,\rangle)$$
will simply be denoted by
$$\langle\,label_1\,|\,label_2\,\rangle$$
and will be called the **bra-(c)-ket** product of the bra $\langle\,label_1\,|$ and the ket $|\,label_2\,\rangle$.

There is a monomorphism (which is an isomorphism if the underlying Hilbert space is finite dimensional)
$$\mathcal{H} \xrightarrow{\dagger} \mathcal{H}^*$$
defined by
$$|\,label\,\rangle \longmapsto (\,|\,label\,\rangle\,, -)$$

---

[3]It should be mentioned that the bra and ket vectors of physics form a more general space than Hilbert space. P.A.M. Dirac [**28**, page 40] writes:

> The space of bra and ket vectors when the vectors are restricted to be of finite length and to have finite scalar products is called by mathematicians a *Hilbert space*. The bra and ket vectors that we now use form a more general space than Hilbert space.

Readers interested in pursuing this further should refer to the the theory of **Gel'fand triplets**, also known as **rigged Hilbert spaces**.

The bra $(\,|\,label\,\rangle,-)$ is denoted by $\langle\,label\,|$.

Hence,
$$\langle\,label_1\,|\,label_2\,\rangle = (|\,label_1\,\rangle,|\,label_2\,\rangle)$$

REMARK 3. *Please note that* $(\lambda\,|\,label\,\rangle)^{\dagger} = \lambda^{*}\,\langle label|$ *and* $\|\,|label\rangle\,\| = \sqrt{\langle\,label\,|\,label\,\rangle}$.

The **tensor product**[4] $\mathcal{H}\otimes\mathcal{K}$ of two Hilbert spaces $\mathcal{H}$ and $\mathcal{K}$ is simply the "simplest" Hilbert space such that

1) $(h_1 + h_2)\otimes k = h_1\otimes k + h_2\otimes k$, for all $h_1, h_2 \in \mathcal{H}$ and for all $k \in \mathcal{K}$, and
2) $h\otimes(k_1 + k_2) = h\otimes k_1 + h\otimes k_2$ for all $h \in \mathcal{H}$ and for all $k_1, k_2 \in \mathcal{K}$.
3) $\lambda(h\otimes k) \equiv (\lambda h)\otimes k = h\otimes(\lambda k)$ for all $\lambda \in \mathbb{C}$, $h \in \mathcal{H}$, $k \in \mathcal{K}$.

It follows that, if $\{\,e_1, e_2, \ldots, e_m\,\}$ and $\{\,f_1, f_2, \ldots, f_n\,\}$ are respectively bases of the Hilbert spaces $\mathcal{H}$ and $\mathcal{K}$, then $\{\,e_i\otimes f_j\,|\,1\leq i\leq m,\,1\leq j\leq n\,\}$ is a basis of $\mathcal{H}\otimes\mathcal{K}$. Hence, the dimension of the Hilbert space $\mathcal{H}\otimes\mathcal{K}$ is the product of the dimensions of the Hilbert spaces $\mathcal{H}$ and $\mathcal{K}$, i.e.,

$$Dim\,(\mathcal{H}\otimes\mathcal{K}) = Dim\,(\mathcal{H})\cdot Dim\,(\mathcal{K})\,.$$

Finally, if $|\,label_1\,\rangle$ and $|\,label_2\,\rangle$ are kets respectively in Hilbert spaces $\mathcal{H}_1$ and $\mathcal{H}_2$, then their tensor product will be written in any one of the following three ways:

$$|\,label_1\,\rangle\otimes|\,label_2\,\rangle$$

$$|\,label_1\,\rangle\,|\,label_2\,\rangle$$

$$|\,label_1\,,\,label_2\,\rangle$$

4.2. Quantum mechanics: Part I. The state of a quantum system.

The states of a quantum system $\mathcal{Q}$ are represented by state kets in a Hilbert space $\mathcal{H}$. Two kets $|\alpha\rangle$ and $|\beta\rangle$ represent the same state of a quantum system $\mathcal{Q}$ if they differ by a non-zero multiplicative constant. In other words, $|\alpha\rangle$ and $|\beta\rangle$ represent the same quantum state $\mathcal{Q}$ if there exists a non-zero $\lambda \in \mathbb{C}$ such that

$$|\alpha\rangle = \lambda\,|\beta\rangle$$

---

[4]Readers well versed in homological algebra will recognize this informal definition as a slightly disguised version of the more rigorous universal definition of the tensor product. For more details, please refer to [16], or any other standard reference on homological algebra.

Hence, quantum states are simply elements of the manifold

$$\mathcal{H}/\tilde{\,} = \mathbb{C}P^{n-1}$$

where n denotes the dimension of $\mathcal{H}$, and $\mathbb{C}P^{n-1}$ denotes **complex projective $(n-1)$-space** .

> **Convention:** Since a quantum mechanical state is represented by a state ket up to a multiplicative constant, we will, unless stated otherwise, choose those kets $|\alpha\rangle$ which are of unit length, i.e., such that
>
> $$\langle \alpha \mid \alpha \rangle = 1 \iff \| \,|\alpha\rangle\| = 1$$

4.3. Polarized light: Part II. The quantum mechanical perspective.

As an illustration of the above concepts, we consider the polarization states of a photon.

The polarization states of a photon are represented as state kets in a two dimensional Hilbert space $\mathcal{H}$. One orthonormal basis of $\mathcal{H}$ consists of the kets

$$|\circlearrowleft\rangle \text{ and } |\circlearrowright\rangle$$

which represent respectively the quantum mechanical states of left- and right-circularly polarized photons[5]. Another orthonormal basis consists of the kets

$$|\updownarrow\rangle \text{ and } |\leftrightarrow\rangle$$

representing respectively vertically and horizontally linearly polarized photons. And yet another orthonormal basis consists of the kets

$$|\nearrow\rangle \text{ and } |\nwarrow\rangle$$

for linearly polarized photons at the angles $\theta = \pi/4$ and $\theta = -\pi/4$ off the vertical, respectively.

These orthonormal bases are related as follows:

$$\begin{cases} |\nearrow\rangle = \frac{1}{\sqrt{2}}(|\updownarrow\rangle + |\leftrightarrow\rangle) \\ |\nwarrow\rangle = \frac{1}{\sqrt{2}}(|\updownarrow\rangle - |\leftrightarrow\rangle) \end{cases} \qquad \begin{cases} |\nearrow\rangle = \frac{1+i}{2}|\circlearrowleft\rangle + \frac{1-i}{2}|\circlearrowright\rangle \\ |\nwarrow\rangle = \frac{1-i}{2}|\circlearrowleft\rangle + \frac{1+i}{2}|\circlearrowright\rangle \end{cases}$$

$$\begin{cases} |\updownarrow\rangle = \frac{1}{\sqrt{2}}(|\nearrow\rangle + |\nwarrow\rangle) \\ |\leftrightarrow\rangle = \frac{1}{\sqrt{2}}(|\nearrow\rangle - |\nwarrow\rangle) \end{cases} \qquad \begin{cases} |\updownarrow\rangle = \frac{1}{\sqrt{2}}(|\circlearrowleft\rangle + |\circlearrowright\rangle) \\ |\leftrightarrow\rangle = \frac{i}{\sqrt{2}}(|\circlearrowleft\rangle - |\circlearrowright\rangle) \end{cases}$$

$$\begin{cases} |\circlearrowleft\rangle = \frac{1}{\sqrt{2}}(|\updownarrow\rangle - i|\leftrightarrow\rangle) \\ |\circlearrowright\rangle = \frac{1}{\sqrt{2}}(|\updownarrow\rangle + i|\leftrightarrow\rangle) \end{cases} \qquad \begin{cases} |\circlearrowleft\rangle = \frac{1-i}{2}|\nearrow\rangle + \frac{1+i}{2}|\nwarrow\rangle \\ |\circlearrowright\rangle = \frac{1+i}{2}|\nearrow\rangle + \frac{1-i}{2}|\nwarrow\rangle \end{cases}$$

---

[5]Please refer to [**84**, pages 9-10] for a discussion in regard to the circular polarization states of light.

The bracket products of the various polarization kets are given in the table below:

	$|\updownarrow\rangle$	$|\leftrightarrow\rangle$	$|\nearrow\rangle$	$|\searrow\rangle$	$|\circlearrowleft\rangle$	$|\circlearrowright\rangle$
$\langle\updownarrow|$	1	0	$\frac{1}{\sqrt{2}}$	$\frac{1}{\sqrt{2}}$	$\frac{1}{\sqrt{2}}$	$\frac{1}{\sqrt{2}}$
$\langle\leftrightarrow|$	0	1	$\frac{1}{\sqrt{2}}$	$-\frac{1}{\sqrt{2}}$	$-\frac{i}{\sqrt{2}}$	$\frac{i}{\sqrt{2}}$
$\langle\nearrow|$	$\frac{1}{\sqrt{2}}$	$\frac{1}{\sqrt{2}}$	1	0	$\frac{1-i}{2}$	$\frac{1+i}{2}$
$\langle\searrow|$	$\frac{1}{\sqrt{2}}$	$-\frac{1}{\sqrt{2}}$	0	1	$\frac{1+i}{2}$	$\frac{1-i}{2}$
$\langle\circlearrowleft|$	$\frac{1}{\sqrt{2}}$	$\frac{i}{\sqrt{2}}$	$\frac{1+i}{2}$	$\frac{1-i}{2}$	1	0
$\langle\circlearrowright|$	$\frac{1}{\sqrt{2}}$	$-\frac{i}{\sqrt{2}}$	$\frac{1-i}{2}$	$\frac{1+i}{2}$	0	1

In terms of the basis $\{|\updownarrow\rangle, |\leftrightarrow\rangle\}$ and the dual basis $\{\langle\updownarrow|, \langle\leftrightarrow|\}$, these kets and bras can be written as matrices as indicated below:

$$\begin{cases} \langle\updownarrow| = \begin{pmatrix} 1 & 0 \end{pmatrix}, & |\updownarrow\rangle = \begin{pmatrix} 1 \\ 0 \end{pmatrix} \\[6pt] \langle\leftrightarrow| = \begin{pmatrix} 0 & 1 \end{pmatrix}, & |\leftrightarrow\rangle = \begin{pmatrix} 0 \\ 1 \end{pmatrix} \\[6pt] \langle\nearrow| = \frac{1}{\sqrt{2}} \begin{pmatrix} 1 & 1 \end{pmatrix}, & |\nearrow\rangle = \frac{1}{\sqrt{2}} \begin{pmatrix} 1 \\ 1 \end{pmatrix} \\[6pt] \langle\searrow| = \frac{1}{\sqrt{2}} \begin{pmatrix} 1 & -1 \end{pmatrix}, & |\searrow\rangle = \frac{1}{\sqrt{2}} \begin{pmatrix} 1 \\ -1 \end{pmatrix} \\[6pt] \langle\circlearrowleft| = \frac{1}{\sqrt{2}} \begin{pmatrix} 1 & i \end{pmatrix}, & |\circlearrowleft\rangle = \frac{1}{\sqrt{2}} \begin{pmatrix} 1 \\ -i \end{pmatrix} \\[6pt] \langle\circlearrowright| = \frac{1}{\sqrt{2}} \begin{pmatrix} 1 & -i \end{pmatrix}, & |\circlearrowright\rangle = \frac{1}{\sqrt{2}} \begin{pmatrix} 1 \\ i \end{pmatrix} \end{cases}$$

In this basis, for example, the tensor product $|\nearrow\circlearrowright\rangle$ is

$$|\nearrow\circlearrowright\rangle = \begin{pmatrix} \frac{1}{\sqrt{2}} \\ \frac{1}{\sqrt{2}} \end{pmatrix} \otimes \begin{pmatrix} \frac{1}{\sqrt{2}} \\ -\frac{i}{\sqrt{2}} \end{pmatrix} = \frac{1}{2} \begin{pmatrix} 1 \\ -i \\ 1 \\ -i \end{pmatrix}$$

and the projection operator $|\circlearrowright\rangle\langle\circlearrowright|$ is:

$$|\circlearrowright\rangle\langle\circlearrowright| = \frac{1}{\sqrt{2}} \begin{pmatrix} 1 \\ i \end{pmatrix} \otimes \frac{1}{\sqrt{2}} \begin{pmatrix} 1 & -i \end{pmatrix} = \frac{1}{2} \begin{pmatrix} 1 & -i \\ i & 1 \end{pmatrix}$$

REMARK 4. *Please note that, if bras and kets are represented as matrices, then the adjoint is nothing more than the conjugate transpose.*

4.4. A Rosetta stone for Dirac notation: Part II. Operators.

An **(linear) operator** or **transformation** $\mathcal{O}$ on a ket space $\mathcal{H}$ is a Hilbert space morphism of $\mathcal{H}$ into $\mathcal{H}$, i.e., is an element of

$$Hom_{\mathbb{C}}(\mathcal{H}, \mathcal{H})$$

The **adjoint** $\mathcal{O}^{\dagger}$ of an operator $\mathcal{O}$ is that operator such that

$$\left(\mathcal{O}^{\dagger} \,|\, label_1 \,\rangle , |\, label_2 \,\rangle\right) = \left(|\, label_1 \,\rangle , \mathcal{O} \,|\, label_2 \,\rangle\right)$$

for all kets $|\, label_1 \,\rangle$ and $|\, label_2 \,\rangle$.

In like manner, an (linear) operator or transformation on a bra space $\mathcal{H}^*$ is an element of

$$Hom_{\mathbb{C}}(\mathcal{H}^*, \mathcal{H}^*)$$

Moreover, each operator $\mathcal{O}$ on $\mathcal{H}$ can be identified with an operator, also denoted by $\mathcal{O}$, on $\mathcal{H}^*$ defined by

$$\langle\, label_1 \,| \longmapsto \langle\, label_1 \,|\, \mathcal{O}$$

where $\langle\, label_1 \,|\, \mathcal{O}$ is the bra defined by

$$(\langle\, label_1 \,|\, \mathcal{O})(|\, label_2 \,\rangle) = \langle\, label_1 \,|\, (\mathcal{O} \,|\, label_2 \,\rangle)$$

(This is sometimes called Dirac's associativity law.) Hence, the expression

$$\langle\, label_1 \,|\, \mathcal{O} \,|\, label_2 \,\rangle$$

is unambiguous.

REMARK 5. *Please note that*

$$(\mathcal{O} \,|\, label \rangle)^{\dagger} = \langle label | \, \mathcal{O}^{\dagger}$$

4.5. Quantum mechanics: Part II. Observables.

In quantum mechanics, an **observable** is simply a **Hermitian** (also called **self-adjoint**) operator on a Hilbert space $\mathcal{H}$, i.e., a linear operator $\mathcal{O}$ such that

$$\mathcal{O}^{\dagger} = \mathcal{O} \,.$$

A class of operators that play a fundamental role in quantum measurement are projection operators. A **projection operator** P on a Hilbert space $\mathcal{H}$ is a linear operator such that $P^2 = P$. Let V_P denote the sub-Hilbert space of all kets lying in the image of P. Then P is called the **projector for the subspace** V_P, and P is said to **project** $\mathcal{H}$ onto V_P. Projection operators are indeed very special observables, for:

PROPOSITION 1. *Every projection operator is an observable.*

Projection operators can be naturally expressed in terms of Dirac notation. For let P be a projection operator for a subspace V_P of the Hilbert space $\mathcal{H}$. Then, if
$$\{|label_0\rangle, |label_1\rangle, \ldots, |label_{n-1}\rangle\}$$
is an orthonormal basis of V_P, the projection operator P can be written as
$$P = \sum_{j=0}^{n-1} |label_j\rangle \langle label_j|,$$
where $|label_j\rangle \langle label_j|$ denotes, for each j, the outer product of the unit length ket $|label_j\rangle$ with the unit length bra $\langle label_j|$. Moreover, $P_j = |label_j\rangle \langle label_j|$ is for each j the projection operator for the subspace spanned by the ket $|label_j\rangle$.

An **eigenvalue** a of an operator A is a complex number for which there exists a ket $|label\rangle$ in $\mathcal{H}$ such that
$$A|label\rangle = a|label\rangle.$$
The ket $|label\rangle$ is called an **eigenket** of A corresponding to the eigenvalue a. The **eigenspace** V_a of the eigenvalue a is the subspace of $\mathcal{H}$ of all eigenkets corresponding to the eigenvalue a, i.e., the space
$$V_a = \{\ |label\rangle \ |\ A|label\rangle = a|label\rangle\ \}$$
We denote the **projection operator for the eigenspace** V_a by P_a.

THEOREM 1. *The eigenvalues a_j of an observable $\mathcal{O}$ are all real numbers. Moreover, eigenkets corresponding to distinct eigenvalues are orthogonal.*

The observables and unitary transformations of quantum mechanics are examples of a larger class of linear operators, called normal operators.

DEFINITION 1. *A linear operator A is said to be **normal** if it commutes with its adjoint, i.e.,*
$$AA^\dagger = A^\dagger A.$$

The normal linear operators form a class of linear operators with many notable properties, one of the most important of which is the spectral decomposition theorem[6].

THEOREM 2 (Spectral Decomposition). *Let $a_0, a_1, \ldots, a_{n-1}$ be the eigenvalues of a linear operator A on the Hilbert space $\mathcal{H}$. Then A is a normal operator if and only if it can be written in the form*
$$A = \sum_{j=0}^{n-1} a_j P_{a_j},$$

---

[6]A proof of the spectral decomposition theorem can be found in almost any suitably advanced book on linear algebra, such as for example [**45**].

where P_{a_j} denotes the projection operator for the eigenspace V_{a_j} corresponding to the eigenvalue a_j. Moreover, a linear operator A is normal if and only if the Hilbert space $\mathcal{H}$ is a direct sum of the eigenspaces of A, i.e., if and only if

$$\mathcal{H} = V_{a_0} \oplus V_{a_1} \oplus \cdots \oplus V_{a_{n-1}},$$

where the eigenspaces are mutually orthogonal.

REMARK 6. *Hence, the projection operators $P_{a_0}, P_{a_1}, \ldots, P_{a_{n-1}}$ are **mutually orthogonal**, i.e., $P_{a_i} P_{a_j} = 0$ if $i \neq j$, and **complete**, i.e.,*

$$\sum_{j=0}^{n-1} P_{a_j} = 1,$$

where "1" denotes the identity linear transformation on the Hilbert space $\mathcal{H}$. Moreover, for all kets $|\psi\rangle$ in $\mathcal{H}$, we have the decomposition

$$|\psi\rangle = \sum_{j=0}^{n-1} P_{a_j} |\psi\rangle.$$

Thus, if A is a normal linear operator, then there is an orthonormal basis of the underlying Hilbert space $\mathcal{H}$ consisting entirely of eigenkets of A. In other words, the operator A can be **diagonalized**, i.e., written as a diagonal matrix as follows:

For each j, let

$$\{e_{j1}, e_{j2}, \ldots, e_{jm_j}\}$$

be an orthonormal basis of the eigenspace V_{a_j}. Then

$$\{e_{01}, e_{02}, \ldots, e_{0m_0}, \quad e_{11}, e_{12}, \ldots, e_{1m_0}, \quad \ldots, \quad e_{(n-1)1}, e_{(n-1)2}, \ldots, e_{(n-1)m_{n-1}}\}$$

is an orthonormal basis of the Hilbert space $\mathcal{H}$. In terms of this basis, the matrix representation of the normal operator A is the $n \times n$ diagonal matrix with the following diagonal:

$$\left(\underbrace{a_0, \ldots, a_0}_{m_0}, \underbrace{a_1, \ldots, a_1}_{m_1}, \ldots, \underbrace{a_{n-1}, \ldots, a_{n-1}}_{m_{n-1}} \right)$$

DEFINITION 2. *Let A be a normal linear operator on a Hilbert space $\mathcal{H}$. Then an eigenvalue a is said to be **degenerate** if the corresponding eigenspace V_a is of dimension greater than 1. Otherwise, the eigenvalue a is said to be **nondegenerate**. A linear operator A is said to be **nondegenerate** it all its eigenvalues are nondegenerate. Otherwise, it is said to be **degenerate**.*

Notational Convention. *Let $a_0, a_1, \ldots, a_{n-1}$ be the eigenvalues of a normal operator A. Let $V_{a_0}, V_{a_1}, \ldots, V_{a_{n-1}}$ denote the corresponding eigenspaces of respective dimensions $n_{a_0}, n_{a_1}, \ldots, n_{a_{n-1}}$. If an eigenvalue a_j is degenerate, then we will frequently denote an orthonormal basis of the eigenspace V_{a_j} simply by*

$$|a_j, 0\rangle, |a_j, 1\rangle, \ldots, |a_j, n_{a_j} - 1\rangle.$$

On the other hand, if a_j is a nondegenerate eigenvalue of A, then we will denote a normalized (i.e., unit length) eigenket corresponding to a_j simply by

$$|a_j\rangle \ .$$

In terms of this notation, the spectral decomposition of A is

$$A = \sum_{j=0}^{n-1} a_j \sum_{k=0}^{n_{a_k}-1} |a_j, k\rangle \langle a_j, k| \ .$$

If A is nondegenerate, the spectral decomposition can be written more simply as

$$A = \sum_{j=0}^{n-1} a_j |a_j\rangle \langle a_j| \ .$$

EXAMPLE 1. Let $\mathcal{H}$ be the Hilbert space with orthonormal basis

$$\{\ |0\rangle, |1\rangle, |2\rangle, |3\rangle\ \} \ ,$$

and let $\mathcal{O}$ be the observable which, when written as a matrix in the above orthonormal basis, is given.

$$\mathcal{O} = \begin{pmatrix} 0 & 0 & -2i & 0 \\ 0 & 0 & 0 & 2i \\ 2i & 0 & 0 & 0 \\ 0 & -2i & 0 & 0 \end{pmatrix}$$

The observable $\mathcal{O}$ is degenerate. Its eigenvalues and corresponding eigenspace orthonormal bases (using the notation convention found on page 13) are given in the table below:

Eigenvalue	Orthonormal Basis of Eigenspace $\mathbf{V}_{a_j}$
$a_0 = 2$	$\|+2, 0\rangle = \frac{1}{\sqrt{2}}(\|0\rangle + i\|2\rangle) = \frac{1}{\sqrt{2}}(1, 0, i, 0)^T$
	$\|+2, 1\rangle = \frac{1}{\sqrt{2}}(\|1\rangle - i\|3\rangle) = \frac{1}{\sqrt{2}}(0, 1, 0, -i)^T$
$a_1 = -2$	$\|-2, 0\rangle = \frac{1}{\sqrt{2}}(\|0\rangle - i\|2\rangle) = \frac{1}{\sqrt{2}}(1, 0, -i, 0)^T$
	$\|-2, 1\rangle = \frac{1}{\sqrt{2}}(\|1\rangle + i\|3\rangle) = \frac{1}{\sqrt{2}}(0, 1, 0, i)^T$

Its spectral decomposition is

$$\mathcal{O} = a_0 P_0 + a_1 P_1$$

$$= (+2)\Big(|+2,0\rangle\langle +2,0| \ + \ |+2,1\rangle\langle +2,1|\Big)$$

$$+ (-2)\Big(|-2,0\rangle\langle -2,0| \ + \ |-2,1\rangle\langle -2,1|\Big)$$

$$= 2 \begin{pmatrix} \frac{1}{2} & 0 & -\frac{i}{2} & 0 \\ 0 & \frac{1}{2} & 0 & \frac{i}{2} \\ \frac{i}{2} & 0 & \frac{1}{2} & 0 \\ 0 & -\frac{i}{2} & 0 & \frac{1}{2} \end{pmatrix} + (-2) \begin{pmatrix} \frac{1}{2} & 0 & \frac{i}{2} & 0 \\ 0 & \frac{1}{2} & 0 & -\frac{i}{2} \\ -\frac{i}{2} & 0 & \frac{1}{2} & 0 \\ 0 & \frac{i}{2} & 0 & \frac{1}{2} \end{pmatrix}$$

In terms of the eigenket orthonormal basis

$$\Big\{ |+2,0\rangle, \ |+2,1\rangle, \ |-2,0\rangle, \ |-2,1\rangle \Big\},$$

the matrix representation of the observable becomes the diagonal matrix

$$\mathcal{O} = \begin{pmatrix} a_0 & 0 & 0 & 0 \\ 0 & a_0 & 0 & 0 \\ 0 & 0 & a_1 & 0 \\ 0 & 0 & 0 & a_1 \end{pmatrix} = \begin{pmatrix} 2 & 0 & 0 & 0 \\ 0 & 2 & 0 & 0 \\ 0 & 0 & -2 & 0 \\ 0 & 0 & 0 & -2 \end{pmatrix}$$

*Finally, as an illustration of **Remark 6**, we have*

$$P_0 + P_1 = \Big(|+2,0\rangle\langle +2,0| \ + \ |+2,1\rangle\langle +2,1|\Big)$$

$$+ \Big(|-2,0\rangle\langle -2,0| \ + \ |-2,1\rangle\langle -2,1|\Big)$$

$$= \begin{pmatrix} 1 & 0 & 0 & 0 \\ 0 & 1 & 0 & 0 \\ 0 & 0 & 1 & 0 \\ 0 & 0 & 0 & 1 \end{pmatrix} = 1,$$

where the last "1" denotes the identity operator.

EXAMPLE 2. *The Pauli spin matrices*

$$\sigma_1 = \begin{pmatrix} 0 & 1 \\ 1 & 0 \end{pmatrix}, \quad \sigma_2 = \begin{pmatrix} 0 & -i \\ i & 0 \end{pmatrix}, \quad \sigma_3 = \begin{pmatrix} 1 & 0 \\ 0 & -1 \end{pmatrix}$$

are examples of observables that frequently appear in quantum mechanics and quantum computation. Their eigenvalues and eigenkets are given in the following table:

Pauli Matrices	Eigenvalue/Eigenket	
$\sigma_1 = \begin{pmatrix} 0 & 1 \\ 1 & 0 \end{pmatrix}$	$+1$	$\frac{\|0\rangle+\|1\rangle}{\sqrt{2}} = \frac{1}{\sqrt{2}}\begin{pmatrix} 1 \\ 1 \end{pmatrix}$
	-1	$\frac{\|0\rangle-\|1\rangle}{\sqrt{2}} = \frac{1}{\sqrt{2}}\begin{pmatrix} 1 \\ -1 \end{pmatrix}$
$\sigma_2 = \begin{pmatrix} 0 & -i \\ i & 0 \end{pmatrix}$	$+1$	$\frac{\|0\rangle+i\|1\rangle}{\sqrt{2}} = \frac{1}{\sqrt{2}}\begin{pmatrix} 1 \\ i \end{pmatrix}$
	-1	$\frac{\|0\rangle-i\|1\rangle}{\sqrt{2}} = \frac{1}{\sqrt{2}}\begin{pmatrix} 1 \\ -i \end{pmatrix}$
$\sigma_3 = \begin{pmatrix} 1 & 0 \\ 0 & -1 \end{pmatrix}$	$+1$	$\|0\rangle = \begin{pmatrix} 1 \\ 0 \end{pmatrix}$
	-1	$\|1\rangle = \begin{pmatrix} 0 \\ 1 \end{pmatrix}$

Thus, the spectral decomposition of the Pauli operator σ_1 is:

$$\sigma_1 = (+1)\left(\frac{|0\rangle+|1\rangle}{\sqrt{2}}\right)\left(\frac{\langle 0|+\langle 1|}{\sqrt{2}}\right) + (-1)\left(\frac{|0\rangle-|1\rangle}{\sqrt{2}}\right)\left(\frac{\langle 0|-\langle 1|}{\sqrt{2}}\right),$$

which, in terms of the notation convention found on page 13, can be written as

$$\sigma_1 = (+1)|+1\rangle\langle +1| \quad + \quad (-1)|-1\rangle\langle -1|.$$

Please note that, as an illustration of **Remark 6,** we have

$$|+1\rangle\langle +1| \quad + \quad |-1\rangle\langle -1| = 1,$$

where " 1 " denotes the identity operator.

4.6. Quantum mechanics: Part III. Quantum measurement — General principles.

According to quantum measurement theory, measurement is defined by the following rubric:

Quantum Measurement Rubric: *The **measurement of an observable** A of a quantum system $\mathcal{Q}$ in the state $|\psi\rangle$ produces the eigenvalue a_i as the measured result with probability*

$$Prob(\text{Value} \quad a_i \quad \text{is} \quad \text{observed}) = \|P_{a_i}|\psi\rangle\|^2 = \langle\psi \mid P_{a_i} \mid \psi\rangle,$$

and forces the quantum system $\mathcal{Q}$ into the state

$$\frac{P_{a_i}|\psi\rangle}{\sqrt{\langle\psi \mid P_{a_i} \mid \psi\rangle}},$$

which lies in the corresponding eigenspace V_{a_i}.

Since quantum measurement is such a hotly debated topic among physicists, we (in self-defense) quote P.A.M. Dirac[28]:

> "A measurement always causes the (quantum mechanical) system to jump into an eigenstate of the dynamical variable that is being measured."

If $\mathcal{Q}$ is a quantum system in state $|\psi\rangle$, then the measurement of the system $\mathcal{Q}$ with respect to the observable A can be diagrammatically represented as follows:

	First Meas. of A		Second Meas. of A	
$\|\psi\rangle = \sum_i P_{a_i}\|\psi\rangle$	$\underset{Prob = \langle\psi\|P_{a_j}\|\psi\rangle}{\Longrightarrow}$	$\dfrac{P_{a_j}\|\psi\rangle}{\sqrt{\langle\psi\|P_{a_j}\|\psi\rangle}}$	$\underset{Prob = 1}{\Longrightarrow}$	$\dfrac{P_{a_j}\|\psi\rangle}{\sqrt{\langle\psi\|P_{a_j}\|\psi\rangle}}$

Please note that the measured value is the eigenvalue a_j with probability $\langle\psi \mid P_{a_j} \mid \psi\rangle$. If the same measurement is repeated on the quantum system $\mathcal{Q}$ after the first measurement, then the result of the second measurement is no longer probabilistic. It produces the previously measured eigenvalue a_j, and the state of $\mathcal{Q}$ remains the same, i.e., $\dfrac{P_{a_j}|\psi\rangle}{\sqrt{\langle\psi|P_{a_j}|\psi\rangle}}$.

Example 1. (Cont.) *Let the Hilbert space $\mathcal{H}$ and the observable $\mathcal{O}$ be as in **Example 1**. Let $\mathcal{Q}$ be a quantum system in the state $|\psi\rangle \in \mathcal{H}$ given by*

$$|\psi\rangle = \frac{1}{3}(2|0\rangle + 2i|1\rangle - |3\rangle) = \frac{1}{3}(2, 2i, 0, -1)^T$$

Then measurement of the quantum system $\mathcal{Q}$ with respect to the observable $\mathcal{O}$ is summarized in the table below:

Index j	Probability $\langle\psi \mid \mathbf{P_{a_j}}\mid \psi\rangle$	Eigenvalue a_j	Resulting State $\dfrac{\mathbf{P_{a_j}}\|\psi\rangle}{\sqrt{\langle\psi\|\mathbf{P_{a_j}}\|\psi\rangle}}$
0	$\frac{5}{18}$	$+2$	$\frac{1}{\sqrt{10}}(\ 2\|0\rangle + i\|1\rangle + 2i\|2\rangle + \|3\rangle\ )$
1	$\frac{13}{18}$	-2	$\frac{1}{\sqrt{26}}(\ 2\|0\rangle + 3i\|1\rangle - 2i\|2\rangle - 3\|3\rangle\ )$

EXAMPLE 3. *Let $\mathcal{H}$ be a Hilbert space with orthonormal basis*

$$\{\ |0\rangle, |1\rangle, |2\rangle, \ldots, |n-1\rangle\ \},$$

and let $\mathcal{O}$ be the observable

$$\mathcal{O} = |0\rangle\langle 0|\ .$$

The operator is a degenerate observable. Its eigenvalues and corresponding eigenkets are given in the table below:

Eigenvalue	Orthonormal Basis of Eigenspace V_{a_j}
$a_0 = 1$	$\|0\rangle$
$a_1 = 0$	$\begin{cases} \|0,0\rangle = \|1\rangle \\ \|0,1\rangle = \|2\rangle \\ \vdots \\ \|0, n-1\rangle = \|n\rangle \end{cases}$

Let $\mathcal{Q}$ be a quantum system in the state

$$|\psi\rangle = \sum_{j=0}^{n-1} \alpha_j |j\rangle ,$$

where

$$\sum_{j=0}^{n-1} |\alpha_j|^2 = 1 ,$$

If the quantum system $\mathcal{Q}$ is measured with respect to the observable

$$\mathcal{O} = |0\rangle \langle 0| ,$$

then the result of this measurement is given in the table below:

Index j	Probability $\langle \psi \| \mathbf{P}_{\mathbf{a_j}} \| \psi \rangle$	Eigenvalue $\mathbf{a_j}$	Resulting State $\dfrac{\mathbf{P}_{\mathbf{a_j}}\|\psi\rangle}{\sqrt{\langle \psi\|\mathbf{P}_{\mathbf{a_j}}\|\psi\rangle}}$
0	$\|\alpha_0^2\|$	+1	$\|0\rangle$
1	$\sum_{j=1}^{n-1} \|\alpha_j\|^2$	0	$\dfrac{\sum_{j=1}^{n-1} \alpha_j \|j\rangle}{\sqrt{\sum_{j=1}^{n-1} \|\alpha_j\|^2}}$

It should be mentioned that Julian Schwinger's **algebra of measurement operators**, a slightly different perspective on quantum measurement, can be useful in describing quantum measurement. It is very briefly described as follows:

There are quantum measuring devices that block all states resulting from a measurement of an observable $\mathcal{O}$ except for states corresponding to a selected eigenvalue a' of $\mathcal{O}$. A measurement, in this instance, is called a **selective measurement** (or **filtration**) of the observable $\mathcal{O}$ with respect to the selected eigenvalue a'.

Let $|\psi\rangle$ be the state of a quantum system $\mathcal{Q}$ before a selective measurement of the observable $\mathcal{O}$ with respect to the eigenvalue a'. After selective measurement, the quantum system $\mathcal{Q}$ is discarded if the resulting state

$$P_a |\psi\rangle / \sqrt{\langle \psi | P_a | \psi \rangle}$$

does not lie in the selected eigenspace $V_{a'}$, i.e., if $a \neq a'$, and accepted if it does, i.e., if $a = a'$. If $a_0, a_1, \ldots, a_{n-1}$ are all the distinct eigenvalues of an observable $\mathcal{O}$, then a selective measurement with respect to a chosen eigenvalue a_j can be diagrammatically depicted as:

```
                          Prob = ⟨ψ|P_{a_j}|ψ⟩
              ┌─────────┐ ================>   P_a|ψ⟩/√⟨ψ|P_a|ψ⟩
              │Select(a_j)│
|ψ⟩  ==>      │ Measure │
              │    O    │
              └─────────┘ ================>   Not Selected
                        Prob = 1 - ⟨ψ|P_{a_j}|ψ⟩
```

REMARK 7. *The above selective measurement is the same as a selective measurement of a_j with respect to the observable P_{a_j}. For this reason, the projection operator P_{a_j} is sometimes called a **selective measurement operator**.*

Readers interested in learning more about selective measurement should refer to the many references on Julian Schwinger's algebra of measurement operators, such as [86] and [36].

In the above discussion, we have focused mainly on von Neumann measure theory, and just briefly touched upon Schwinger's theory of measurement. The most general theory of quantum measurement is that of **positive operator valued measures (POVMs)** (also known as, **probability operator valued measures**). For more information on POVM theory, please refer to such books as, for example, [23], [48], [77], and [81].

4.7. Polarized light: Part III. Three examples of quantum measurement.

We can now apply the above general principles of quantum measurement to polarized light. Three examples are given below:[7]

---

[7]The last two examples can easily be verified experimentally with at most three pair of polarized sunglasses.

EXAMPLE 4.

$$
\begin{array}{l}
\text{Rt. Circularly polarized photon} \\
|\circlearrowright\rangle = \tfrac{1}{\sqrt{2}}(|\updownarrow\rangle + i|\leftrightarrow\rangle)
\end{array}
\Longrightarrow
\begin{array}{c}
\text{Vertical} \\
\text{Polaroid} \\
\text{filter} \\
\\
Select\,(\updownarrow) \\
Meas. \\
|\updownarrow\rangle\langle\updownarrow|
\end{array}
\begin{array}{l}
\overset{Prob=\frac{1}{2}}{\Longrightarrow} \quad \begin{array}{c}\text{Vertically}\\ \text{polarized}\\ \text{photon}\\ |\updownarrow\rangle\end{array} \\[1em]
\overset{Prob=\frac{1}{2}}{\Longrightarrow} \quad \begin{array}{c}\text{No}\\ \text{Photon}\end{array}
\end{array}
$$

EXAMPLE 5. *A vertically polarized filter followed by a horizontally polarized filter.*

$$
\begin{array}{c}
\text{photon}\\
\alpha|\updownarrow\rangle + \beta|\leftrightarrow\rangle \\
|\alpha|^2 + |\beta|^2 = 1
\end{array}
\Longrightarrow
\begin{array}{c}
\text{Vert.}\\ \text{polar.}\\ \text{filter}\\ \\
Select\,(\updownarrow)\\ Meas.\\ |\updownarrow\rangle\langle\updownarrow|
\end{array}
\overset{Prob=|\alpha|^2}{\Longrightarrow}
\begin{array}{c}\text{Vert.}\\ \text{polar.}\\ \text{photon}\\ \\ |\updownarrow\rangle\end{array}
\Longrightarrow
\begin{array}{c}
\text{Horiz.}\\ \text{polar.}\\ \text{filter}\\ \\
Select\,(\leftrightarrow)\\ Meas.\\ |\leftrightarrow\rangle\langle\leftrightarrow|
\end{array}
\overset{Prob=1}{\Longrightarrow}
\begin{array}{c}\text{No}\\ \text{Photon}\end{array}
$$

EXAMPLE 6. *But if we insert a diagonally polarized filter (by 45° off the vertical) between the two polarized filters in the above example, we have:*

$$
\begin{array}{c}\\ Select\,(\updownarrow)\\ Meas.\\ |\updownarrow\rangle\langle\updownarrow|\end{array}
\overset{|\alpha|^2}{\Longrightarrow} |\updownarrow\rangle = \tfrac{1}{\sqrt{2}}(|\nearrow\rangle + |\nwarrow\rangle)
\begin{array}{c}\\ Select\,(\nearrow)\\ Meas.\\ |\nearrow\rangle\langle\nearrow|\end{array}
\overset{\frac{1}{2}}{\Longrightarrow} |\nearrow\rangle = \tfrac{1}{\sqrt{2}}(|\updownarrow\rangle + |\leftrightarrow\rangle)
\begin{array}{c}\\ Select\,(\leftrightarrow)\\ Meas.\\ |\leftrightarrow\rangle\langle\leftrightarrow|\end{array}
\overset{\frac{1}{2}}{\Longrightarrow} |\leftrightarrow\rangle
$$

where the input to the first filter is $\alpha|\updownarrow\rangle + \beta|\leftrightarrow\rangle$.

4.8. A Rosetta stone for Dirac notation: Part III. Expected values.

In this section we prove the following proposition which is an almost immediate consequence of the spectral decomposition theorem:

PROPOSITION 2. *Let $|\psi\rangle$ be the state of a quantum system $\mathcal{Q}$, and let A be an observable of $\mathcal{Q}$. Then The **average value (expected value)** of a measurement with respect to an observable A of a quantum system in a state $|\psi\rangle$ is:*

$$\langle A \rangle = \langle \psi | A | \psi \rangle$$

PROOF. Let $a_0, a_1, \ldots, a_{n-1}$ be all the distinct eigenvalues of the observable A, and let $P_{a_0}, P_{a_1}, \ldots, P_{a_{n-1}}$ be the corresponding projection operators. Then by the spectral decomposition theorem, we have

$$A = \sum_{j=0}^{n-1} a_j P_{a_j} .$$

So,

$$\langle A \rangle = \langle \psi | A | \psi \rangle = \left\langle \psi \left| \sum_{j=0}^{n-1} a_j P_{a_j} \right| \psi \right\rangle = \sum_{j=0}^{n-1} a_j \langle \psi | P_{a_j} | \psi \rangle .$$

But

$$\langle \psi | P_{a_j} | \psi \rangle = Prob\,(\text{Observing } a_j \text{ on input } |\psi\rangle) .$$

Hence,

$$\langle A \rangle = \sum_{j=0}^{n-1} a_j Prob\,(\text{Observing } a_j \text{ on input } |\psi\rangle)$$

is indeed the average observed eigenvalue. □

4.9. Quantum Mechanics: Part IV. The Heisenberg uncertainty principle.

There is, surprisingly enough, a limitation of what we can observe in the quantum world.

From classical probability theory, we know that one yardstick of uncertainty is the **standard deviation**, which measures the average fluctuation about the mean. Thus, the **uncertainty** involved in the measurement of a quantum observable A is defined as the standard deviation of the observed eigenvalues. This standard deviation is given by the expression

$$Uncertainty(A) = \sqrt{\left\langle (\triangle A)^2 \right\rangle}$$

where

$$\triangle A = A - \langle A \rangle$$

Two observables A and B are said to be **compatible** if they commute, i.e., if

$$AB = BA.$$

Otherwise, they are said to be **incompatible**.

Let $[A, B]$, called the **commutator** of A and B, denote the expression
$$[A, B] = AB - BA$$
In this notation, two operators A and B are compatible if and only if $[A, B] = 0$.

The following principle is one expression of how quantum mechanics places limits on what can be observed:

Heisenberg's Uncertainty Principle[8]
$$\left\langle (\triangle A)^2 \right\rangle \left\langle (\triangle B)^2 \right\rangle \geq \frac{1}{4} |\langle [A, B] \rangle|^2$$

Thus, if A and B are incompatible, i.e., do not commute, then, by measuring A more precisely, we are forced to measure B less precisely, and vice versa. We can not simultaneously measure both A and B to unlimited precision. Measurement of A somehow has an impact on the measurement of B, and vice versa.

4.10. Quantum mechanics: Part V. Dynamics of closed quantum systems: Unitary transformations, the Hamiltonian, and Schrödinger's equation.

An operator U on a Hilbert space $\mathcal{H}$ is **unitary** if
$$U^\dagger = U^{-1} .$$
Unitary operators are of central importance in quantum mechanics for many reasons. We list below only two:
- Closed quantum mechanical systems transform only via unitary transformations
- Unitary transformations preserve quantum probabilities

Let $|\psi(t)\rangle$ denote the state as a function of time t of a closed quantum mechanical system $\mathcal{Q}$. Then the dynamical behavior of the state of $\mathcal{Q}$ is determined by the **Schrödinger equation**
$$i\hbar \frac{\partial}{\partial t} |\psi(t)\rangle = H |\psi(t)\rangle ,$$
where $\hbar$ denotes **Planck's constant** divided by 2π, and where H denotes an observable of $\mathcal{Q}$ called the **Hamiltonian**. The Hamiltonian is the quantum mechanical analog of the Hamiltonian of classical mechanics. In classical physics, the Hamiltonian is the total energy of the system.

REMARK 8. *The dynamical behavior of non-closed quantum systems is much more complex. (See [**77**, Chapter 8].)*

---

[8]We have assumed units have been chosen such that $\hbar = 1$.

4.11. The mathematical perspective.

From the mathematical perspective, Schrödinger's equation is written as:

$$\frac{\partial}{\partial t} U(t) = -\frac{i}{\hbar} H(t) U(t),$$

where

$$|\psi(t)\rangle = U |\psi(0)\rangle,$$

and where $-\frac{i}{\hbar} H(t)$ is a skew-Hermitian operator lying in the Lie algebra of the unitary group. The solution is given by a multiplicative integral, called the **path-ordered integral**,

$$U(t) = {}_t\!\!\int\!\!{}_0\, e^{-\frac{i}{\hbar} H(t) dt},$$

which is taken over the path $-\frac{i}{\hbar} H(t)$ in the Lie algebra of the unitary group. The path-ordered integral is given by:

$$\begin{aligned}{}_t\!\!\int\!\!{}_0\, e^{-\frac{i}{\hbar} H(t) dt} &= \lim_{n \to \infty} \prod_{k=n}^{0} e^{-\frac{i}{\hbar} H(k \frac{t}{n}) \frac{t}{n}} \\ &= \lim_{n \to \infty} \left[e^{-\frac{i}{\hbar} H(n \cdot \frac{t}{n}) \frac{t}{n}} \cdot e^{-\frac{i}{\hbar} H((n-1) \cdot \frac{t}{n}) \frac{t}{n}} \cdot \ldots \cdot e^{-\frac{i}{\hbar} H(1 \cdot \frac{t}{n}) \frac{t}{n}} \cdot e^{-\frac{i}{\hbar} H(0 \cdot \frac{t}{n}) \frac{t}{n}} \right]\end{aligned}$$

REMARK 9. *The standard notation for the above path-ordered integral is*

$$\mathbf{P} \exp\left(-\frac{i}{\hbar} \int_0^t H(t) dt \right)$$

We prefer the elongated "P" notation for multiplicative integrals because it is similar to the elongated "S" notation for additive integrals.

If the Hamiltonian $H(t) = H$ is independent of time, then all matrices commute and the above path-ordered integral simplifies to

$${}_t\!\!\int\!\!{}_0\, e^{-\frac{i}{\hbar} H dt} = e^{\int_0^t -\frac{i}{\hbar} H dt} = e^{-\frac{i}{\hbar} H t}$$

Thus, in this case, $U(t)$ is nothing more than a one parameter subgroup of the unitary group.

5. The Density Operator

5.1. Introducing the density operator.

John von Neumann suggested yet another way of representing the state of a quantum system.

Let $|\psi\rangle$ be a unit length ket (i.e., $\langle \psi | \psi \rangle = 1$) in the Hilbert space $\mathcal{H}$ representing the state of a quantum system[9]. The **density operator** ρ associated with the state ket $|\psi\rangle$ is defined as the outer product of the ket $|\psi\rangle$ (which can be thought of as a column vector) with the bra $\langle\psi|$ (which can be thought of as a row vector), i.e.,

$$\rho = |\psi\rangle\langle\psi|$$

The density operator formalism has a number of advantages over the ket state formalism. One advantage is that the density operator can also be used to represent hybrid quantum/classical states, i.e., states which are a classical statistical mixture of quantum states. Such hybrid states may also be thought of as quantum states for which we have incomplete information.

For example, consider a quantum system which is in the states (each of unit length)

$$|\psi_1\rangle, |\psi_2\rangle, \ldots, |\psi_n\rangle$$

with probabilities

$$p_1, p_2, \ldots, p_n$$

respectively, where

$$p_1 + p_2 + \ldots + p_n = 1$$

(Please note that the states $|\psi_1\rangle, |\psi_2\rangle, \ldots, |\psi_n\rangle$ need not be orthogonal.) Then the density operator representation of this state is defined as

$$\rho = p_1 |\psi_1\rangle\langle\psi_1| + p_2 |\psi_2\rangle\langle\psi_2| + \ldots + p_n |\psi_n\rangle\langle\psi_n|$$

If a density operator ρ can be written in the form

$$\rho = |\psi\rangle\langle\psi|,$$

it is said to represent a **pure ensemble**. Otherwise, it is said to represent a **mixed ensemble**.

---

[9]Please recall that each of the kets in the set $\{ \lambda|\psi\rangle \mid \lambda \in \mathbb{C}, \lambda \neq 0 \}$ represent the same state of a quantum system. Hence, we can always (and usually do) represent the state of a quantum system as a unit normal ket, i.e., as a ket such that $\langle \psi | \psi \rangle = 1$.

5.2. Properties of density operators.

It can be shown that all density operators are positive semi-definite Hermitian operators of trace 1, and vice versa. As a result, we have the following crisp mathematical definition:

DEFINITION 3. *An linear operator on a Hilbert space $\mathcal{H}$ is a **density operator** if it is a positive semi-definite Hermitian operator of trace 1.*

It can be shown that a density operator represents a pure ensemble if and only if $\rho^2 = \rho$, or equivalently, if and only if $Trace(\rho^2) = 1$. For all ensembles, both pure and mixed, $Trace(\rho^2) \leq 1$.

From standard theorems in linear algebra, we know that, for every density operator ρ, there exists a unitary matrix U which **diagonalizes** ρ, i.e., such that $U\rho U^\dagger$ is a diagonal matrix. The diagonal entries in this matrix are, of course, the eigenvalues of ρ. These are non-negative real numbers which all sum to 1.

Finally, if we let $\mathcal{D}$ denote the set of all density operators for a Hilbert space $\mathcal{H}$, then $i\mathcal{D}$ is a convex subset of the Lie algebra of the unitary group associated with $\mathcal{H}$.

5.3. Quantum measurement in terms of density operators.

Let $a_0, a_1, \ldots, a_{n-1}$ denote all the distinct eigenvalues of an observable A, and let $P_{a_0}, P_{a_1}, \ldots, P_{a_{n-1}}$ be the corresponding projection operators. Finally, let $\mathcal{Q}$ be a quantum system with state given by the density operator ρ.

If the quantum system $\mathcal{Q}$ is measured with respect to the observable A, then with probability
$$p_i = Trace\left(P_{a_i}\rho\right)$$
the resulting measured eigenvalue is a_i, and the resulting state of $\mathcal{Q}$ is given by the density operator
$$\rho_i = \frac{P_{a_i}\rho P_{a_i}}{Trace\left(P_{a_i}\rho\right)} \;.$$

Moreover, for an observable A, the averaged observed eigenvalue expressed in terms of the density operator is:
$$\langle A \rangle = trace(\rho A)$$
Thus, we have extended the definition of $\langle A \rangle$ so that it applies to mixed as well as pure ensembles, i.e., generalized the following formula to mixed ensembles:
$$\langle A \rangle = \langle \psi \mid A \mid \psi \rangle = trace\left(|\psi\rangle\langle\psi| A\right) = trace(\rho A) \;.$$

EXERCISE 1. *Let the Hilbert space $\mathcal{H}$ and the observable $\mathcal{O}$ be as defined in **Example 1** on page 14. Let $\mathcal{Q}$ be a quantum system with state given by the density operator*

$$\rho = \begin{pmatrix} \frac{3}{8} & 0 & 0 & -\frac{3}{8} \\ 0 & \frac{1}{4} & 0 & 0 \\ 0 & 0 & 0 & 0 \\ -\frac{3}{8} & 0 & 0 & \frac{3}{8} \end{pmatrix}.$$

Assume that the quantum system $\mathcal{Q}$ is measured with respect to the observable $\mathcal{O}$.

For $j = 0, 1$, find the probability $Prob\,(Observing\,a_j)$ of observing the eigenvalue a_j, and find the corresponding state ρ_j of the measured $\mathcal{Q}$.

5.4. Some examples of density operators.

For example, consider the following mixed ensemble of the polarization state of a photon:

EXAMPLE 7.

Ket	$\vert\updownarrow\rangle$	$\vert\nearrow\rangle$
Prob.	$\frac{3}{4}$	$\frac{1}{4}$

In terms of the basis $\vert\leftrightarrow\rangle, \vert\updownarrow\rangle$ of the two dimensional Hilbert space $\mathcal{H}$, the density operator ρ of the above mixed ensemble can be written as:

$$\begin{aligned}
\rho &= \tfrac{3}{4}\vert\updownarrow\rangle\langle\updownarrow\vert + \tfrac{1}{4}\vert\nearrow\rangle\langle\nearrow\vert \\
&= \tfrac{3}{4}\begin{pmatrix}1\\0\end{pmatrix}\begin{pmatrix}1 & 0\end{pmatrix} + \tfrac{1}{4}\begin{pmatrix}1/\sqrt{2}\\1/\sqrt{2}\end{pmatrix}\begin{pmatrix}1/\sqrt{2} & 1/\sqrt{2}\end{pmatrix} \\
&= \tfrac{3}{4}\begin{pmatrix}1 & 0\\0 & 0\end{pmatrix} + \tfrac{1}{8}\begin{pmatrix}1 & 1\\1 & 1\end{pmatrix} = \begin{pmatrix}\frac{7}{8} & \frac{1}{8}\\ \frac{1}{8} & \frac{1}{8}\end{pmatrix}
\end{aligned}$$

EXAMPLE 8. *The following two **preparations** produce mixed ensembles with the same density operator:*

Ket	$\vert\updownarrow\rangle$	$\vert\leftrightarrow\rangle$
Prob.	$\frac{1}{2}$	$\frac{1}{2}$

and

Ket	$\vert\nearrow\rangle$	$\vert\nwarrow\rangle$
Prob.	$\frac{1}{2}$	$\frac{1}{2}$

For, for the left preparation, we have

$$\begin{aligned}
\rho &= \tfrac{1}{2}\vert\updownarrow\rangle\langle\updownarrow\vert + \tfrac{1}{2}\vert\leftrightarrow\rangle\langle\leftrightarrow\vert \\
&= \tfrac{1}{2}\begin{pmatrix}1\\0\end{pmatrix}\begin{pmatrix}1 & 0\end{pmatrix} + \tfrac{1}{2}\begin{pmatrix}0\\1\end{pmatrix}\begin{pmatrix}0 & 1\end{pmatrix} \\
&= \tfrac{1}{2}\begin{pmatrix}1 & 0\\0 & 1\end{pmatrix}
\end{aligned}$$

And for the right preparation, we have

$$\begin{aligned}\rho &= \tfrac{1}{2}|\nearrow\rangle\langle\nearrow| + \tfrac{1}{2}|\nwarrow\rangle\langle\nwarrow| \\ &= \tfrac{1}{2}\tfrac{1}{\sqrt{2}}\begin{pmatrix}1\\1\end{pmatrix}\tfrac{1}{\sqrt{2}}\begin{pmatrix}1 & 1\end{pmatrix} + \tfrac{1}{2}\tfrac{1}{\sqrt{2}}\begin{pmatrix}1\\-1\end{pmatrix}\tfrac{1}{\sqrt{2}}\begin{pmatrix}1 & -1\end{pmatrix} \\ &= \tfrac{1}{4}\begin{pmatrix}1 & 1\\1 & 1\end{pmatrix} + \tfrac{1}{4}\begin{pmatrix}1 & -1\\-1 & 1\end{pmatrix} = \tfrac{1}{2}\begin{pmatrix}1 & 0\\0 & 1\end{pmatrix}\end{aligned}$$

There is no way of physically distinguishing the above two mixed ensembles which were prepared in two entirely different ways. For the density operator represents all that can be known about the state of the quantum system.

5.5. The partial trace of a linear operator.

In order to deal with a quantum system composed of many quantum subsystems, we need to define the partial trace.

Let
$$\mathcal{O}: \mathcal{H} \longrightarrow \mathcal{H} \in Hom_\mathbb{C}(\mathcal{H},\mathcal{H})$$
be a linear operator on the Hilbert space $\mathcal{H}$.

Since Hilbert spaces are free algebraic objects, it follows from standard results in abstract algebra[10] that
$$Hom_\mathbb{C}(\mathcal{H},\mathcal{H}) \cong \mathcal{H} \otimes \mathcal{H}^*,$$
where we recall that
$$\mathcal{H}^* = Hom_\mathbb{C}(\mathcal{H},\mathbb{C}).$$

It follows that such an operator $\mathcal{O}$ can be written in the form
$$\mathcal{O} = \sum_\alpha a_\alpha |h_\alpha\rangle \otimes \langle k_\alpha|,$$
where the kets $|h_\alpha\rangle$ lie in $\mathcal{H}$ and the bras $\langle k_\alpha|$ lie in $\mathcal{H}^\dagger$.

Thus, the standard **trace** of a linear operator
$$Trace: Hom_\mathbb{C}(\mathcal{H},\mathcal{H}) \longrightarrow \mathbb{C}$$
is nothing more than a contraction, i.e.,
$$Trace(\mathcal{O}) = \sum_\alpha a_\alpha \langle k_\alpha | h_\alpha \rangle,$$
i.e., a replacement of each outer product $|h_\alpha\rangle \otimes \langle k_\alpha|$ by the corresponding bracket

---
[10] See for example [**61**].

$\langle k_\alpha | h_\alpha \rangle$.

We can generalize the *Trace* as follows:

Let $\mathcal{H}$ now be the tensor product of Hilbert spaces $\mathcal{H}_1, \mathcal{H}_2, \ldots, \mathcal{H}_n$, i.e.,

$$\mathcal{H} = \bigotimes_{j=1}^{n} \mathcal{H}_j .$$

It follows once again from standard results in abstract algebra that

$$Hom_{\mathbb{C}}(\mathcal{H}, \mathcal{H}) \cong \bigotimes_{j=1}^{n} \left(\mathcal{H}_j \otimes \mathcal{H}_j^* \right) .$$

Hence, the operator $\mathcal{O}$ can be written in the form

$$\mathcal{O} = \sum_{\alpha} a_\alpha \bigotimes_{j=1}^{n} |h_{\alpha,j}\rangle \otimes \langle k_{\alpha,j}| ,$$

where, for each j, the kets $|h_{\alpha,j}\rangle$ lie in $\mathcal{H}_j$ and the bras $\langle k_{\alpha,j}|$ lie in $\mathcal{H}_j^*$ for all α.

Next we note that, for every subset $\mathcal{I}$ of the set of indices $\mathcal{J} = \{1, 2, \ldots, n\}$, we can define the **partial trace** over $\mathcal{I}$, written

$$Trace_{\mathcal{I}} : Hom_{\mathbb{C}} \left(\bigotimes_{j \in \mathcal{J}} \mathcal{H}_j, \bigotimes_{j \in \mathcal{J}} \mathcal{H}_j \right) \longrightarrow Hom_{\mathbb{C}} \left(\bigotimes_{j \in \mathcal{J}-\mathcal{I}} \mathcal{H}_j, \bigotimes_{j \in \mathcal{J}-\mathcal{I}} \mathcal{H}_j \right) ,$$

as the contraction on the indices $\mathcal{I}$, i.e.,

$$Trace_{\mathcal{I}}(\mathcal{O}) = \sum_{\alpha} a_\alpha \left(\prod_{j \in \mathcal{I}} \langle k_{\alpha,j} | h_{\alpha,j} \rangle \right) \bigotimes_{j \in \mathcal{J}-\mathcal{I}} |h_{\alpha,j}\rangle \langle k_{\alpha,j}| .$$

For example, let $\mathcal{H}_1$ and $\mathcal{H}_0$ be two dimensional Hilbert spaces with selected orthonormal bases $\{|0_1\rangle, |1_1\rangle\}$ and $\{|0_0\rangle, |1_0\rangle\}$, respectively. Thus, $\{|0_1 0_0\rangle, |0_1 1_0\rangle, |1_1 0_0\rangle, |1_1 1_0\rangle\}$ is an orthonormal basis of $\mathcal{H} = \mathcal{H}_1 \otimes \mathcal{H}_0$.

Let $\rho \in Hom_{\mathbb{C}}(\mathcal{H}, \mathcal{H})$ be the operator

$$\rho = \left(\frac{|0_1 0_0\rangle - |1_1 1_0\rangle}{\sqrt{2}} \right) \otimes \left(\frac{\langle 0_1 0_0| - \langle 1_1 1_0|}{\sqrt{2}} \right)$$

$$= \frac{1}{2} \left(|0_1 0_0\rangle \langle 0_1 0_0| - |0_1 0_0\rangle \langle 1_1 1_0| - |1_1 1_0\rangle \langle 0_1 0_0| + |1_1 1_0\rangle \langle 1_1 1_0| \right)$$

which in terms of the basis $\{|0_1 0_0\rangle, |0_1 1_0\rangle, |1_1 0_0\rangle, |1_1 1_0\rangle\}$ can be written as the matrix

$$\rho = \frac{1}{2} \begin{pmatrix} 1 & 0 & 0 & -1 \\ 0 & 0 & 0 & 0 \\ 0 & 0 & 0 & 0 \\ -1 & 0 & 0 & 1 \end{pmatrix} ,$$

where the rows and columns are listed in the order $|0_1 0_0\rangle, |0_1 1_0\rangle, |1_1 0_0\rangle, |1_1 1_0\rangle$

The partial trace $Trace_0$ with respect to $\mathcal{I} = \{0\}$ of ρ is

$$\rho_1 = Trace_0(\rho)$$
$$= \frac{1}{2} Trace_0 \left(|0_1 0_0\rangle \langle 0_1 0_0| - |0_1 0_0\rangle \langle 1_1 1_0| - |1_1 1_0\rangle \langle 0_1 0_0| + |1_1 1_0\rangle \langle 1_1 1_0| \right)$$
$$= \frac{1}{2} \left(\langle 0_0 | 0_0 \rangle |0_1\rangle \langle 0_1| - \langle 1_0 | 0_0 \rangle |0_1\rangle \langle 1_1| - \langle 0_0 | 1_0 \rangle |1_1\rangle \langle 0_1| + \langle 1_0 | 1_0 \rangle |1_1\rangle \langle 1_1| \right)$$
$$= \frac{1}{2} \left(|0_1\rangle \langle 0_1| - |0_1\rangle \langle 1_1| \right)$$

which in terms of the basis $\{|0_1\rangle, |1_1\rangle\}$ becomes

$$\rho_1 = Trace_0(\rho) = \frac{1}{2} \begin{pmatrix} 1 & 0 \\ 0 & 1 \end{pmatrix},$$

where the rows and columns are listed in the order $|0_1\rangle, |1_1\rangle$.

5.6. Multipartite quantum systems.

One advantage density operators have over kets is that they provide us with a means for dealing with multipartite quantum systems.

DEFINITION 4. *Let $\mathcal{Q}_1, \mathcal{Q}_2, \ldots, \mathcal{Q}_n$ be quantum systems with underlying Hilbert spaces $\mathcal{H}_1, \mathcal{H}_2, \ldots, \mathcal{H}_n$, respectively. The global quantum system $\mathcal{Q}$ consisting of the quantum systems $\mathcal{Q}_1, \mathcal{Q}_2, \ldots, \mathcal{Q}_n$ is called a **multipartite quantum system**. Each of the quantum systems $\mathcal{Q}_j$ ($j = 1, 2, \ldots, n$) is called a **constituent "part"** of $\mathcal{Q}$. The underlying Hilbert space $\mathcal{H}$ of $\mathcal{Q}$ is the tensor product of the Hilbert spaces of the constituent "parts," i.e.,*

$$\mathcal{H} = \bigotimes_{j=1}^{n} \mathcal{H}_j .$$

If the density operator ρ is the state of a multipartite quantum system $\mathcal{Q}$, then the state of each constituent "part" $\mathcal{Q}_j$ is the density operator ρ_j given by the partial trace

$$\rho_j = Trace_{\mathcal{J} - \{j\}} (\rho) ,$$

where $\mathcal{J} = \{1, 2, \ldots, n\}$ is the set of indices.

Obviously, much more can be said about the states of multipartite systems and their constituent parts. However, we will forego that discussion until after we have had an opportunity introduce the concepts of quantum entanglement and von Neumann entropy.

5.7. Quantum dynamics in density operator formalism.

Under a unitary transformation U, a density operator ρ transforms according to the rubric:
$$\rho \longmapsto U\rho U^{\dagger}$$
Moreover, in terms of the density operator, Schrödinger's equation[11] becomes:
$$i\hbar \frac{\partial \rho}{\partial t} = [H, \rho] ,$$
where $[H, \rho]$ denotes the **commutator** of H and ρ, i.e.,
$$[H, \rho] = H\rho - \rho H$$

5.8. The mathematical perspective.

From the mathematical perspective, one works with $i\rho$ instead of ρ because $i\rho$ lies in the Lie algebra of the unitary group. Thus, the density operator transforms under a unitary transformation U according to the rubric:
$$i\rho \longmapsto Ad_U(i\rho) ,$$
where Ad_U denotes the **big adjoint representation**[12], i.e., the representation
$$i\rho \longmapsto U(i\rho) U^{-1}$$

From the mathematical perspective, Schrödinger's equation is in this case more informatively written as:
$$\frac{\partial (i\rho)}{\partial t} = ad_{-\frac{i}{\hbar}H}(i\rho) ,$$
where $ad_{-\frac{i}{\hbar}H}$ denotes the **little adjoint representation**[13] , i.e.,
$$ad_{-\frac{i}{\hbar}H}(i\rho) = \left[-\frac{i}{\hbar}H, i\rho\right] = \left(-\frac{i}{\hbar}H\right)(i\rho) - (i\rho)\left(-\frac{i}{\hbar}H\right) .$$
Thus, the solution to the above form of Schrödinger's equation is given by the path ordered integral:
$$\rho = \left({}_t\mathcal{S}_0 \, e^{-\frac{1}{\hbar}\left(ad_{iH(t)}\right)dt} \right) \rho_0$$
where ρ_0 denotes the density operator at time $t = 0$.

---

[11]Schrödinger's equation determines the dynamics of closed quantum systems. However, non-closed quantum systems are also of importance in quantum computation and quantum information theory. See for example the Schumacher's work on superoperators [**85**], or [**77**, Chapter 8].

[12]For a more in depth discussion of the big adjoint representation, see[**67**].

[13]For a more in depth discussion of the little adjoint representation, see [**67**].

6. The Heisenberg model of quantum mechanics

Consider a computing device with inputs and outputs. Assume we have no knowledge of the internal workings of the device. We are allowed to probe the device with inputs and observe the corresponding outputs. But we are given no information as to how the device performs its calculation. We call such a device a **blackbox** computing device.

For such blackboxes, we say that two theoretical models for blackboxes are **equivalent** provided both predict the same input/output behavior. We may prefer one model over the other for various reasons, such as simplicity, aesthetics, or whatever meets our fancy. But the basic fact is that each of the two equivalent models is just as "correct" as the other.

In like manner, two theoretical models of the quantum world are said to be **equivalent** if they both predict the same results in regard to quantum measurements.

Up to this point, we have been describing the Schrödinger model of quantum mechanics, frequently called the **Schrödinger picture**. Heisenberg proposed yet another model, called the **Heisenberg picture**. Both models have been proven to be equivalent.

In the Heisenberg picture, state kets remain stationary with time, but observables move with time. While state kets, and hence density operators, remain fixed with respect to time, the observables A change dynamically as:

$$A \longmapsto U^\dagger A U$$

under a unitary transformation $U = U(t)$, where the unitary transformation is determined by the equation

$$i\hbar \frac{\partial U}{\partial t} = HU$$

It follows that the equation of motion of observables is according to the following equation

$$i\hbar \frac{\partial A}{\partial t} = [A, H]$$

One advantage the Heisenberg picture has over the Schrödinger picture is that the equations appearing in it are similar to those found in classical mechanics.

In summary, we have the following table which contrasts the two pictures:

	Schrödinger Picture	Heisenberg Picture
State ket	Moving $$|\psi_0\rangle \longmapsto |\psi\rangle = U|\psi_0\rangle$$	Stationary $$|\psi_0\rangle$$
Density Operator	Moving $$\rho_0 \longmapsto \rho = U\rho_0 U^\dagger = Ad_U(\rho_0)$$	Stationary $$\rho_0$$
Observable	Stationary $$A_0$$	Moving $$A_0 \longmapsto A = U^\dagger A_0 U = Ad_{U^\dagger}(A_0)$$
Observable Eigenvalues	Stationary $$a_j$$	Stationary $$a_j$$
Observable Frame	Stationary $$A_0 = \sum_j a_j |a_j\rangle_0 \langle a_j|_0$$	Moving $$A_0 = \sum_j a_j |a_j\rangle_0 \langle a_j|_0$$ $$\longmapsto$$ $$A_t = \sum_j a_j |a_j\rangle_t \langle a_j|_t$$ where $|a_j\rangle_t = U^\dagger |a_j\rangle_0$
Dynamical Equations	$$i\hbar \frac{\partial U}{\partial t} = H^{(S)} U$$ $$i\hbar \frac{\partial}{\partial t} |\psi\rangle = H^{(S)} |\psi\rangle$$	$$i\hbar \frac{\partial U}{\partial t} = H^{(H)} U$$ $$i\hbar \frac{\partial A}{\partial t} = [A, H^{(H)}]$$
Measurement	Measurement of observable A_0 produces eigenvalue a_j with probability $$\left|(\langle a_j|_0)|\psi\rangle\right|^2 = \left|(\langle a_j|_0)|\psi\rangle\right|^2$$	Measurement of observable A produces eigenvalue a_j with probability $$\left|(\langle a_j|_t)|\psi_0\rangle\right|^2 = \left|(\langle a_j|_0)|\psi\rangle\right|^2$$

where
$$H^{(H)} = U^\dagger H^{(S)} U$$
It follows that the Schrödinger Hamiltonian $H^{(S)}$ and the Heisenberg Hamiltonian are related as follows:
$$\frac{\partial H^{(S)}}{\partial t} = U \frac{\partial H^{(H)}}{\partial t} U^\dagger,$$
where terms containing $\frac{\partial U}{\partial t}$ and $\frac{\partial U^\dagger}{\partial t}$ have cancelled out as a result of the Schrödinger equation.

We should also mention that the Schrödinger and Heisenberg pictures can be transformed into one another via the mappings:

$S \longrightarrow H$	$H \longrightarrow S$
$\left|\psi^{(S)}\right\rangle \longmapsto \left|\psi^{(H)}\right\rangle = U^\dagger \left|\psi^{(S)}\right\rangle$	$\left|\psi^{(H)}\right\rangle \longmapsto \left|\psi^{(S)}\right\rangle = U \left|\psi^{(H)}\right\rangle$
$\rho^{(S)} \longmapsto \rho^{(H)} = U^\dagger \rho^{(S)} U$	$\rho^{(H)} \longmapsto \rho^{(S)} = U \rho^{(H)} U^\dagger$
$A^{(S)} \longmapsto A^{(H)} = U^\dagger A^{(S)} U$	$A^{(H)} \longmapsto A^{(S)} = U A^{(H)} U^\dagger$
$A^{(S)} \longmapsto A^{(H)} = U^\dagger A^{(S)} U$	$A^{(H)} \longmapsto A^{(S)} = U A^{(H)} U^\dagger$

Obviously, much more could be said on this topic.

For quantum computation from the perspective of the Heisenberg model, please refer to the work of Deutsch and Hayden[26], and also to Gottesman's "study of the ancient Hittites" :-) [34].

7. Quantum entanglement

7.1. The juxtaposition of two quantum systems.

Let $\mathcal{Q}_1$ and $\mathcal{Q}_2$ be two quantum systems that have been separately prepared respectively in states $|\psi_1\rangle$ and $|\psi_2\rangle$, and that then have been united without interacting. Because $\mathcal{Q}_1$ and $\mathcal{Q}_2$ have been separately prepared without interacting, their states $|\psi_1\rangle$ and $|\psi_2\rangle$ respectively lie in distinct Hilbert spaces $\mathcal{H}_1$ and $\mathcal{H}_2$. Moreover, because of the way in which $\mathcal{Q}_1$ and $\mathcal{Q}_2$ have been prepared, no physical prediction relating to one of these quantum systems depends in any way whatsoever on the other quantum system.

The global quantum system $\mathcal{Q}$ consisting of the two quantum systems $\mathcal{Q}_1$ and $\mathcal{Q}_2$ as prepared above is called a **juxtaposition** of the quantum systems $\mathcal{Q}_1$ and $\mathcal{Q}_2$. The state of the global quantum system $\mathcal{Q}$ is the tensor product of the states $|\psi_1\rangle$ and $|\psi_2\rangle$. In other words, the state of $\mathcal{Q}$ is:

$$|\psi_1\rangle \otimes |\psi_2\rangle \in \mathcal{H}_1 \otimes \mathcal{H}_2$$

7.2. An example: An n-qubit register $\mathcal{Q}$ consisting of the juxtaposition of n qubits.

Let $\mathcal{H}$ be a two dimensional Hilbert space, and let $\{|0\rangle, |1\rangle\}$ denote an arbitrarily selected orthonormal basis[14]. Let $\mathcal{H}_{n-1}, \mathcal{H}_{n-2}, \ldots, \mathcal{H}_0$ be distinct Hilbert spaces, each isomorphic to $\mathcal{H}$, with the obvious induced orthonormal bases

$$\{|0_{n-1}\rangle, |1_{n-1}\rangle\}, \{|0_{n-2}\rangle, |1_{n-2}\rangle\}, \ldots, \{|0_0\rangle, |1_0\rangle\}$$

respectively.

---

[14]We obviously have chosen to label the basis elements in a suggestive way.

Consider n qubits $\mathcal{Q}_{n-1}, \mathcal{Q}_{n-2}, \ldots, \mathcal{Q}_0$ separately prepared in the states
$$\frac{1}{\sqrt{2}}\left(|0_{n-1}\rangle + |1_{n-1}\rangle\right), \frac{1}{\sqrt{2}}\left(|0_{n-2}\rangle + |1_{n-2}\rangle\right), \ldots, \frac{1}{\sqrt{2}}\left(|0_0\rangle + |1_0\rangle\right),$$

respectively. Let $\mathcal{Q}$ denote the global system consisting of the separately prepared (without interacting) qubits $\mathcal{Q}_{n-1}, \mathcal{Q}_{n-2}, \ldots, \mathcal{Q}_0$. Then the state $|\psi\rangle$ of $\mathcal{Q}$ is the tensor product:

$$\begin{aligned}|\psi\rangle &= \frac{1}{\sqrt{2}}\left(|0_{n-1}\rangle + |1_{n-1}\rangle\right) \otimes \frac{1}{\sqrt{2}}\left(|0_{n-2}\rangle + |1_{n-2}\rangle\right) \otimes \ldots \otimes \frac{1}{\sqrt{2}}\left(|0_0\rangle + |1_0\rangle\right) \\ &= \left(\frac{1}{\sqrt{2}}\right)^n \left(|0_{n-1}0_{n-2}\ldots0_10_0\rangle + |0_{n-1}0_{n-2}\ldots0_11_0\rangle + \ldots + |1_{n-1}1_{n-2}\ldots1_11_0\rangle\right)\end{aligned}$$

which lies in the Hilbert space
$$\mathcal{H} = \mathcal{H}_{n-1} \otimes \mathcal{H}_{n-2} \otimes \ldots \otimes \mathcal{H}_0.$$

Notational Convention: We will usually omit subscripts whenever they can easily be inferred from context.

Thus, the global system $\mathcal{Q}$ consisting of the n qubits $\mathcal{Q}_{n-1}, \mathcal{Q}_{n-2}, \ldots, \mathcal{Q}_0$ is in the state

$$|\psi\rangle = \left(\frac{1}{\sqrt{2}}\right)^n \left(|00\ldots00\rangle + |00\ldots01\rangle + \ldots + |11\ldots11\rangle\right) \in \bigotimes_{0}^{n-1} \mathcal{H}$$

The reader should note that the n-qubit register $\mathcal{Q}$ is a superposition of kets with labels consisting of all the binary n-tuples. If each binary n-tuple $b_{n-1}b_{n-2}\ldots b_0$ is identified with the integer

$$b_{n-1}2^{n-1} + b_{n-2}2^{n-2} + \ldots + b_0 2^0 ,$$

i.e., if we interpret each binary n-tuple as the radix 2 representation of an integer, then we can rewrite the state as

$$|\psi\rangle = \left(\frac{1}{\sqrt{2}}\right)^n \left(|0\rangle + |1\rangle + |2\rangle + \ldots + |2^n - 1\rangle\right).$$

In other words, this n-qubit register contains all the integers from 0 to $2^n - 1$ in superposition. But most importantly, it contains all the integers 0 to $2^n - 1$ *simultaneously*!

This is an example of the massive parallelism that is possible within quantum computation. However, there is a downside. If we observe (measure) the register, then all the massive parallelism disappears. On measurement, the quantum world selects for us one and only one of the 2^n integers. The probability of observing any particular one of the integers is $\left|(1/\sqrt{2})^n\right|^2 = (\frac{1}{2})^n$. The selection of which integer is observed is unfortunately not made by us, but by the quantum world.

Thus, harnessing the massive parallelism of quantum mechanics is no easy task! As we will see, a more subtle approach is required.

7.3. An example of the dynamic behavior of a 2-qubit register.

We now consider the previous n-qubit register for $n = 2$. In terms of the bases described in the previous section, we have:

$$\begin{cases} |0\rangle = |00\rangle = \begin{pmatrix} 1 \\ 0 \end{pmatrix} \otimes \begin{pmatrix} 1 \\ 0 \end{pmatrix} = \begin{pmatrix} 1 \\ 0 \\ 0 \\ 0 \end{pmatrix} \\ \\ |1\rangle = |01\rangle = \begin{pmatrix} 1 \\ 0 \end{pmatrix} \otimes \begin{pmatrix} 0 \\ 1 \end{pmatrix} = \begin{pmatrix} 0 \\ 1 \\ 0 \\ 0 \end{pmatrix} \\ \\ = \\ \\ |2\rangle = |10\rangle = \begin{pmatrix} 0 \\ 1 \end{pmatrix} \otimes \begin{pmatrix} 1 \\ 0 \end{pmatrix} = \begin{pmatrix} 0 \\ 0 \\ 1 \\ 0 \end{pmatrix} \\ \\ |3\rangle = |11\rangle = \begin{pmatrix} 0 \\ 1 \end{pmatrix} \otimes \begin{pmatrix} 0 \\ 1 \end{pmatrix} = \begin{pmatrix} 0 \\ 0 \\ 0 \\ 1 \end{pmatrix} \end{cases}$$

Let us assume that the initial state $|\psi\rangle_{t=0}$ of our 2-qubit register is

$$|\psi\rangle_{t=0} = \left(\frac{|0\rangle - |1\rangle}{\sqrt{2}}\right) \otimes |0\rangle = \frac{1}{\sqrt{2}}(|00\rangle - |10\rangle) = \frac{1}{\sqrt{2}}(|0\rangle - |2\rangle) = \frac{1}{\sqrt{2}}\begin{pmatrix} 1 \\ 0 \\ -1 \\ 0 \end{pmatrix}$$

Let us also assume that from time $t = 0$ to time $t = 1$ the dynamical behavior of the above 2-qubit register is determined by a constant Hamiltonian H, which when written in terms of the basis $\{|00\rangle, |01\rangle, |10\rangle, |11\rangle\} = \{|0\rangle, |1\rangle, |2\rangle, |3\rangle\}$ is given by

$$H = \frac{\pi \hbar}{2} \begin{pmatrix} 0 & 0 & 0 & 0 \\ 0 & 0 & 0 & 0 \\ 0 & 0 & 1 & -1 \\ 0 & 0 & -1 & 1 \end{pmatrix},$$

where the rows and the columns are listed in the order $|00\rangle, |01\rangle, |10\rangle, |11\rangle$, i.e., in the order $|0\rangle, |1\rangle, |2\rangle, |3\rangle$.

Then, as a consequence of Schrödinger's equation, the Hamiltonian H determines a unitary transformation

$$U_{CNOT} = {}_t\mathcal{S}_0 \, e^{-\frac{i}{\hbar}Hdt} = e^{\int_0^1 -\frac{i}{\hbar}Hdt} = e^{-\frac{i}{\hbar}H}$$

$$= \begin{pmatrix} 1 & 0 & 0 & 0 \\ 0 & 1 & 0 & 0 \\ 0 & 0 & 0 & 1 \\ 0 & 0 & 1 & 0 \end{pmatrix} = |0\rangle\langle 0| + |1\rangle\langle 1| + |2\rangle\langle 3| + |3\rangle\langle 2|$$

which moves the 2-qubit register from the initial state $|\psi\rangle_{t=0}$ at time $t = 0$ to $|\psi\rangle_{t=1} = U_{CNOT}|\psi\rangle_{t=0}$ at time $t = 1$. Thus,

$$|\psi\rangle_{t=1} = U_{CNOT}|\psi\rangle_{t=0} = \begin{pmatrix} 1 & 0 & 0 & 0 \\ 0 & 1 & 0 & 0 \\ 0 & 0 & 0 & 1 \\ 0 & 0 & 1 & 0 \end{pmatrix} \cdot \frac{1}{\sqrt{2}} \begin{pmatrix} 1 \\ 0 \\ -1 \\ 0 \end{pmatrix}$$

$$= \frac{1}{\sqrt{2}} \begin{pmatrix} 1 \\ 0 \\ 0 \\ -1 \end{pmatrix} = \frac{1}{\sqrt{2}}(|00\rangle - |11\rangle) = \frac{1}{\sqrt{2}}(|0\rangle - |3\rangle)$$

The resulting state (called an **EPR pair** of qubits for reasons we shall later explain) can no longer be written as a tensor product of two states. Consequently, we no longer have the juxtaposition of two qubits.

Somehow, the resulting two qubits have in some sense "lost their separate identities." Measurement of any one of the qubits immediately impacts the other.

For example, if we measure the 0-th qubit (i.e., the right-most qubit), the EPR state in some sense "jumps" to one of two possible states. Each of the two possibilities occurs with probability $\frac{1}{2}$, as indicated in the table below:

$\frac{1}{\sqrt{2}}(|0_1 0_0\rangle - |1_1 1_0\rangle)$	
↙↙↙ Meas. 0-th Qubit ↘↘↘	
$Prob = \frac{1}{2}$ $\|0_1 0_0\rangle$	$Prob = \frac{1}{2}$ $\|1_1 1_0\rangle$

So we see that a measurement of one of the qubits causes a change in the other.

7.4. Definition of quantum entanglement.

The above mentioned phenomenon is so unusual and so non-classical that it warrants a name.

DEFINITION 5. *Let $Q_1, Q_2, \ldots, Q_n$ be quantum systems with underlying Hilbert spaces $\mathcal{H}_1, \mathcal{H}_2, \ldots, \mathcal{H}_n$, respectively. Then the global quantum system Q consisting of the quantum systems $Q_1, Q_2, \ldots, Q_n$ is said to be **entangled** if its state $|\psi\rangle \in \mathcal{H} = \bigotimes_{j=1}^{n} \mathcal{H}_j$ can not be written in the form*

$$|\psi\rangle = \bigotimes_{j=1}^{n} |\psi_j\rangle ,$$

*where each ket $|\psi_j\rangle$ lies in the Hilbert space $\mathcal{H}_j$ for, $j = 1, 2, \ldots, n$. We also say that such a state $|\psi\rangle$ is **entangled**.*

Thus, the state

$$|\psi\rangle_{t=1} = \frac{1}{\sqrt{2}} (|00\rangle - |11\rangle)$$

of the 2-qubit register of the previous section is entangled.

REMARK 10. *In terms of density operator formalism, a pure ensemble ρ is entangled if it can not be written in the form*

$$\rho = \bigotimes_{j=1}^{n} \rho_j ,$$

where the ρ_j's denote density operators.

Please note that we have defined entanglement only for pure ensembles. For mixed ensembles, entanglement is not well understood[15]. As a result, the "right" definition of entanglement of mixed ensembles is still unresolved. We give one definition below:

DEFINITION 6. *A density operator ρ on a Hilbert space $\mathcal{H}$ is said to be entangled with respect to the Hilbert space decomposition*

$$\mathcal{H} = \bigotimes_{j=1}^{n} \mathcal{H}_j$$

if it can not be written in the form

$$\rho = \sum_{k=1}^{\ell} \lambda_k \left(\bigotimes_{j=1}^{n} \rho_{(j,k)} \right) ,$$

for some positive integer ℓ, where the λ_k's are positive real numbers such that

$$\sum_{k=1}^{\ell} \lambda_k = 1 .$$

and where each $\rho_{(j,k)}$ is a density operator on the Hilbert space $\mathcal{H}_j$.

---

[15]Quantum entanglement is not even well understood for pure ensembles.

Readers interested in pursuing this topic further should refer to the works of Bennett, the Horodecki's, Nielsen, Smolin, Wootters, and others[8], [51], [65], [76], [1].

7.5. Einstein, Podolsky, Rosen's (EPR's) grand challenge to quantum mechanics.

Albert Einstein was skeptical of quantum mechanics, so skeptical that he together with Podolsky and Rosen wrote a joint paper[29] appearing in 1935 challenging the very foundations of quantum mechanics. Their paper hit the scientific community like a bombshell. For it delivered a direct frontal attack at the very heart and center of quantum mechanics.

At the core of their objection was quantum entanglement. Einstein and his colleagues had insightfully recognized the central importance of this quantum phenomenon.

Their argument centered around the fact that quantum mechanics violated either the **principle of non-locality**[16] or the **principle of reality**[17]. They argued that, as a result, quantum mechanics must be incomplete, and that quantum entanglement could be explained by missing **hidden variables**.

For many years, no one was able to conceive of an experiment that could determine which of the two theories, i.e., quantum mechanics or EPR's hidden variable theory, was correct. In fact, many believed that the two theories were not distinguishable on physical grounds.

It was not until Bell developed his famous inequalities [5],[6], [15], that a physical criterion was found to distinguish the two theories. Bell developed inequalities which, if violated, would clearly prove that quantum mechanics is correct, and hidden variable theories are not. Many experiments were performed[18]. Each emphatically supported quantum mechanics, and clearly demonstrated the incorrectness of hidden variable theory. Quantum mechanics was the victor!

7.6. Why did Einstein, Podolsky, Rosen (EPR) object?

But why did Einstein and his colleagues object so vehemently to quantum entanglement?

As a preamble to our answer to this question, we note that Einstein and his colleagues were convinced of the validity of the following two physical principles:
1) The **principle of local interactions**, i.e., that all the known forces of nature are local interactions,
2) The **principle of non-locality**, i.e., that spacelike separated regions of spacetime are physically independent of one another.

---

[16] We will later explain the principle of non-locality. See also [15].

[17] For an explanation of the principle of reality as well as the principle of non-localty, please refer, for example, to [81], [15].

[18] See for example [2].

Their conviction in regard to principle 1) was based on the fact that all four known forces of nature, i.e., gravitational, electromagnetic, weak, and strong forces, are **local interactions**. By this we mean:

i) They are mediated by another entity, e.g., graviton, photon, etc.
ii) They propagate no faster than the speed c of light
iii) Their strength drops off with distance

Their conviction in regard to principle 2) was based on the following reasoning:

Two points in spacetime $P_1 = (x_1, y_1, z_1, t_1)$ and $P_2 = (x_2, y_2, z_2, t_2)$ are separated by a **spacelike distance** provided the distance between (x_1, y_1, z_1) and (x_2, y_2, z_2) is greater than $c|t_2 - t_1|$, i.e.,

$$Distance\left((x_1, y_1, z_1), (x_2, y_2, z_2)\right) > c|t_2 - t_1|,$$

where c denotes the speed of light. In other words, no signal can travel between points that are said to be separated by a spacelike distance unless the signal travels faster than the speed of light. But because of the basic principles of relativity, such superluminal communication is not possible.

Hence we have:

The principle of non-locality: Spacelike separated regions of spacetime are physically independent. In other words, spacelike separated regions can not influence one another.

7.7. EPR's objection.

We now are ready to explain why Einstein and his colleagues objected so vehemently to quantum entanglement. We explain Bohm's simplified version of their argument.

Consider a two qubit quantum system that has been prepared by **Alice**[19] in her laboratory in the state

$$|\psi\rangle = \frac{1}{\sqrt{2}}\left(|0_1 0_0\rangle - |1_1 1_0\rangle\right).$$

After the preparation, she decides to keep qubit #1 in her laboratory, but enlists Captain James T. Kirk of the Starship Enterprise to transport qubit #0 to her friend **Bob**[20] who is at some far removed distant part of the universe, such as at a Federation outpost orbiting about the double star Alpha Centauri in the constellation Centaurus.

After Captain Kirk has delivered qubit #0, Alice's two qubits are now separated by a spacelike distance. Qubit #1 is located in her Earth based laboratory. Qubits #0 is located with Bob at a Federation outpost orbiting Alpha Centauri. But the

---

[19] Alice is a well known personality in quantum computation, quantum cryptography, and quantum information theory.

[20] Bob is another well known personality in quantum computation, quantum cryptography, and quantum information theory.

two qubits are still entangled, even in spite of the fact that they are separated by a spacelike distance.

If Alice now measures qubit #1 (which is located in her Earth based laboratory), then the principles of quantum mechanics force her to conclude that instantly, without any time lapse, both qubits are "effected." As a result of the measurement, both qubits will be either in the state $|0_1 0_0\rangle$ or the state $|1_1 1_0\rangle$, each possibility occurring with probability 1/2.

This is a non-local "interaction." For,

- The "interaction" occurred without the presence of any force. It was not mediated by anything.
- The measurement produced an instantaneous change, which was certainly faster than the speed of light.
- The strength of the "effect" of the measurement did not drop off with distance.

No wonder Einstein was highly skeptical of quantum entanglement. Yet puzzlingly enough, since no information is exchanged by the process, the principles of general relativity are not violated. As a result, such an "effect" can not be used for superluminal communication.

For a more in-depth discussion of the EPR paradox and the foundations of quantum mechanics, the reader should refer to [**15**].

7.8. Quantum entanglement: The Lie group perspective.

Many aspects of quantum entanglement can naturally be captured in terms of Lie groups and their Lie algebras.

Let
$$\mathcal{H} = \mathcal{H}_{n-1} \otimes \mathcal{H}_{n-2} \otimes \ldots \otimes \mathcal{H}_0 = \bigotimes_{0}^{n-1} \mathcal{H}_j$$
be a decomposition of a Hilbert space $\mathcal{H}$ into the tensor product of the Hilbert spaces $\mathcal{H}_{n-1}, \mathcal{H}_{n-2}, \ldots, \mathcal{H}_0$. Let $\mathbb{U} = \mathbb{U}(\mathcal{H})$, $\mathbb{U}_{n-1} = \mathbb{U}(\mathcal{H}_{n-1})$, $\mathbb{U}_{n-2} = \mathbb{U}(\mathcal{H}_{n-2})$, $\ldots, \mathbb{U}_0 = \mathbb{U}(\mathcal{H}_0)$, denote respectively the Lie groups of all unitary transformations on $\mathcal{H}, \mathcal{H}_{n-1}, \mathcal{H}_{n-2}, \ldots, \mathcal{H}_0$. Moreover, let $\mathbf{u} = \mathbf{u}(\mathcal{H})$, $\mathbf{u}_{n-1} = \mathbf{u}_{n-1}(\mathcal{H}_{n-1})$, $\mathbf{u}_{n-2} = \mathbf{u}_{n-2}(\mathcal{H}_{n-2}), \ldots, \mathbf{u}_0 = \mathbf{u}_0(\mathcal{H}_0)$ denote the corresponding Lie algebras.

DEFINITION 7. *The **local subgroup*** $\mathbb{L} = \mathbb{L}(\mathcal{H})$ *of* $\mathbb{U} = \mathbb{U}(\mathcal{H})$ *is defined as the subgroup*
$$\mathbb{L} = \mathbb{U}_{n-1} \otimes \mathbb{U}_{n-2} \otimes \ldots \otimes \mathbb{U}_0 = \bigotimes_{0}^{n-1} \mathbb{U}_j \ .$$
The elements of $\mathbb{L}$ *are called **local unitary transformations**. Unitary transformations which are in* $\mathbb{U}$ *but not in* $\mathbb{L}$ *are called **global unitary transformations**. The corresponding lie algebra*
$$\ell = \mathbf{u}_{n-1} \boxplus \mathbf{u}_{n-2} \boxplus \ldots \boxplus \mathbf{u}_0$$

is called the **local Lie algebra**, where '⊞' denotes the **Kronecker sum**[21].

Local unitary transformations can not entangle quantum systems with respect to the above tensor product decomposition. However, global unitary transformations are those unitary transformations which can and often do produce interactions which entangle quantum systems. This leads to the following definition:

DEFINITION 8. *Two states $|\psi_1\rangle$ and $|\psi_2\rangle$ in $\mathcal{H}$ are said to be **locally equivalent** (or, of the **same entanglement type**) , written*

$$|\psi_1\rangle \underset{local}{\sim} |\psi_2\rangle \ ,$$

if there exists a local unitary transformation $U \in \mathbb{L}$ such that

$$U|\psi_1\rangle = |\psi_2\rangle \ .$$

*The equivalence classes of local equivalence $\underset{local}{\sim}$ are called the **entanglement classes of $\mathcal{H}$**. Two density operators ρ_1 and ρ_2, (and hence, the corresponding two skew Hermitian operators $i\rho_1$ and $i\rho_2$ lying in $\mathbf{u}$) are said to be **locally equivalent** (or, of the **same entanglement type**), written*

$$\rho_1 \underset{local}{\sim} \rho_2 \ ,$$

if there exists a local unitary transformation $U \in \mathbb{L}$ such that

$$Ad_U(\rho_1) = \rho_2 \ ,$$

*where Ad_U denotes the big adjoint representation, i.e., $Ad_U(i\rho) = U(i\rho)U^\dagger$. The equivalence classes under this relation are called **entanglement classes** of the Lie algebra $\mathbf{u}(\mathcal{H})$.*

Thus, the entanglement classes of the Hilbert space $\mathcal{H}$ are just the **orbits** of the group action of $\mathbb{L}(\mathcal{H})$ on $\mathcal{H}$. In like manner, the entanglement classes of the Lie algebra $\mathbf{u}(\mathcal{H})$ are the **orbits** of the big adjoint action of $\mathbb{L}(\mathcal{H})$ on $\mathbf{u}(\mathcal{H})$. Two states are entangled in the same way if and only if they lie in the same entanglement class, i.e., the same orbit.

For example, let us assume that Alice and Bob collectively possess two qubits $\mathcal{Q}_{AB}$ which are in the entangled state

$$|\psi_1\rangle = \frac{|0_B 0_A\rangle + |1_B 1_A\rangle}{\sqrt{2}} = \frac{1}{\sqrt{2}} \begin{pmatrix} 1 \\ 0 \\ 0 \\ 1 \end{pmatrix} \ ,$$

and moreover that Alice possesses the qubit labeled A, but not the qubit labeled B, and that Bob holds qubit B, but not qubit A. Let us also assume that Alice and Bob are also separated by a spacelike distance. As a result, they can only apply local unitary transformations to the qubits that they possess.

---

[21]The Kronecker sum $A \boxplus B$ is defined as

$$A \boxplus B = A \otimes \mathbf{1} + \mathbf{1} \otimes B \ ,$$

where $\mathbf{1}$ denotes the identity transformation.

Alice could, for example, apply the local unitary transformation

$$U_A = \begin{pmatrix} 0 & 1 \\ -1 & 0 \end{pmatrix} \otimes \begin{pmatrix} 1 & 0 \\ 0 & 1 \end{pmatrix} = \begin{pmatrix} 0 & 0 & 1 & 0 \\ 0 & 0 & 0 & 1 \\ -1 & 0 & 0 & 0 \\ 0 & -1 & 0 & 0 \end{pmatrix}$$

to her qubit to move Alice's and Bob's qubits A and B respectively into the state

$$|\psi_2\rangle = \frac{|0_B 1_A\rangle - |1_B 0_A\rangle}{\sqrt{2}} = \frac{1}{\sqrt{2}} \begin{pmatrix} 0 \\ 1 \\ -1 \\ 0 \end{pmatrix},$$

Bob also could accomplish the same by applying the local unitary transformation

$$U_B = \begin{pmatrix} 1 & 0 \\ 0 & 1 \end{pmatrix} \otimes \begin{pmatrix} 0 & -1 \\ 1 & 0 \end{pmatrix} = \begin{pmatrix} 0 & -1 & 0 & 0 \\ 1 & 0 & 0 & 0 \\ 0 & 0 & 0 & -1 \\ 0 & 0 & 1 & 0 \end{pmatrix}$$

to his qubit.

By local unitary transformations, Alice and Bob can move the state of their two qubits to any other state within the same entanglement class. But with local unitary transformations, there is no way whatsoever that Alice and Bob can transform the two qubits into a state lying in a different entanglement class (i.e., a different orbit), such as

$$|\psi_3\rangle = |0_B 0_A\rangle.$$

The only way Alice and Bob could transform the two qubits from state $|\psi_1\rangle$ to the state $|\psi_3\rangle$ is for Alice and Bob to come together, and make the two qubits interact with one another via a global unitary transformation such as

$$U_{AB} = \frac{1}{\sqrt{2}} \begin{pmatrix} 1 & 0 & 0 & 1 \\ 0 & 1 & 1 & 0 \\ 0 & -1 & 1 & 0 \\ -1 & 0 & 0 & 1 \end{pmatrix}$$

The main objective of this approach to quantum entanglement is to determine when two states lie in the same orbit or in different orbits? In other words, what is needed is a complete set of invariants, i.e., invariants that completely specify all the orbits (i.e., all the entanglement classes). We save this topic for another lecture[67].

At first it would seem that state kets are a much better vehicle than density operators for the study of quantum entanglement. After all, state kets are much simpler mathematical objects. So why should one deal with the additional mathematical luggage of density operators?

Actually, density operators have a number of advantages over state kets. The most obvious advantage is that density operators certainly have an upper hand over state kets when dealing with mixed ensembles. But their most important advantage is that the orbits of the adjoint action are actually manifolds, which have a very rich and pliable mathematical structure. Needless to say, this topic is beyond the scope of this paper.

REMARK 11. *It should also be mentioned that the mathematical approach discussed in this section by no means captures every aspect of the physical phenomenon of quantum entanglement. The use of ancilla and of classical communication have not been considered. For an in-depth study of the relation between quantum entanglement and classical communication (including catalysis), please refer to* [**1**, *Chapter 5*], [**76**], *and to the article by Popescu and Rohrlich in* [**66**].

In regard to describing the locality of unitary operations, we will in Section 10 of this paper have need for a little less precision than that given in the above definitions. So we give the following (unfortunately rather technical) definitions:

DEFINITION 9. *Let $\mathcal{H}$, $\mathcal{H}_{n-1}$, $\mathcal{H}_{n-2}$, ..., $\mathcal{H}_0$ be as stated above. Let $\mathcal{P} = \{B_\alpha\}$ be a **partition** of the set of indices $\{0, 1, 2, \ldots, n-1\}$, i.e., $\mathcal{P}$ is a collection of disjoint subsets B_α of $\{0, 1, 2, \ldots, n-1\}$, called **blocks**, such that $\bigcup_\alpha B_\alpha = \{0, 1, 2, \ldots, n-1\}$. Then the $\mathcal{P}$-**tensor product decomposition** of $\mathcal{H}$ is defined as*

$$\mathcal{H} = \bigotimes_{B_\alpha \in \mathcal{P}} \mathcal{H}_{B_\alpha},$$

where

$$\mathcal{H}_{B_\alpha} = \bigotimes_{j \in B_\alpha} \mathcal{H}_j,$$

*for each block B_α in $\mathcal{P}$. Also the subgroup of $\mathcal{P}$-**local unitary transformations** $\mathbb{L}_\mathcal{P}(\mathcal{H})$ is defined as the subgroup of local unitary transformations of $\mathcal{H}$ corresponding to the $\mathcal{P}$-tensor decomposition of $\mathcal{H}$.*

*We define the **fineness of a partition** $\mathcal{P}$, written $fineness(\mathcal{P})$, as the maximum number of indices in a block of $\mathcal{P}$. We say that a unitary transformation U of $\mathcal{H}$ is **sufficiently local** if there exists a partition $\mathcal{P}$ with sufficiently small $fineness(\mathcal{P})$ (e.g., $fineness(\mathcal{P}) \leq 3$) such that $U \in \mathbb{L}_\mathcal{P}(\mathcal{H})$.*

REMARK 12. *The above lack of precision is needed because there is no way to know what kind (if any) of quantum computing devices will be implemented in the future. Perhaps we will at some future date be able to construct quantum computing devices that locally manipulate more than 2 or 3 qubits at a time?*

8. Entropy and quantum mechanics

8.1. Classical entropy, i.e., Shannon Entropy.

Let $\mathcal{S}$ be a probability distribution on a finite set $\{s_1, s_2, \ldots, s_n\}$ of elements called **symbols** given by

$$\text{Prob}(s_j) = p_j,$$

where $\sum_{j=1}^n p_j = 1$. Let s denote the random variable (i.e., **finite memoryless stochastic source**) that produces the value s_j with probability p_j.

DEFINITION 10. *The **classical entropy** (also called the **Shannon entropy**) $H(S)$ of a probability distribution S (or of the source s) is defined as:*

$$H(\mathcal{S}) = H(s) = -\sum_{j=1}^{n} p_j \lg(p_j) ,$$

where 'lg' denotes the log to the base 2 .

Classical entropy $H(\mathcal{S})$ is a measure of the uncertainty inherent in the probability distribution $\mathcal{S}$. Or in other words, it is the measure of the uncertainty of an observer before the source s "outputs" a symbol s_j.

One property of such classical **stochastic sources** we often take for granted is that the output symbols s_j are completely distinguishable from one another. We will see that this is not necessarily the case in the strange world of the quantum.

8.2. Quantum entropy, i.e., Von Neumann entropy.

Let $\mathcal{Q}$ be a quantum system with state given by the density operator ρ.

Then there are many **preparations**

Preparation			
$\|\psi_1\rangle$	$\|\psi_2\rangle$	$\ldots$	$\|\psi_n\rangle$
p_1	p_2	$\ldots$	p_n

which will produce the same state ρ. These preparations are classical stochastic sources with classical entropy given by

$$H = -\sum p_j \lg(p_j) .$$

Unfortunately, the classical entropy H of a preparation does not necessarily reflect the uncertainty in the resulting state ρ. For two different preparations $\mathcal{P}_1$ and $\mathcal{P}_2$, having different entropies $H(\mathcal{P}_1)$ and $H(\mathcal{P}_2)$, can (and often do) produce the same state ρ. The problem is that the states of the preparation may not be completely physically distinguishable from one another. This happens when the states of the preparation are not orthogonal. (Please refer to the Heisenberg uncertainty principle.)

John von Neumann found that the true measure of quantum entropy can be defined as follows:

DEFINITION 11. *Let $\mathcal{Q}$ be a quantum system with state given by the density operator ρ. Then the **quantum entropy** (also called the **von Neumann entropy**) of $\mathcal{Q}$, written $S(\mathcal{Q})$, is defined as*

$$S(\mathcal{Q}) = -Trace\,(\rho \lg \rho) ,$$

where '$\lg \rho$' denotes the log to the base 2 of the operator ρ.

REMARK 13. *The operator $\lg \rho$ exists and is an analytic map $\rho \longmapsto \lg \rho$ given by the power series*

$$\lg \rho = \frac{1}{\ln 2} \sum_{n=1}^{\infty} (-1)^{n+1} \frac{(\rho - I)^n}{n}$$

provided that ρ is sufficiently close to the identity operator I, i.e., provided

$$\|\rho - I\| < 1 ,$$

where

$$\|A\| = \sup_{v \in \mathcal{H}} \frac{\|Av\|}{\|v\|} .$$

It can be shown that this is the case for all positive definite Hermitian operators of trace 1.

For Hermitian operators ρ of trace 1 which are not positive definite, but only positive semi-definite (i.e., which have a zero eigenvalue), the logarithm $\lg(\rho)$ does not exist. However, there exists a sequence $\rho_1, \rho_2, \rho_3, \ldots$ of positive definite Hermitian operators of trace 1 which converges to ρ, i.e., such that

$$\rho = \lim_{k \longrightarrow \infty} \rho_k$$

It can then be shown that the limit

$$\lim_{k \longrightarrow \infty} \rho_k \lg \rho_k$$

exists.

Hence, $S(\rho)$ is defined and exists for all density operators ρ.

Quantum entropy is a measure of the uncertainty at the quantum level. As we shall see, it is very different from the classical entropy that arises when a measurement is made.

One important feature of quantum entropy $S(\rho)$ is that it is invariant under the big adjoint action of unitary transformations, i.e.,

$$S(\ Ad_U(\rho)\ ) = S\left(U\rho U^\dagger\right) = S(\rho) .$$

It follows that, for closed quantum systems, it is a **dynamical invariant.** As the state ρ moves according to Schrödinger's equation, the quantum entropy $S(\rho)$ of ρ remains constant. It does not change unless measurement is made, or, as we shall see, unless we ignore part of the quantum system.

Because of unitary invariance, the quantum entropy can be most easily computed by first diagonalizing ρ with a unitary transformation U, i.e.,

$$U\rho U^\dagger = \triangle(\vec{\lambda}) ,$$

where $\triangle(\vec{\lambda})$ denotes the diagonal matrix with diagonal $\vec{\lambda} = (\lambda_1, \lambda_2, \ldots, \lambda_n)$.

Once ρ has been diagonalized, we have

$$S(\rho) = -Trace\left(\triangle(\vec{\lambda})\lg\triangle(\vec{\lambda})\right)$$
$$= -Trace\left(\triangle(\lambda_1\lg\lambda_1,\ \lambda_2\lg\lambda_2,\ \ldots\ ,\ \lambda_n\lg\lambda_n)\right)$$
$$= -\sum_{j=1}^{n}\lambda_j\lg\lambda_j\ ,$$

where the λ_j's are the eigenvalues of ρ, and where $0\lg 0 \equiv 0$.

Please note that, because ρ is positive semi-definite Hermitian of trace 1, all the eigenvalues of ρ are non-negative real numbers such that

$$\sum_{j=1}^{n}\lambda_j = 1\ .$$

As an immediate corollary we have that the quantum entropy of a pure ensemble must be zero, i.e.,

$$\boxed{\rho\ \text{pure ensemble}\ \Longrightarrow\ S(\rho) = 0}$$

There is no quantum uncertainty in a pure ensemble. However, as expected, there is quantum uncertainty in mixed ensembles.

8.3. How is quantum entropy related to classical entropy?

But how is classical entropy H related to quantum entropy S?

Let A be an observable of the quantum system $\mathcal{Q}$. Then a measurement with respect to A of $\mathcal{Q}$ produces an eigenvalue a_i with probability

$$p_i = Trace\left(P_{a_i}\rho\right)\ ,$$

where P_{a_i} denotes the projection operator for the eigenspace of the eigenvalue a_i. For example, if a_i is a non-degenerate eigenvalue, then $P_{a_i} = |a_i\rangle\langle a_i|$.

In other words, measurement with respect to A of a quantum system $\mathcal{Q}$ in state ρ can be identified with a classical stochastic source with the eigenvalues a_i as output symbols occurring with probability p_i. We denote this classical stochastic source simply by (ρ, A).

The two entropies $S(\rho)$ and $H(\rho, A)$ are by no means the same. One is a measure of quantum uncertainty before measurement, the other a measure of the classical uncertainty that results from measurement. The quantum entropy $S(\rho)$ is usually a lower bound for the classical entropy, i.e.,

$$S(\rho) \leq H(\rho, A)\ .$$

If A is a complete observable (hence, non-degenerate), and if A is compatible with ρ, i.e., $[\rho, A] = 0$, then $S(\rho) = H(\rho, A)$.

8.4. When a part is greater than the whole, then Ignorance = uncertainty.

Let $\mathcal{Q}$ be a multipartite quantum system with constituent parts $\mathcal{Q}_{n-1}, \ldots, \mathcal{Q}_1, \mathcal{Q}_0$, and let the density operator ρ denote the state of $\mathcal{Q}$. Then from section 5.6 of this paper we know that the state ρ_j of each constituent "part" $\mathcal{Q}_j$ is given by the partial trace over all degrees of freedom except $\mathcal{Q}_j$, i.e., by

$$\rho_j = \mathop{Trace}_{\substack{0 \leq k \leq n-1 \\ k \neq j}} (\rho) \ .$$

By applying the above partial trace, we are focusing only on the quantum system $\mathcal{Q}_j$, and literally ignoring the remaining constituent "parts" of $\mathcal{Q}$. By taking the partial trace, we have done nothing physical to the quantum system. We have simply ignored parts of the quantum system.

What is surprising is that, by intentionally ignoring "part" of the quantum system, we can in some cases create more quantum uncertainty. This happens when the constituent "parts" of $\mathcal{Q}$ are quantum entangled.

For example, let $\mathcal{Q}$ denote the bipartite quantum system consisting of two qubits $\mathcal{Q}_1$ and $\mathcal{Q}_0$ in the entangled state

$$|\Psi_{\mathcal{Q}}\rangle = \frac{|0_1 0_0\rangle - |1_1 1_0\rangle}{\sqrt{2}} \ .$$

The corresponding density operator $\rho_{\mathcal{Q}}$ is

$$\rho_{\mathcal{Q}} = \frac{1}{2} \left(|0_1 0_0\rangle\langle 0_1 0_0| - |0_1 0_0\rangle\langle 1_1 1_0| - |1_1 1_0\rangle\langle 0_1 0_0| + |1_1 1_0\rangle\langle 1_1 1_0| \right)$$

$$= \frac{1}{2} \begin{pmatrix} 1 & 0 & 0 & -1 \\ 0 & 0 & 0 & 0 \\ 0 & 0 & 0 & 0 \\ -1 & 0 & 0 & 1 \end{pmatrix}$$

Since $\rho_{\mathcal{Q}}$ is a pure ensemble, there is no quantum uncertainty, i.e.,

$$S(\rho_{\mathcal{Q}}) = 0 \ .$$

Let us now focus on qubit #0 (i.e., $\mathcal{Q}_0$). The resulting density operator ρ_0 for qubit #0 is obtained by tracing over $\mathcal{Q}_1$, i.e.,

$$\rho_0 = Trace_1(\rho_{\mathcal{Q}}) = \frac{1}{2} \left(|0\rangle\langle 0| + |1\rangle\langle 1| \right) = \frac{1}{2} \begin{pmatrix} 1 & 0 \\ 0 & 1 \end{pmatrix} \ .$$

Hence, the quantum uncertainty of qubit #0 is

$$S(\rho_0) = 1 \ .$$

Something most unusual, and non-classical, has happened. Simply by ignoring part of the quantum system, we have increased the quantum uncertainty. The quantum uncertainty of the constituent "part" Q_0 is greater than that of he whole quantum system Q. This is not possible in the classical world, i.e., not possible for Shannon entropy. (For more details, see [**17**].)

9. There is much more to quantum mechanics

There is much more to quantum mechanics. For more in-depth overviews, there are many outstanding books. Among such books are [**15**], [**18**], [**28**], [**31**], [**43**], [**47**], [**48**], [**52**], [**54**], [**72**], [**78**], [**75**], [**81**], [**83**], [**84**], and many more.

Part 3. Part of a Rosetta Stone for Quantum Computation

10. The Beginnings of Quantum Computation – Elementary Quantum Computing Devices

We begin this section with some examples of quantum computing devices. By a **quantum computing device**[22] we mean a unitary transformation U that is the composition of finitely many sufficiently local unitary transformations, i.e.,

$$U = U_{n-1}U_{n-2}\ldots U_1 U_0,$$

where $U_{n-1}, U_{n-2}, \ldots, U_1, U_0$ are sufficiently local[23] unitary transformations. Each U_j is called a **computational step** of the device.

Our first examples will be constructed by embedding classical computing devices within the realm of quantum mechanics. We will then look at some other quantum computing devices that are not the embeddings of classical devices.

10.1. Embedding classical (memoryless) computation in quantum mechanics.

One objective in this section is to represent[24] classical computing devices as unitary transformations. Since unitary transformations are invertible, i.e., reversible, it follows that the only classical computing devices that can be represented as such transformations must of necessity be reversible devices. Hence, the keen interest in reversible computation[25].

For a more in depth study of reversible computation, please refer to the work of Bennett and others.

---

[22]Unfortunately, Physicists have "stolen" the akronym QCD. :-)
[23]See Definition 9 in Section 7.8 of this paper for a definition of the term 'sufficiently local'.
[24]Double meaning is intended.
[25]For references on reversible computation, see [**49**, Chapter 5] and [**77**, Chapter 3].

10.2. Classical reversible computation without memory.

$$\text{Input} \begin{cases} x_{n-1} & \longrightarrow \\ x_{n-2} & \longrightarrow \\ \vdots & \vdots \\ x_1 & \longrightarrow \\ x_0 & \longrightarrow \end{cases} \boxed{\text{CRCD}_n} \begin{matrix} \longrightarrow & y_{n-1} \\ \longrightarrow & y_{n-2} \\ \vdots & \vdots \\ \longrightarrow & y_1 \\ \longrightarrow & y_0 \end{matrix} \Bigg\} \text{Output}$$

Each **classical n-input/n-output (binary memoryless) reversible computing device** (**CRCD$_n$**) can be identified with a bijection

$$\pi : \{0,1\}^n \longrightarrow \{0,1\}^n$$

on the set $\{0,1\}^n$ of all binary n-tuples. Thus, we can in turn identify each CRCD$_n$ with an element of the permutation group S_{2^n} on the 2^n symbols

$$\{\, |\vec{a}\rangle \quad | \quad \vec{a} \in \{0,1\}^n \,\} \, .$$

Let

$$\mathcal{B}_n = \mathcal{B}\langle x_0, x_1, \ldots, x_{n-1}\rangle$$

denote the **free Boolean ring** on the symbols $x_0, x_1, \ldots, x_{n-1}$. Then the binary n-tuples $\vec{a} \in \{0,1\}^n$ are in one-to-one correspondence with the **minterms** of $\mathcal{B}_n$, i.e.,

$$\vec{a} \longleftrightarrow x^{\vec{a}} = \prod_{j=0}^{n-1} x_j^{a_j} \, ,$$

where

$$\begin{cases} x_j^0 & = \overline{x}_j \\ x_j^1 & = x_j \end{cases}$$

Since there is a one-to-one correspondence between the automorphisms of $\mathcal{B}_n$ and the permutations on the set of minterms, it follows that CRCD$_n$'s can also be identified with the **automorphism group** $Aut(\mathcal{B}_n)$ of the free Boolean ring $\mathcal{B}_n$.

Moreover, since the set of binary n-tuples $\{0,1\}^n$ is in one-to-one correspondence with the set of integers $\{0, 1, 2, \ldots, 2^n - 1\}$ via the radix 2 representation of integers, i.e.,

$$(b_{n-1}, b_{n-2}, \ldots, b_1, b_0) \longleftrightarrow \sum_{j=0}^{n-1} b_j 2^j \, ,$$

we can, and frequently do, identify binary n-tuples with integers.

For example, consider the Controlled-NOT gate, called **CNOT**, which is defined by the following **wiring diagram**:

$$\mathbf{CNOT} = \boxed{\begin{matrix} c & \longrightarrow \oplus \longrightarrow & b+c \\ & | & \\ b & \longrightarrow \bullet \longrightarrow & b \\ & & \\ a & \longrightarrow \longrightarrow \longrightarrow & a \end{matrix}} \, ,$$

where '•' and '⊕' denote respectively a **control bit** and a **target bit**, and where '$a+b$' denotes the exclusive 'or' of bits a and b. This corresponds to the permutation $\pi = (26)(37)$, i.e.,

$$\begin{cases} |0\rangle = & |000\rangle & \longmapsto & |000\rangle & = |0\rangle \\ |1\rangle = & |001\rangle & \longmapsto & |001\rangle & = |1\rangle \\ |2\rangle = & |010\rangle & \longmapsto & |110\rangle & = |6\rangle \\ |3\rangle = & |011\rangle & \longmapsto & |111\rangle & = |7\rangle \\ \\ |4\rangle = & |100\rangle & \longmapsto & |100\rangle & = |4\rangle \\ |5\rangle = & |101\rangle & \longmapsto & |101\rangle & = |5\rangle \\ |6\rangle = & |110\rangle & \longmapsto & |010\rangle & = |2\rangle \\ |7\rangle = & |111\rangle & \longmapsto & |011\rangle & = |3\rangle \end{cases},$$

where we have used the following indexing conventions:

$$\begin{cases} \text{First=Right=Bottom} \\ \text{Last=Left=Top} \end{cases}$$

As another example, consider the **Toffoli** gate, which is defined by the following wiring diagram:

$$\textbf{Toffoli} = \begin{array}{|ccc|} \hline c & \longrightarrow \oplus \longrightarrow & c+ab \\ & | & \\ b & \longrightarrow \bullet \longrightarrow & b \\ & | & \\ a & \longrightarrow \bullet \longrightarrow & a \\ \hline \end{array},$$

where 'ab' denotes the logical 'and' of a and b. As before, '+' denotes exclusive 'or'. This gate corresponds to the permutation $\pi = (67)$.

In summary, we have:

$$\boxed{\{\ CRCD_n\ \} = S_2^n = Aut\,(\mathcal{B}_n)}$$

10.3. Embedding classical irreversible computation within classical reversible computation.

A classical 1-input/n-output (binary memoryless) irreversible computing device can be thought of as a Boolean function $f = f(x_{n-2}, \ldots, x_1, x_0)$ in $\mathcal{B}_{n-1} = \mathcal{B}\langle x_0, x_1, \ldots, x_{n-2}\rangle$. Such irreversible computing devices can be transformed into reversible computing devices via the monomorphism

$$\iota : \mathcal{B}_{n-1} \longrightarrow Aut(\mathcal{B}_n),$$

where $\iota(f)$ is the automorphism in $Aut(\mathcal{B}_n)$ defined by

$$(x_{n-1}, x_{n-2}, \ldots, x_1, x_0) \longmapsto (x_{n-1} \oplus f, x_{n-2}, \ldots, x_1, x_0),$$

and where '⊕' denotes exclusive 'or'. Thus, the image of each Boolean function f is a product of disjoint transpositions in S_{2^n}.

As an additive group (ignoring ring structure), $\mathcal{B}_{n-1}$ is the abelian group $\bigoplus_{j=0}^{2^{(n-1)}-1} \mathbb{Z}_2$, where $\mathbb{Z}_2$ denotes the cyclic group of order two.

Classical Binary Memoryless Computation is summarized in the table below:

Summary
Classical Binary Memoryless Computation
$\mathcal{B}_{n-1} = \bigoplus_{j=0}^{2^{(n-1)}-1} \mathbb{Z}_2 \xrightarrow{\iota} S_{2^n} = Aut(\mathcal{B}_n)$

10.4. The unitary representation of reversible computing devices.

It is now a straight forward task to represent $CRCD_n$'s as unitary transformations. We simply use the **standard unitary representation**

$$\boxed{\nu : S_2^n \longrightarrow \mathbb{U}(2^n; \mathbb{C})}$$

of the symmetric group S_{2^n} into the group of $2^n \times 2^n$ unitary matrices $\mathbb{U}(2^n; \mathbb{C})$. This is the representation defined by

$$\pi \longmapsto (\delta_{k, \pi k})_{2^n \times 2^n} ,$$

where $\delta_{k\ell}$ denotes the Kronecker delta, i.e.,

$$\delta_{k\ell} = \begin{cases} 1 & \text{if } k = \ell \\ 0 & \text{otherwise} \end{cases}$$

We think of such unitary transformations as quantum computing devices.

For example, consider the controlled-NOT gate **CNOT'** $= (45)(67) \in S_8$ given by the wiring diagram

$$\mathbf{CNOT'} = \boxed{\begin{array}{ccc} c & \longrightarrow \bullet \longrightarrow & c \\ & | & \\ b & \longrightarrow \longrightarrow \longrightarrow & b \\ & | & \\ a & \longrightarrow \oplus \longrightarrow & a+c \end{array}}$$

This corresponds to the unitary transformation

$$U_{\mathbf{CNOT'}} = \nu(\mathbf{CNOT'}) = \begin{pmatrix} 1 & 0 & 0 & 0 & 0 & 0 & 0 & 0 \\ 0 & 1 & 0 & 0 & 0 & 0 & 0 & 0 \\ 0 & 0 & 1 & 0 & 0 & 0 & 0 & 0 \\ 0 & 0 & 0 & 1 & 0 & 0 & 0 & 0 \\ 0 & 0 & 0 & 0 & 0 & 1 & 0 & 0 \\ 0 & 0 & 0 & 0 & 1 & 0 & 0 & 0 \\ 0 & 0 & 0 & 0 & 0 & 0 & 0 & 1 \\ 0 & 0 & 0 & 0 & 0 & 0 & 1 & 0 \end{pmatrix}$$

Moreover, consider the Toffoli gate **Toffoli'** $= (57) \in S_8$ given by the wiring diagram

$$\textbf{Toffoli'} = \begin{array}{|ccc|} \hline c & \longrightarrow \bullet \longrightarrow & c \\ & | & \\ b & \longrightarrow \oplus \longrightarrow & b + ac \\ & | & \\ a & \longrightarrow \bullet \longrightarrow & a \\ \hline \end{array}$$

This corresponds to the unitary transformation

$$U_{\textbf{Toffoli'}} = \nu(\textbf{Toffoli'}) = \begin{pmatrix} 1 & 0 & 0 & 0 & 0 & 0 & 0 & 0 \\ 0 & 1 & 0 & 0 & 0 & 0 & 0 & 0 \\ 0 & 0 & 1 & 0 & 0 & 0 & 0 & 0 \\ 0 & 0 & 0 & 1 & 0 & 0 & 0 & 0 \\ 0 & 0 & 0 & 0 & 1 & 0 & 0 & 0 \\ 0 & 0 & 0 & 0 & 0 & 0 & 0 & 1 \\ 0 & 0 & 0 & 0 & 0 & 0 & 1 & 0 \\ 0 & 0 & 0 & 0 & 0 & 1 & 0 & 0 \end{pmatrix}$$

Abuse of Notation and a Caveat: Whenever it is clear from context, we will use the name of a $CRCD_n$ to also refer to the unitary transformation corresponding to the $CRCD_n$. For example, we will denote $\nu(CNOT)$ and $\nu(Toffoli)$ simply by $CNOT$ and $Toffoli$. Moreover we will also use the wiring diagram of a $CRCD_n$ to refer to the unitary transformation corresponding to the $CRCD_n$. For quantum computation beginners, this can lead to some confusion. Be careful!

10.5. Some other simple quantum computing devices.

After $CRCD_n$'s are embedded as quantum computing devices, they are no longer classical computing devices. After the embedding, they suddenly have acquired much more computing power. Their inputs and outputs can be a superposition of many states. They can entangle their outputs. It is misleading to think of their input qubits as separate, for they could be entangled.

As an illustration of this fact, please note that the quantum computing device **CNOT''** given by the wiring diagram

$$\textbf{CNOT''} = \begin{array}{|ccc|} \hline b & \longrightarrow \bullet \longrightarrow & a + b \\ & | & \\ a & \longrightarrow \oplus \longrightarrow & a \\ \hline \end{array} = \begin{pmatrix} 1 & 0 & 0 & 0 \\ 0 & 1 & 0 & 0 \\ 0 & 0 & 0 & 1 \\ 0 & 0 & 1 & 0 \end{pmatrix}$$

is far from classical. It is more than a permutation. It is a linear operator that respects quantum superposition.

For example, **CNOT''** can take two non-entangled qubits as input, and then produce two entangled qubits as output. This is something no classical computing device can do. For example,

$$\frac{|0\rangle - |1\rangle}{\sqrt{2}} \otimes |0\rangle = \frac{1}{\sqrt{2}} (|00\rangle - |10\rangle) \longmapsto \frac{1}{\sqrt{2}} (|00\rangle - |11\rangle)$$

For completeness, we list two other quantum computing devices that are embeddings of CRCD$_n$'s, **NOT** and **SWAP**:

$$\mathbf{NOT} = \boxed{a \longrightarrow \boxed{\text{NOT}} \longrightarrow a+1} = \begin{pmatrix} 0 & 1 \\ 1 & 0 \end{pmatrix} = \sigma_1$$

and

$$\mathbf{SWAP} = \boxed{\begin{array}{c} b \longrightarrow \bullet \longrightarrow \longrightarrow \oplus \longrightarrow \longrightarrow \bullet \longrightarrow a \\ \quad | | | \\ a \longrightarrow \oplus \longrightarrow \longrightarrow \bullet \longrightarrow \longrightarrow \oplus \longrightarrow b \end{array}} = \begin{pmatrix} 1 & 0 & 0 & 0 \\ 0 & 0 & 1 & 0 \\ 0 & 1 & 0 & 0 \\ 0 & 0 & 0 & 1 \end{pmatrix}$$

10.6. Quantum computing devices that are not embeddings.

We now consider quantum computing devices that are not embeddings of CRCD$_n$'s.

The **Hadamard** gate **H** is defined as:

$$\mathbf{H} = \boxed{\longrightarrow \boxed{\mathbf{H}} \longrightarrow} = \frac{1}{\sqrt{2}} \begin{pmatrix} 1 & 1 \\ 1 & -1 \end{pmatrix}.$$

Another quantum gate is the **square root of NOT**, i.e., $\sqrt{\mathbf{NOT}}$, which is given by

$$\sqrt{\mathbf{NOT}} = \boxed{\longrightarrow \boxed{\sqrt{\text{NOT}}} \longrightarrow} = \frac{1-i}{2} \begin{pmatrix} i & 1 \\ 1 & i \end{pmatrix} = \frac{1+i}{2} \begin{pmatrix} 1 & -i \\ -i & 1 \end{pmatrix}.$$

There is also the **square root of swap** $\sqrt{\mathbf{SWAP}}$ which is defined as:

$$\sqrt{\mathbf{SWAP}} = \boxed{\longrightarrow \boxed{\sqrt{\text{SWAP}}} \longrightarrow} = \begin{pmatrix} 1 & 0 & 0 & 0 \\ 0 & \frac{1+i}{2} & \frac{1-i}{2} & 0 \\ 0 & \frac{1-i}{2} & \frac{1+i}{2} & 0 \\ 0 & 0 & 0 & 1 \end{pmatrix}.$$

Three frequently used unary quantum gates are the rotations:

$$\boxed{\longrightarrow \boxed{e^{i\theta\sigma_1}} \longrightarrow} = \begin{pmatrix} \cos\theta & i\sin\theta \\ i\sin\theta & \cos\theta \end{pmatrix} = e^{i\theta\sigma_1}$$

$$\boxed{\longrightarrow \boxed{e^{i\theta\sigma_2}} \longrightarrow} = \begin{pmatrix} \cos\theta & \sin\theta \\ -\sin\theta & \cos\theta \end{pmatrix} = e^{i\theta\sigma_2}$$

$$\boxed{\longrightarrow \boxed{e^{i\theta\sigma_3}} \longrightarrow} = \begin{pmatrix} e^{i\theta} & 0 \\ 0 & e^{-i\theta} \end{pmatrix} = e^{i\theta\sigma_3}$$

10.7. The implicit frame of a wiring diagram.

Wiring diagrams have the advantage of being a simple means of describing some rather complicated unitary transformations. However, they do have their drawbacks, and they can, if not used with care, be even misleading.

One problem with wiring diagrams is that they are not frame (i.e., basis) independent descriptions of unitary transformations. Each wiring diagram describes a unitary transformation using an implicitly understood basis.

For example, consider **CNOT″** given by the wiring diagram:

$$\mathbf{CNOT''} = \boxed{\begin{array}{ccccc} b & \longrightarrow & \bullet & \longrightarrow & a+b \\ & & | & & \\ a & \longrightarrow & \oplus & \longrightarrow & a \end{array}}.$$

The above wiring diagram defines **CNOT″** in terms of the implicitly understood basis

$$\left\{ |0\rangle = \begin{pmatrix} 1 \\ 0 \end{pmatrix}, \ |1\rangle = \begin{pmatrix} 0 \\ 1 \end{pmatrix} \right\}.$$

This wiring diagram suggests that qubit #1 controls qubit #0, and that qubit #1 is not effected by qubit #0. But this is far from the truth. For, **CNOT″** transforms

$$\frac{|0\rangle + |1\rangle}{\sqrt{2}} \otimes \frac{|0\rangle - |1\rangle}{\sqrt{2}}$$

into

$$\frac{|0\rangle - |1\rangle}{\sqrt{2}} \otimes \frac{|0\rangle - |1\rangle}{\sqrt{2}},$$

where we have used our indexing conventions

$$\begin{cases} \text{First=Right=Bottom} \\ \text{Last=Left=Top} \end{cases}.$$

In fact, in the basis

$$\left\{ |0'\rangle = \frac{|0\rangle + |1\rangle}{\sqrt{2}}, \ |1'\rangle = \frac{|0\rangle - |1\rangle}{\sqrt{2}} \right\}$$

the wiring diagram of the same unitary transformation **CNOT″** is:

$$\boxed{\begin{array}{ccccc} b & \longrightarrow & \oplus & \longrightarrow & a+b \\ & & | & & \\ a & \longrightarrow & \bullet & \longrightarrow & a \end{array}}$$

The roles of the target and control qubits appeared to have switched!

11. The No-Cloning Theorem

In this section, we prove the no-cloning theorem of Dieks[24], Wootters and Zurek [93]. The theorem states that there can be no device that produces exact replicas or copies of a quantum state.

In mathematical terms, a device which replicates quantum states, i.e., a quantum replicator, is defined as follows:

DEFINITION 12. *Let $\mathcal{H}$ be a Hilbert space. Then a **quantum replicator** for $\mathcal{H}$ consists of an auxiliary Hilbert space $\mathcal{H}_A$, a fixed state $|\psi_0\rangle \in \mathcal{H}_A$ (called the **initial state of replicator**), and a unitary transformation*

$$U : \mathcal{H}_A \otimes \mathcal{H} \otimes \mathcal{H} \longrightarrow \mathcal{H}_A \otimes \mathcal{H} \otimes \mathcal{H}$$

such that, for some fixed state $|blank\rangle \in \mathcal{H}$,

$$U |\psi_0\rangle |a\rangle |blank\rangle = |\psi_a\rangle |a\rangle |a\rangle \ ,$$

*for all states $|a\rangle \in \mathcal{H}$, where $|\psi_a\rangle \in \mathcal{H}_A$ (called the **replicator state after replication** of $|a\rangle$) depends on $|a\rangle$.*

THEOREM 3 (No-Cloning). *Let $\mathcal{H}$ be a Hilbert space of dimension greater than one. Then a quantum replicator for $\mathcal{H}$ does not exist.*

The proof of the no-cloning theorem, i.e., the proof that quantum replicators do not exist, is an amazingly simple application of the linearity of quantum mechanics. The key idea is that copying is an inherently nonlinear transformation, while the unitary transformations of quantum mechanics are inherently linear. Ergo, copying can not be a unitary transformation.

More specifically, the proof goes as follows:

PROOF. Since a quantum state is determined by a ket up to a multiplicative non-zero complex number, we can without loss of generality assume that $|\psi_0\rangle$, $|a\rangle$, $|blank\rangle$ are all of unit length. From unitarity, it follows that $|\psi_a\rangle$ is also of unit length.

Let $|a\rangle$, $|b\rangle$ be two kets of unit length in $\mathcal{H}$ such that

$$0 < |\langle a | b \rangle| < 1 \ .$$

Then

$$\begin{cases} U |\psi_0\rangle |a\rangle |blank\rangle &= |\psi_a\rangle |a\rangle |a\rangle \\ U |\psi_0\rangle |b\rangle |blank\rangle &= |\psi_b\rangle |b\rangle |b\rangle \end{cases}$$

Hence,

$$\langle blank| \langle a| \langle \psi_0| U^\dagger U |\psi_0\rangle |b\rangle |blank\rangle = \langle blank| \langle a| \langle \psi_0 | \psi_0 \rangle |b\rangle |blank\rangle$$
$$= \langle a | b \rangle$$

On the other hand,

$$\langle blank| \langle a| \langle \psi_0| U^\dagger U |\psi_0\rangle |b\rangle |blank\rangle = \langle a| \langle a| \langle \psi_a | \psi_b \rangle |b\rangle |b\rangle$$
$$= \langle a | b \rangle^2 \langle \psi_a | \psi_b \rangle$$

Thus,
$$\langle a | b \rangle^2 \langle \psi_a | \psi_b \rangle = \langle a | b \rangle .$$
And so,
$$\langle a | b \rangle \langle \psi_a | \psi_b \rangle = 1 .$$

But this equation can not be satisfied since
$$|\langle a | b \rangle| < 1$$
and
$$|\langle \psi_a | \psi_b \rangle| \leq \| |\psi_a\rangle \| \, \| |\psi_b\rangle \| = 1$$

Hence, a quantum replicator cannot exist. □

EXERCISE 2. *Although it is not possible to clone **all** states in $\mathcal{H}$ (emphasis on key word "all"), it is nonetheless still possible to clone all states of a subset of $\mathcal{H}$ consisting of mutually orthogonal states.*

Let $\{|0\rangle, \ldots, |n-1\rangle\}$ be an orthonormal basis of the Hilbert space $\mathcal{H}$. Construct a unitary transformation U such that $U : |k\rangle |blank\rangle \longmapsto |k\rangle |k\rangle$, for $0 \leq k < n$.

EXERCISE 3. *Is cloning possible on a one dimensional Hilbert space $\mathcal{H}$? Please explain your answer.*

12. Quantum teleportation

We now give a brief description of quantum teleportation, a means possibly to be used by future quantum computers to bus qubits from one location to another.

The no-cloning theorem emphatically states that qubits cannot be copied. However, ... qubits can be teleported, as has been demonstrated in laboratory settings. Such a mechanism could be used to bus qubits from one computer location to another. It could be used to create devices called **quantum repeaters**.

But what do we mean by teleportation?

Teleportation is the transferring of an object from one location to another by a process that:

1) Firstly dissociates (i.e., destroys) the object to obtain information. – The object to be teleported is first scanned to extract sufficient information to reassemble the original object.
2) Secondly transmits the acquired information from one location to another.
3) Lastly reconstructs the object at the new location from the received information. – An exact replica is re-assembled at the destination out of locally available materials.

Two key effects of teleportation should be noted:
1) The original object is destroyed during the process of teleportation. Hence, the no-cloning theorem is not violated.
2) An exact replica of the original object is created at the intended destination.

Scotty of the Starship Enterprise was gracious enough to loan me the following teleportation manual. So I am passing it on to you.

Quantum Teleportation Manual

Step. 1 .(Location A) Preparation: At location A, construct an EPR pair of qubits (qubits #2 and #3) in $\mathcal{H}_2 \otimes \mathcal{H}_3$.

$$|00\rangle \longmapsto \boxed{\text{Unitary Matrix}} \longmapsto \frac{|01\rangle - |10\rangle}{\sqrt{2}}$$

$$\mathcal{H}_2 \otimes \mathcal{H}_3 \longrightarrow \mathcal{H}_2 \otimes \mathcal{H}_3$$

Step 2. Transport: Physically transport entangled qubit #3 from location A to location B.

Step 3. : The qubit to be teleported, i.e., qubit #1, is delivered to location A in an unknown state

$$a|0\rangle + b|1\rangle$$

As a result of Steps 1 to 3, we have:
- Locations A and B share an EPR pair, i.e.
  - The qubit to be teleported, i.e., qubit #1, is at Location A
  - Qubit #2 is at Location A
  - Qubit #3 is at Location B
  - Qubits #2 & #3 are entangled
- The current state $|\Phi\rangle$ of all three qubits is:

$$|\Phi\rangle = (a|0\rangle + b|1\rangle)\left(\frac{|01\rangle - |10\rangle}{\sqrt{2}}\right) \in \mathcal{H}_1 \otimes \mathcal{H}_2 \otimes \mathcal{H}_3$$

To better understand what is about to happen, we re-express the state $|\Phi\rangle$ of the three qubits in terms of the following basis (called the **Bell basis**) of $\mathcal{H}_1 \otimes \mathcal{H}_2$:

$$\begin{cases} |\Psi_A\rangle &= \frac{|10\rangle - |01\rangle}{\sqrt{2}} \\ |\Psi_B\rangle &= \frac{|10\rangle + |01\rangle}{\sqrt{2}} \\ |\Psi_C\rangle &= \frac{|00\rangle - |11\rangle}{\sqrt{2}} \\ |\Psi_D\rangle &= \frac{|00\rangle + |11\rangle}{\sqrt{2}} \end{cases}$$

The result is:

$$|\Phi\rangle = \frac{1}{2}[\ |\Psi_A\rangle(-a|0\rangle - b|1\rangle) \\ + |\Psi_B\rangle(-a|0\rangle + b|1\rangle) \\ + |\Psi_C\rangle(a|1\rangle + b|0\rangle) \\ + |\Psi_D\rangle(a|1\rangle - b|0\rangle)\ ],$$

where, as you might have noticed, we have written the expression in a suggestive way.

REMARK 14. *Please note that since the completion of Step 3, we have done nothing physical. We have simply performed some algebraic manipulations of the expression representing the state $|\Phi\rangle$ of the three qubits.*

Let $U : \mathcal{H}_1 \otimes \mathcal{H}_2 \longrightarrow \mathcal{H}_1 \otimes \mathcal{H}_2$ be the unitary transformation defined by

$$\begin{cases} |\Psi_A\rangle &\longmapsto |00\rangle \\ |\Psi_B\rangle &\longmapsto |01\rangle \\ |\Psi_C\rangle &\longmapsto |10\rangle \\ |\Psi_D\rangle &\longmapsto |11\rangle \end{cases}$$

Step 4. (Location A): [26]Apply the local unitary transformation $U \otimes I :$ $\mathcal{H}_1 \otimes \mathcal{H}_2 \otimes \mathcal{H}_3 \longrightarrow \mathcal{H}_1 \otimes \mathcal{H}_2 \otimes \mathcal{H}_3$ to the three qubits (actually more precisely, to qubits #1 and #2). Thus, under $U \otimes I$ the state $|\Phi\rangle$ of all

---

[26]Actually, there is no need to apply the unitary transformation U. We could have instead made a complete Bell state measurement, i.e., a measurement with respect to the compatible observables $|\Psi_A\rangle\langle\Psi_A|, |\Psi_B\rangle\langle\Psi_B|, |\Psi_C\rangle\langle\Psi_C|, |\Psi_D\rangle\langle\Psi_D|$. We have added the additional step 4 to make quantum teleportation easier to understand for quantum computation beginners.

three qubits becomes

$$|\Phi'\rangle = \tfrac{1}{2}[\ |00\rangle(-a|0\rangle - b|1\rangle)$$
$$+ |01\rangle(-a|0\rangle + b|1\rangle)$$
$$+ |10\rangle(a|1\rangle + b|0\rangle)$$
$$+ |11\rangle(a|1\rangle - b|0\rangle)\ ]$$

Step 5. (Location A): Measure qubits #1 and #2 to obtain two bits of classical information. The result of this measurement will be one of the bit pairs $\{00, 01, 10, 11\}$.

Step 6.: Send from location A to location B (via a classical communication channel) the two classical bits obtained in Step 5.

As an intermediate summary, we have:
1) Qubit #1 has been disassembled, and
2) The information obtained during disassembly (two classical bits) has been sent to location B.

Step 7. (Location B): The two bits (i, j) received from location A are used to select from the following table a unitary transformation $U^{(i,j)}$ of $\mathcal{H}_3$, (i.e., a local unitary transformation $I_4 \otimes U^{(i,j)}$ on $\mathcal{H}_1 \otimes \mathcal{H}_2 \otimes \mathcal{H}_3$)

Rec. Bits	$U^{(i,j)}$	Future effect on qubit #3
00	$U^{(00)} = \begin{pmatrix} -1 & 0 \\ 0 & -1 \end{pmatrix}$	$-a|0\rangle - b|1\rangle \longmapsto a|0\rangle + b|1\rangle$
01	$U^{(01)} = \begin{pmatrix} -1 & 0 \\ 0 & 1 \end{pmatrix}$	$-a|0\rangle + b|1\rangle \longmapsto a|0\rangle + b|1\rangle$
10	$U^{(10)} = \begin{pmatrix} 0 & 1 \\ 1 & 0 \end{pmatrix}$	$a|1\rangle + b|0\rangle \longmapsto a|0\rangle + b|1\rangle$
11	$U^{(11)} = \begin{pmatrix} 0 & 1 \\ -1 & 0 \end{pmatrix}$	$a|1\rangle - b|0\rangle \longmapsto a|0\rangle + b|1\rangle$

Step 8. (Location B): The unitary transformation $U^{(i,j)}$ selected in Step 7 is applied to qubit #3.

As a result, qubit #3 is at location B and has the original state of qubit #1 when qubit #1 was first delivered to location A, i.e., the state

$$a|0\rangle + b|1\rangle$$

It is indeed amazing that no one knows the state of the quantum teleported qubit except possibly the individual that prepared the qubit. Knowledge of the actual state of the qubit is not required for teleportaton. If its state is unknown before the teleportation, it remains unknown after the teleportation. All that we know is that the states before and after the teleportation are the same.

13. There is much more to quantum computation

Needles to say, there is much more to quantum computation. I hope that you found this introductory paper useful. For further reading on quantum computation, we refer the reader to the many informative books on subject, such as [**1**], [**41**], [**50**], [**66**], and [**77**].

References

[1] Alber, G., T. Beth, M. Horodecki, P. Horodecki, R. Horodecki, M. Rotteler, H. Weinfurther, R. Werner, and A. Zeilinger, **"Quantum Information: An Introduction to Basic Theorectical Concepts and Experiments,"** Springer-Verlag, (2001).
[2] Aspect, A., J. Dalibard, and G. Roger, Phys. Rev. Lett., 49, (1982), p 1804.
[3] Barenco, A, C.H. Bennett, R. Cleve, D.P. DiVincenzo, N. Margolus, P. Shor, T. Sleator, J.A. Smolin, and H. Weinfurter, **Elementary gates for quantum computation**, Phys. Rev. A, **52**, (1995), pp 3475 - 3467.
[4] Beardon, Alan F., "The Geometry of Discrete Groups," Springer-Verlag, (1983).
[5] Bell, J.S., Physics, 1, (1964), pp 195 - 200.
[6] Bell, J.S., **"Speakable and Unspeakable in Quantum Mechanics,"** Cambridge University Press (1987).
[7] Bennett, C.H. et al., Phys. Rev. Lett. **70**, (1995), pp 1895.
[8] Bennett, C.H., D.P. DiVincenzo, J.A. Smolin, and W.K. Wootters, Ohys. Rev. A, **54**, (1996), pp 3824.
[9] Berman, Gennady, Gary D. Doolen, Ronnie Mainieri, and Vladimir I. Tsifrinovich, "**Introduction to Quantum Computation**," World Scientific (1999).
[10] Bernstein, Ethan, and Umesh Vazirani, **Quantum complexity theory**, Siam J. Comput., Vol. 26, No.5 (1997), pp 1411 - 1473.
[11] Brandt, Howard E., **Qubit devices and the issue of quantum decoherence**, Progress in Quantum Electronics, Vol. 22, No. 5/6, (1998), pp 257 - 370.
[12] Brandt, Howard E., **Qubit devices**, in "Quantum Computation," this AMS Proceedings of Symposia in Applied Mathematics (PSAPM).
[13] Brassard, Gilles, and Paul Bratley, **"Algorithmics: Theory and Practice,"** Printice-Hall, (1988).
[14] Brooks, Michael (Ed.), **"Quantum Computing and Communication**s," Springer-Verlag (1999).
[15] Bub, Jeffrey, **"Interpreting the Quantum World,"** Cambridge University Press (1997).
[16] Cartan, Henri, and Samuel Eilenberg, **"Homological Algebra,"** Princeton University Press, Princeton, New Jersey, (1956)
[17] Cerf, Nicholas J. and Chris Adami, "**Quantum information theory of entanglement and measurement**," in **Proceedings of Physics and Computation, PhysComp'96**, edited by J. Leao T. Toffoli, pp 65 - 71. See also http://xxx.lanl.gov/abs/quant-ph/9605039 .

[18] Cohen-Tannoudji, Claude, Bernard Diu, and Frank Laloë, "**Quantum Mechanics**," Volumes 1 & 2, John Wiley & Sons (1977)
[19] D'Espagnat, Bernard, "**Veiled Reality: Analysis of Present Day Quantum Mechanical Concepts**," Addison-Wesley (1995)
[20] D'Espagnat, Bernard, "**Conceptual Foundations of Quantum Mechanics**," (Second Edition), Addison-Wesley (1988)
[21] Cormen, Thomas H., Charles E. Leiserson, and Ronald L. Rivest, "**Introduction to Algorithms**," McGraw-Hill, (1990).
[22] Cox, David, John Little, and Donal O'Shea, "**Ideals, Varieties, and Algorithms**," (second edition), Springer-Verlag, (1996).
[23] Davies, E.B., "**Quantum Theory of Open Systems**," Academic Press, (1976).
[24] Dieks, D., Phys. Lett., **92**, (1982), p 271.
[25] Deutsch, David, "**The Fabric of Reality**," Penguin Press, New York (1997).
[26] Deutsch, David, and Patrick Hayden, **Information flow in entangled quantum systems**, http://xxx.lanl.gov/abs/quant-ph/9906007.
[27] Deutsch, David, **Quantum theory, the Church-Turing principle and the universal quantum computer**, Proc. Royal Soc. London A, **400**, (1985), pp 97 - 117.
[28] Dirac, P.A.M., "**The Principles of Quantum Mechanics**," (Fourth edition). Oxford University Press (1858).
[29] Einstein, A., B. Podolsky, N. Rosen, **Can quantum, mechanical description of physical reality be considered complete?**, Phys. Rev. **47**, 777 (1935); D. Bohm "**Quantum Theory**", Prentice-Hall, Englewood Cliffs, NJ (1951).
[30] Ekert, Artur K.and Richard Jozsa, **Quantum computation and Shor's factoring algorithm**, Rev. Mod. Phys., 68,(1996), pp 733-753.
[31] Feynman, Richard P., Robert B. Leighton, and Matthew Sands, "**The Feyman Lectures on Physics: Vol. III. Quantum Mechanics**," Addison-Wesley Publishing Company, Reading, Massachusetts (1965).
[32] Feynman, Richard P., "**Feynman Lectures on Computation**," (Edited by Anthony J.G. Hey and Robin W. Allen), Addison-Wesley, (1996).
[33] Gilmore, Robert, "**Alice in Quantumland**," Springer-Verlag (1995).
[34] Gottesman, Daniel, **The Heisenberg representation of quantum computers**, http://xxx.lanl.gov/abs/quant-ph/9807006.
[35] Gottesman, Daniel, **An introduction to quantum error correction**, in "**Quantum Computation**," this AMS Proceedings of the Symposia in Applied Mathematics (PSAPM). (http://xxx.lanl.gov/abs/quant-ph/0004072)
[36] Gottfried, "**Quantum Mechanics: Volume I. Fundamentals**," Addison-Wesley (1989).
[37] Grover, Lov K., **Quantum computer can search arbitrarily large databases by a single querry**, Phys. Rev. Letters (1997), pp 4709-4712.
[38] Grover, Lov K., **A framework for fast quantum mechanical algorithms**, http://xxx.lanl.gov/abs/quant-ph/9711043.
[39] Grover, L., Proc. 28th Annual ACM Symposium on the Theory of Computing, ACM Press, New Yorkm (1996), pp 212 - 219.
[40] Grover, L., Phys. Rev. Lett. 78, (1997), pp 325 - 328.
[41] Gruska, Jozef, "**Quantum Computing**," McGraw-Hill, (1999)
[42] Gunther, Ludwig, "An Axiomatic Basis for Quantum Mechanics: Volume I. Derivation of Hilbert Space Structure," Springer-Verlag (1985).
[43] Haag, R., "**Local Quantum Physics: Fields, Particles, Algebras**," (2nd revised edition), Springer-Verlag.
[44] Halmos, Paul R., "**Lectures on Boolean Algebras**," Van Nostrand, (1967).
[45] Halmos, Paul R., "**Finite-Dimensional Vector Spaces**," Van Nostrand, (1958).
[46] Hardy, G.H., and E.M. Wright, "**An Introduction to the Theory of Numbers**," Oxford Press, (1965).
[47] Heisenberg, Werner, "**The Physical Principles of Quantum Theory**," translated by Eckart and Hoy, Dover.
[48] Helstrom, Carl W., "**Quantum Detection and Estimation Theory**," Academic Press (1976).
[49] Hey, Anthony J.G. (editor), "**Feynman and Computation**," Perseus Books, Reading, Massachusetts, (1998).

[50] Hirvensalo, Mika, **"Quantum Computing,"** Springer-Verlag, (2001).
[51] Horodecki, O., M. Horodecki, and R. Horodecki, Phys. Rev. Lett. **82**, (1999), pp 1056.
[52] Holevo, A.S., **"Probabilistic and Statistical Aspects of Quantum Theory,"** North-Holland, (1982).
[53] Hoyer, Peter, **Efficient quantum transforms**, http://xxx.lanl.gov/abs/quant-ph/9702028.
[54] Jauch, Josef M., **"Foundations of Quantum Mechanics,"** Addison-Wesley Publishing Company, Reading, Massachusetts (1968).
[55] Jozsa, Richard, **Searching in Grover's Algorithm**, http://xxx.lanl.gov/abs/quant-ph/9901021.
[56] Jozsa, Richard, **Quantum algorithms and the Fourier transform**, quant-ph preprint archive 9707033 17 Jul 1997.
[57] Jozsa, Richard, Proc. Roy. Soc. London Soc., Ser. A, 454, (1998), 323 - 337.
[58] Kauffman, Louis H., **Quantum topology and quantum computing**, in "Quantum Computation," this AMS Proceedings of the Symposia in Applied Mathematics (PSAPM).
[59] Kitaev, A., **Quantum measurement and the abelian stabiliser problem,** (1995), http://xxx.lanl.gov/abs/quant-ph/9511026.
[60] Kitaev, Alexei, **Quantum computation with anyons**, this AMS Proceedings of the Symposia in Applied Mathematics (PSAPM).
[61] Lang, Serge, **"Algebra,"** Addison- Wesley (1971).
[62] Lenstra, A.K., and H.W. Lenstra, Jr., eds., **"The Development of the Number Field Sieve,"** Lecture Notes in Mathematics, Vol. 1554, Springer-Velag, (1993).
[63] Lenstra, A.K., H.W. Lenstra, Jr., M.S. Manasse, and J.M. Pollard, **The number field sieve**. Proc. 22nd Annual ACM Symposium on Theory of ComputingACM, New York, (1990), pp 564 - 572. (See exanded version in Lenstra & Lenstra, (1993), pp 11 - 42.)
[64] LeVeque, William Judson, **"Topics in Number Theory: Volume I,"** Addison-Wesley, (1958).
[65] Linden, N., S. Popescu, and A. Sudbery, **Non-local properties of multi-particle density matrices**, http://xxx.lanl.gov/abs/quant-ph/9801076.
[66] Lo, Hoi-Kwong, Tim Spiller & Sandu Popescu(editors), **"Introduction to Quantum Computation & Information,"** World Scientific (1998).
[67] Lomonaco, Samuel J., Jr., **"A entangled tale of quantum entanglement,** in "Quantum Computation," this AMS Proceedings of the Symposia in Applied Mathematics (PSAPM). (http://xxx.lanl.gov/abs/quant-ph/0101120)
[68] Lomonaco, Samuel J., Jr., **A quick glance at quantum cryptography**, Cryptologia, Vol. 23, No. 1, January,1999, pp 1-41. (http://xxx.lanl.gov/abs/quant-ph/9811056)
[69] Lomonaco, Samuel J., Jr., **A talk on quantum cryptography: How Alice Outwits Eve**, in "Coding Theory and Cryptography: From Enigma and Geheimsschreiber to Quantum Theory," edited by David Joyner, Springer-Verlag, (2000), pp 144 - 174. Also in this AMS Proceedings of the Symposia in Applied Mathematics (PSAPM).
[70] Lomonaco, Samuel J., Jr., **Shor's quantum factoring algorithm**, in "Quantum Computation," this AMS Proceedings of the Symposia in Applied Mathematics (PSAPM). (http://xxx.lanl.gov/abs/quant-ph/0010034)
[71] Lomonaco, Samuel J., Jr., **Grover's quantum search algorithm**, in "Quantum Computation," this AMS Proceedings of the Symposia in Applied Mathematics (PSAPM). (http://xxx.lanl.gov/abs/quant-ph/0010040)
[72] Mackey, George W., **"Mathematical Foundations of Quantum Mechanics,"** Addison-Wesley (1963).
[73] Milburn, Gerald J., **"The Feynman Processor,"** Perseus Books, Reading, Massachusetts (1998)
[74] Miller, G.L., **Riemann's hypothesis and tests for primality**, J. Comput. System Sci., 13, (1976), pp 300 - 317.
[75] von Neumann, John, **"Mathematical Foundations of Quantum Mechanics,"** Princeton University Press.
[76] Nielsen, M.A., **Continuity bounds on entanglement**, Phys. Rev. A, Vol. 61, 064301, pp 1-4.
[77] Nielsen, Michael A., and Isaac L. Chuang, **"Quantum Computation and Quantum Information,"** Cambridge University Press, (2000).

[78] Omnès, Roland, **"An Interpretation of Quantum Mechanics,"** Princeton University Press, Princeton, New Jersey, (1994).
[79] Omnès, Roland, "**Understanding Quantum Mechanics**," Princeton University Press (1999).
[80] Penrose, Roger, "**The Large, the Small and the Human Mind**," Cambridge University Press, (1997).
[81] Peres, Asher, **"Quantum Theory: Concepts and Methods,"** Kluwer Academic Publishers, Boston, (1993).
[82] Raymond, Pierre, "**Field Theory: A Modern Primer**," Addison-Wesley (1989).
[83] Piron, C., "**Foundations of Quantum Physics**," Addison-Wesley, (1976).
[84] Sakurai, J.J., "**Modern Quantum Mechanics**," (Revised edition), Addison-Wesley Publishing Company, Reading, Massachusetts (1994).
[85] Schumacher, Benjamin, **Sending entanglement through noisy quantum channels**, (22 April 1996), http://xxx.lanl.gov/abs/quant-ph/9604023.
[86] Schwinger, Julian, **"Quantum Mechanics: Symbolism of Quantum Measurement,"** Springer-Verlag, (2001).
[87] Shor, Peter W., **Polynomial time algorithms for prime factorization and discrete logarithms on a quantum computer**, SIAM J. on Computing, 26(5) (1997), pp 1484 - 1509. (http://xxx.lanl.gov/abs/quant-ph/9508027)
[88] Shor, Peter W., **Introduction to quantum algorithms**, in "**Quantum Computation**," this AMS Proceedings of the Symposia in Applied Mathematics (PSAPM). (http://xxx.lanl.gov/abs/quant-ph/0005003)
[89] Stinson, Douglas R., "**Cryptography: Theory and Practice**," CRC Press, Boca Raton, (1995).
[90] Vazirani, Umesh, **Quantum complexity theory**, in "**Quantum Computation**," this AMS Proceedings of the Symposia in Applied Mathematics (PSAPM).
[91] Williams, Collin P., and Scott H. Clearwater, "**Explorations in Quantum Computation**," Springer-Verlag (1997).
[92] Williams, Colin, and Scott H. Clearwater, "**Ultimate Zero and One**," Copernicus, imprint by Springer-Verlag, (1998).
[93] Wootters, W.K., and W.H. Zurek, **A single quantum cannot be cloned,** Nature, Vol. 299, 28 October 1982, pp 982 - 983.

UNIVERSITY OF MARYLAND BALTIMORE COUNTY, BALTIMORE, MD 21250
E-mail address: Lomonaco@UMBC.EDU
URL: http://www.csee.umbc.edu/~lomonaco

Qubit Devices

Howard E. Brandt

ABSTRACT. This lecture presents brief mathematical descriptions of a variety of potential qubit devices. The qubit devices examined include an interaction-free detector, a quantum key receiver, quantum games, quantum gates, qubit entanglers, Bell state synthesizers, Bell state analyzers, quantum dense coders, entanglement swappers, quantum teleporters, quantum copiers, quantum error correctors, quantum computers, and quantum robots.

Contents

1. Introduction
2. Interaction-free detector
3. Quantum key receiver
4. Quantum decoherence
5. Quantum games
6. Quantum gates
7. Single-photon balanced Mach-Zehnder interferometer
8. Qubit entanglers
9. EPR-pair sources
10. Bell-state synthesizer
11. Quantum dense coder
12. Bell-state analyzer
13. Entanglement swappers
14. Quantum teleporter
15. Quantum copiers
16. All-optical quantum information processors
17. Universal quantum computer
18. Quantum simulators
19. Quantum factorizer
20. Quantum register
21. Quantum error correctors

2000 *Mathematics Subject Classification.* [2000]Primary 81P68, 81V80, 68-01, 68-02, 68Q05, 94-02, 81V45.

Key words and phrases. Quantum information processing, quantum computation, quantum communication.

22. Quantum computers
23. Quantum robots
24. Conclusions
25. Acknowledgement
References

1. Introduction

A *qubit device* [1] is a physical implementation of a set of quantum bits, or qubits, as they are popularly known [2,3]. A *qubit* is a quantum system with a two-dimensional Hilbert space, capable of existing in a superposition of Boolean states, and also capable of being entangled with the states of other qubits [4]. Qubit devices include quantum computers, quantum gates, quantum key receivers, entanglement swappers, quantum teleporters, quantum dense coders, interaction-free detectors, quantum robots, quantum games, quantum copiers, etc.

The exciting new interdisciplinary fields of quantum information processing, quantum computing, quantum communication, and quantum cryptography are rich with a plethora of potentially useful qubit devices [5]. The major obstacle to the successful development of these devices is the phenomenon of quantum decoherence, in which even weak interactions of the qubits with noncomputational environmental degrees of freedom can rapidly attenuate elements of the density matrix off-diagonal in the computational basis, obliterating essential quantum coherence and quantum entanglement. This lecture is an abbreviated version of recent work by the author [5], with some additions; it presents brief mathematical descriptions of a variety of potential qubit devices, and includes expository discussions of the issue of quantum decoherence as it relates to the possible practical development of these devices.

2. Interaction-free detector

The interaction-free detector [6–8] provides a simple example of the practical use of path qubits. (The two-dimensional Hilbert space of a path qubit represents two possible quantum-interfering paths of a photon in spacetime [5].) In this photonic device, the presence of an opaque object inside one arm of an interferometer destroys the interference of an incident photon, sometimes signaling the presence of the object, even though the photon could not have taken a path intersecting the object.

A simple example of an interaction-free detector [8] is schematized in Fig. 1. Here, a single photon in polarization state $|u\rangle$ enters a Michelson interferometer consisting of a 50/50 beamsplitter (BS), a 90° phase shifter (ϕ), two mirrors (M_1 and M_2), and a single-photon photodetector (D). All optical elements are here assumed to be ideal and not to affect the polarization u. An object with transmission coefficient T may or may not lie in the path between the beamsplitter and mirror M_1. If it is not present, its absence is effectively described by $T = 1$. The quantum mechanical probability amplitude [9–12] that the photon reflects at the beamsplitter BS toward mirror M_1 is $2^{-1/2}i$. The factor of $2^{-1/2}$ is due to the 50/50 beamsplitter (with transmission coefficient $1/2$), and the factor of i is due to the reflection at the beamsplitter, which produces a phase shift $e^{i\pi/2} = i$ [13–15]. Then the probability amplitude A_1 that the photon also passes through the object with transmission coefficient T, reflects from mirror M_1, returns through the object and the beamsplitter, and goes on to the detector D is given by

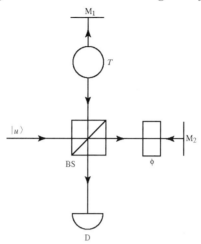

Figure 1. Interaction-free detector.

(1) $$A_1 = \left(2^{-1/2}\right)\left(T^{1/2}\right)\left(T^{1/2}\right)\left(2^{-1/2}i\right) = 2^{-1}iT.$$

The factor $\left(2^{-1/2}i\right)$ is due to the reflection at the beamsplitter, the two factors of $T^{1/2}$ are due to the two passages through the object characterized by its transmission coefficient T, and the factor $\left(2^{-1/2}\right)$ is due to the passage through the 50/50 beamsplitter with transmission coefficient $1/2$. Similarly, the probability amplitude A_2 that the photon initially passes along the other possible path directly through the beamsplitter BS, passes through the 90° phase shifter ϕ to mirror M_2, returns back again to the beamsplitter, and is then reflected to the detector D is given by

(2) $$A_2 = \left(2^{-1/2}i\right)\left(e^{i\pi/2}\right)\left(e^{i\pi/2}\right)\left(2^{-1/2}\right) = -2^{-1}i.$$

Here, reading from right to left, the factor $\left(2^{-1/2}\right)$ is due to the photon passing through the 50/50 beamsplitter with transmission coefficient $1/2$, the factors of $e^{i\pi/2}$ are due to the two traversals of the 90° phase shifter, and the factor $\left(2^{-1/2}i\right)$ is due to the reflection at the beamsplitter. It follows that the total quantum mechanical probability amplitude A that the photon triggers the detector D is given by the sum of the amplitudes for both possible paths of the photon through the interferometer to the detector, namely,

(3) $$A \equiv A_1 + A_2.$$

This is single-particle quantum interference. The beamsplitter effectively converts the state of the photon into a two-state system, corresponding to the two possible paths 1 and 2 of the photon through the interferometer to the detector. This is a *path qubit* [5]. Substituting Eqs. (1) and (2) in Eq. (3), one obtains for the total amplitude

(4) $$A = 2^{-1}i(T - 1).$$

The probability that the detector D is triggered is then given by

(5) $$P_D(T) = |A|^2 = \frac{1}{4}(T - 1)^2.$$

If the object is not present, then effectively $T = 1$, and according to Eq. (5), the probability that the detector is triggered is vanishing, namely,

(6) $$P_D(1) = 0.$$

Therefore, if the object is not there, the detector D cannot be triggered. If the object is present and opaque, namely $T = 0$, then according to Eq. (5),

(7) $$P_D(0) = \frac{1}{4}.$$

Thus, if the object is there, the detector D will be triggered 25% of the time (probability 1/4 for each single incident photon). The probability that the photon is absorbed if the object is present is clearly 1/2, because that is the probability that the photon entering the interferometer is reflected to the object by the 50/50 beamsplitter. The probability that the photon instead returns to the entrance port is clearly 1/4, and the probabilities add to unity as they must. Thus, with this particular interaction-free detector, one can detect the presence of the object 25% of the time. Clearly, when the object is present and the detector is triggered, the photon cannot have taken the path intersecting the object, because if it did it would be absorbed before reaching the detector. This is the basis for saying that the photon does not "interact" with the object. However, there must be an interaction term in any Hamiltonian describing the system [8].

3. Quantum key receiver

Another simple example of a photonic qubit device is a quantum key receiver based on a positive operator valued measure (POVM). Specifically, a POVM is a set of nonnegative Hermitian operators A_μ that act in the Hilbert space of a quantum system and sum to the identity operator, namely [16–18],

(8) $$\sum_\mu A_\mu = 1.$$

The index μ labels the various possible outcomes of a measurement implementing the POVM. The probability P_μ of outcome μ, if the system is in a state described by the density matrix ρ, is given by

(9) $$P_\mu = Tr\left(A_\mu \rho\right).$$

The advantage of a POVM is that it may have a lower inconclusive rate and may allow the extraction of more information than can the usual von Neumann-type projective measurement [16,20].

Because of the noncommutativity of nonorthogonal photon polarization projective measurement operators, a simple von Neumann-type projective measurement of a particular polarization state cannot conclusively distinguish between that state and one to which it is nonorthogonal. If one wants to be able to distinguish conclusively between two nonorthogonal photon states $|u\rangle$ and $|v\rangle$ at least some of the time, it is useful to consider a POVM used in quantum cryptography [16,19–25]. Quantum cryptography [26–31] provides a number of practical applications for qubit devices.

In quantum cryptography, one is able, in principle, to produce a shared key whose security is guaranteed by the laws of quantum mechanics. The key is a random bit sequence which, when added to the message (encoded in binary), forms the encrypted message, and which, when subtracted from the encrypted message, yields the decrypted message. Qubit devices can be used in transmitting and receiving the key, and also in eavesdropping.

A POVM used in quantum cryptography consists of the following set of three nonnegative Hermitian operators:

$$A_u = (1 + \langle u|v\rangle)^{-1} [1 - |v\rangle \langle v|], \tag{10}$$

$$A_v = (1 + \langle u|v\rangle)^{-1} [1 - |u\rangle \langle u|], \tag{11}$$

$$A_? = 1 - A_u - A_v, \tag{12}$$

in which kets $|u\rangle$ and $|v\rangle$ represent nonorthogonal single-photon states. The POVM operators, Eqs. (10) to (12), clearly satisfy Eq. (8). Specifically in the present work, the states $|u\rangle$ and $|v\rangle$ are taken to be linear-polarization states with the Dirac bracket $\langle u|v\rangle$ given by

$$\langle u|v\rangle = \cos\theta, \tag{13}$$

where θ is the angle between the two polarization states. A general qubit state is given by

$$|\psi\rangle = \alpha |u\rangle + \beta |v\rangle, \tag{14}$$

where α and β are complex numbers. The operators (10) to (12) are clearly Hermitian; however, they are not projection operators, since the product of any one of them with itself does not yield itself. Also, for nonorthogonal states they do not commute.

The probability that an arbitrary qubit $|\psi\rangle$ given by Eq. (14) is measured to be in the u-polarization state can be calculated with Eqs. (9), (10), (13), and (14); the calculation yields

$$P_u = Tr(A_u |\psi\rangle \langle\psi|) = \langle\psi|A_u|\psi\rangle = |\alpha|^2 (1 - \cos\theta). \tag{15}$$

The first equality holds, since for a pure state $|\psi\rangle$, the density matrix is $|\psi\rangle \langle\psi|$. A_u is a positive operator, as it must be. This follows since $|\psi\rangle$ can represent any state in the two-dimensional Hilbert space of states, and the right-hand side of Eq. (15) is nonnegative. This must be so, since P_u is a probability. Analogously, one obtains

$$P_v = \langle\psi|A_v|\psi\rangle = |\beta|^2 (1 - \cos\theta), \tag{16}$$

and

$$P_? = \langle\psi|A_?|\psi\rangle = |\alpha + \beta|^2 \cos\theta, \tag{17}$$

both of which are also clearly nonnegative. In Eq. (16), the qubit is measured to be in the state $|v\rangle$. In Eq. (17), $P_?$ is the probability of an inconclusive measurement, meaning that it is undecided whether the qubit is in state $|u\rangle$ or $|v\rangle$. If the state $|\psi\rangle$, Eq. (14), is normalized to unity, then

$$1 = \langle\psi|\psi\rangle = |\alpha|^2 + (\alpha^*\beta + \alpha\beta^*)\cos\theta + |\beta|^2, \tag{18}$$

and it then follows from Eqs. (15) to (17) that the probabilities sum to unity, as they must:

$$P_u + P_v + P_? = 1. \tag{19}$$

When the incident photon is in the state $|u\rangle$, one has $(\alpha, \beta) = (1, 0)$, and Eq. (16) becomes $\langle u| A_v |u\rangle = 0$. When the incident photon is in the state $|v\rangle$, one has $(\alpha, \beta) = (0, 1)$, and Eq. (15) becomes $\langle v| A_u |v\rangle = 0$. Therefore, when an ideal detector representing the operator A_u responds positively, it follows that a photon with a v-polarization state cannot have been received. Likewise, when an ideal detector representing the operator A_v responds, a photon with a u-polarization state cannot have been received. The operator $A_?$ represents inconclusive responses, since Eq. (17) is nonvanishing for $(\alpha, \beta) = (0,1)$ or $(1,0)$. Thus a u-polarized photon can result in a nonzero expectation value (and the associated response) only for detectors representing the A_u or $A_?$ operators. A v-polarized photon excites only the A_v or $A_?$ detectors. It follows that the POVM of Eqs. (10) to (12) distinguishes conclusively between two nonorthogonal states $|u\rangle$ and $|v\rangle$ at least some of the time.

For the purpose of secure key generation in quantum cryptography, one can employ a train of single photons having two possible equally likely nonorthogonal polarization states $|u\rangle$ and $|v\rangle$, which encode 0 and 1, respectively, to securely communicate a random bit sequence between a sender (Alice) and a receiver (Bob). To detect a photon, Bob can use a *quantum key receiver*. If Bennett's two-state protocol [25] is employed, it can be advantageous for the receiver to be based on a POVM [16,19,20]. In the two-state protocol, a positive response of the receiver (the reception of a photon in a u- or v-polarization state) is publicly communicated by Bob to Alice, without revealing which polarization was detected, and the corresponding bits then constitute the preliminary key shared by Alice and Bob. Bits corresponding to photons that do not excite the u- or v-polarization state detectors in the receiver are excluded from the key. The ideal receiver must be such that if its u-polarization detector is excited, then Bob is certain that the u-polarization state was received. Also, if the v-polarization detector is excited, then a v-polarization state was received. Because of the noncommutativity of nonorthogonal photon polarization-measurement operators representing nonorthogonal photon polarization states [5], and also because arbitrary quantum states cannot be cloned [32,33], any attempt to eavesdrop can, in principle, be detected by Bob and Alice.

A quantum key receiver based on a POVM is shown schematically in Fig. 2. It is an all-optical implementation of the POVM given in Eqs. (10) to (12) [21–23]. The straight lines with arrows represent possible optical pathways for a photon to move through the device. The path labeled $|\psi\rangle$ is the incoming path for a photonic qubit represented by an arbitrary polarization state, given by Eq. (14). Also in Fig. 2, D_u, D_v, and $D_?$ designate photodetectors representing the measurement operators A_u, A_v, and $A_?$, respectively. Shown also is a Wollaston prism W [34], which is aligned so that an incident photon with polarization vector $\hat{e}_{u+v}$ takes the path labeled by the state $|\psi_1\rangle$ and $\hat{e}_{u+v}$, and not the path labeled by the polarization vector $\hat{e}_{u-v}$ and the state $|\psi_2\rangle$. Here $\hat{e}_{u+v}$ denotes a unit polarization vector corresponding to polarization state $|u+v\rangle = |u\rangle + |v\rangle$, and is perpendicular to the unit polarization vector $\hat{e}_{u-v}$ corresponding to the polarization state $|u-v\rangle = |u\rangle - |v\rangle$. It follows from Eq. (13) that the states $|u+v\rangle$ and $|u-v\rangle$ are orthogonal, namely,

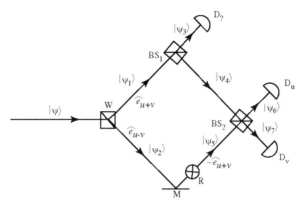

Figure 2. Quantum key receiver.

(20) $$\langle u+v|u-v\rangle = 0.$$

Also, clearly,

(21) $$\hat{e}_{u+v} \cdot \hat{e}_{u-v} = 0,$$

which is consistent with Eqs. (20) and (13). In accordance with the property of a Wollaston prism of separating orthogonal polarization states, an incident photon with polarization vector $\hat{e}_{u-v}$ takes the path labeled by the state $|\psi_2\rangle$.

The device has two beamsplitters [34], designated BS_1 and BS_2 in Fig. 2. Beamsplitter BS_2 is a 50/50 beamsplitter for a photon entering either of its entrance ports. Beamsplitter BS_1 has transmission and reflection coefficients specified in Eqs. (34) and (35) below. The two paths from the Wollaston prism to beamsplitter BS_2 have equal optical path lengths. Also shown in Fig. 2 is a 90° polarization rotator [34], designated R, which transforms a photon with polarization vector $\hat{e}_{u-v}$ into one with polarization vector $-\hat{e}_{u+v}$. There is also a single mirror M, as shown in Fig. 2. All optical elements are here taken to be ideal.

The state of a photon taking the path designated by the state $|\psi_1\rangle$ in Fig. 2 is given by

(22) $$|\psi_1\rangle = |\hat{e}_{u+v}\rangle\langle\hat{e}_{u+v}|\psi\rangle,$$

where $|\hat{e}_{u+v}\rangle$ represents a unit ket corresponding to polarization vector $\hat{e}_{u+v}$. Clearly,

(23) $$|\hat{e}_{u+v}\rangle = \frac{|u\rangle + |v\rangle}{[(\langle u| + \langle v|)(|u\rangle + |v\rangle)]^{1/2}}.$$

Substituting Eq. (23) in Eq. (22), and using Eqs. (14) and (13), one obtains

(24) $$|\psi_1\rangle = 2^{-1/2}(\alpha+\beta)(1+\cos\theta)^{1/2}|\hat{e}_{u+v}\rangle.$$

One also has

(25) $$|\psi_2\rangle = |\hat{e}_{u-v}\rangle\langle\hat{e}_{u-v}|\psi\rangle,$$

where

(26) $$|\hat{e}_{u-v}\rangle = \frac{|u\rangle - |v\rangle}{[(\langle u| - \langle v|)(|u\rangle - |v\rangle)]^{1/2}}$$

is a unit ket corresponding to polarization vector $\hat{e}_{u-v}$. Next, using Eqs. (25), (26), (14), and (13), one obtains

$$|\psi_2\rangle = 2^{-1/2}(\alpha - \beta)(1 - \cos\theta)^{1/2}|\hat{e}_{u-v}\rangle. \tag{27}$$

This device exploits the entanglement of path and polarization qubits. For example, the state $|\psi_{12}\rangle$ of the photon exiting the Wollaston prism is

$$|\psi_{12}\rangle = |\psi_1\rangle + |\psi_2\rangle = 2^{-1/2}(\alpha + \beta)(1 + \cos\theta)^{1/2}|1\rangle \otimes |\hat{e}_{u+v}\rangle$$
$$+ 2^{-1/2}(\alpha - \beta)(1 - \cos\theta)^{1/2}|2\rangle \otimes |\hat{e}_{u-v}\rangle, \tag{28}$$

in which the kets $|1\rangle$ and $|2\rangle$ are unit kets corresponding to the upper and lower paths, respectively (directly to the right of W in Fig. 2). The polarization qubit $\{|u\rangle, |v\rangle\}$ is entangled with the path qubit $\{|1\rangle, |2\rangle\}$.

For ideal photodetectors D_u, D_v, and $D_?$, it is evident from Fig. 2 and Eq. (9) that one must require

$$P_u = |\psi_6|^2 = \langle\psi_6|\psi_6\rangle = \langle\psi|A_u|\psi\rangle, \tag{29}$$

$$P_v = |\psi_7|^2 = \langle\psi_7|\psi_7\rangle = \langle\psi|A_v|\psi\rangle, \tag{30}$$

and

$$P_? = |\psi_3|^2 = \langle\psi_3|\psi_3\rangle = \langle\psi|A_?|\psi\rangle, \tag{31}$$

respectively, in order that the expectation values of A_u, A_v, and $A_?$, measured by detectors D_u, D_v, and $D_?$, respectively, equal the probabilities P_u, P_v, and $P_?$ that a photon is incident. From Eqs. (31), (17), and (14), it follows that, up to an irrelevant phase factor, one has

$$|\psi_3\rangle = (\alpha + \beta)(\cos\theta)^{1/2}|\hat{e}_{u+v}\rangle. \tag{32}$$

For a *single* photon incident on beamsplitter BS_1, one can effectively ignore the unused vacuum port of the beamsplitter (see Ref. 23 and also p. 9 of Ref. 35). The transmission coefficient T_1 of beamsplitter BS_1 must then be given by

$$T_1 = \frac{\langle\psi_3|\psi_3\rangle}{\langle\psi_1|\psi_1\rangle}. \tag{33}$$

Therefore, substituting Eqs. (32) and (24) in Eq. (33), one obtains

$$T_1 = 1 - \tan^2(\theta/2). \tag{34}$$

The corresponding reflection coefficient R_1 is

$$R_1 = 1 - T_1 = \tan^2(\theta/2). \tag{35}$$

From Fig. 2, it is also evident that

$$\langle\psi_4|\psi_4\rangle = R_1\langle\psi_1|\psi_1\rangle; \tag{36}$$

substituting Eqs. (24) and (35) in Eq. (36), one obtains

$$\langle\psi_4|\psi_4\rangle = \frac{1}{2}|\alpha + \beta|^2(1 - \cos\theta). \tag{37}$$

Since the reflection at BS_1 results in a $\pi/2$-phase shift [13–15], resulting in a factor of $\exp(i\pi/2) = i$, it follows from Eqs. (37) and (24) that

$$|\psi_4\rangle = i2^{-1/2}(\alpha + \beta)(1 - \cos\theta)^{1/2}|\hat{e}_{u+v}\rangle. \tag{38}$$

Since the polarization rotator R converts polarization in the direction $\hat{e}_{u-v}$ into that in the direction $-\hat{e}_{u+v}$, it follows from Eq. (27) that

(39) $$|\psi_5\rangle = -2^{-1/2}(\alpha - \beta)(1-\cos\theta)^{1/2}|\hat{e}_{u+v}\rangle.$$

Since the beamsplitter BS_2 is a 50/50 beamsplitter, its reflection coefficient is

(40) $$R_2 = \frac{1}{2}$$

and its transmission coefficient is

(41) $$T_2 = \frac{1}{2}.$$

It then follows that

(42) $$|\psi_6\rangle = 2^{-1/2}|\psi_5\rangle + i2^{-1/2}|\psi_4\rangle,$$

and

(43) $$|\psi_7\rangle = 2^{-1/2}|\psi_4\rangle + i2^{-1/2}|\psi_5\rangle.$$

Therefore, substituting Eqs. (38) and (39) in Eqs. (42) and (43), one obtains

(44) $$|\psi_6\rangle = -\alpha(1-\cos\theta)^{1/2}|\hat{e}_{u+v}\rangle,$$

and

(45) $$|\psi_7\rangle = i\beta(1-\cos\theta)^{1/2}|\hat{e}_{u+v}\rangle.$$

For an ideal detector, Eq. (29) holds, and by substituting Eq. (44), one gets

(46) $$\langle\psi|A_u|\psi\rangle = |\alpha|^2(1-\cos\theta),$$

consistent with Eq. (15). Similarly, from Eqs. (30) and (45), it follows that

(47) $$\langle\psi|A_v|\psi\rangle = |\beta|^2(1-\cos\theta),$$

consistent with Eq. (16). Thus, the device depicted in Fig. 2 satisfies all the appropriate statistics, Eqs. (29) to (31), and is a faithful all-optical implementation of the POVM given by Eqs. (10) to (12).

Since the quantum key receiver is all-optical, all optical path lengths are only of the order of tens of centimeters, and decohering interactions with the photonic qubits are negligible, it follows that, generally, decoherence is not an issue for this device. However, since it is a very serious issue for many other prospective qubit devices, the following section briefly reviews the physics of quantum decoherence.

4. Quantum decoherence

Consider a two-state system in the absence of environmental interactions. The state vector $|\psi\rangle$ for such a two-state closed system lies in a two-dimensional Hilbert space and is given by

(48) $$|\psi\rangle = \alpha_0|0\rangle + \alpha_1|1\rangle,$$

where $|0\rangle$ and $|1\rangle$ are kets representing the two states, here also serving as orthonormal basis vectors, and α_0 and α_1 are complex numbers. Thus, one has

(49) $$\langle 0|0\rangle = \langle 1|1\rangle = 1, \quad \langle 0|1\rangle = 0.$$

The corresponding density operator is given by [36,16]

(50) $$\rho = |\psi\rangle\langle\psi|.$$

Substituting Eq. (48) in Eq. (50), one obtains for the density operator of this two-state closed system

$$\rho = |\alpha_0|^2 |0\rangle\langle 0| + \alpha_0 \alpha_1^* |0\rangle\langle 1| + \alpha_0^* \alpha_1 |1\rangle\langle 0| + |\alpha_1|^2 |1\rangle\langle 1|. \tag{51}$$

The corresponding density matrix is

$$[\rho_{mn}] = [\langle m|\rho|n\rangle] = \begin{bmatrix} |\alpha_0|^2 & \alpha_0 \alpha_1^* \\ \alpha_0^* \alpha_1 & |\alpha_1|^2 \end{bmatrix}. \tag{52}$$

The diagonal components are the populations, and the off-diagonal components are the coherences [36]. The populations measure the probabilities that the system is in either state, and the coherences measure the amount of interference between the states. The expectation value of any observable represented by an operator A for the two-state system is given by

$$\langle\psi|A|\psi\rangle = Tr(\rho A) = \sum_{mn} \rho_{mn} A_{nm}, \tag{53}$$

and it is clear that, in general, the coherences are as important as the populations in determining expectation values of observables.

Generally, a system does not exist in absolute isolation, and possible interactions with both the external and internal environments must be taken into account. If the two states of interest are part of an object containing other degrees of freedom, the latter constitute the internal environment, and the external environment is external to the object. For complex systems, the two states of interest might themselves represent two collective observables [37]. Consider now, therefore, a two-state system with state vector $|\psi(t)\rangle$ at time t, including environmental interactions:

$$|\psi(t)\rangle = \alpha_0 |0\rangle \otimes |e_0\rangle + \alpha_1 |1\rangle \otimes |e_1\rangle, \tag{54}$$

in which now the two possible states of the system, $|0\rangle$ and $|1\rangle$, through unitary evolution, have become entangled with the corresponding normalized environmental states $|e_0\rangle$ and $|e_1\rangle$, respectively. The environmental states are, in general, nonorthogonal. Here $\otimes$ denotes the tensor product. The density operator becomes

$$\begin{aligned}\rho(t) &= |\psi(t)\rangle\langle\psi(t)| \\ &= |\alpha_0|^2 |0\rangle \otimes |e_0\rangle\langle 0| \otimes \langle e_0| + \alpha_0 \alpha_1^* |0\rangle \otimes |e_0\rangle\langle 1| \otimes \langle e_1| \\ &\quad + \alpha_0^* \alpha_1 |1\rangle \otimes |e_1\rangle\langle 0| \otimes \langle e_0| + |\alpha_1|^2 |1\rangle \otimes |e_1\rangle\langle 1| \otimes \langle e_1|.\end{aligned} \tag{55}$$

If we are interested only in what the two-state system is doing, and not the environment, one need only know the reduced density matrix of the two-state system, with the environmental states traced out [36,37]. For this purpose, choose as environmental basis vectors $|e_0\rangle$ and $|e_0^\perp\rangle$, where

$$\langle e_0^\perp | e_0\rangle = 0, \quad \langle e_0 | e_1\rangle \equiv \cos\theta, \quad \langle e_0^\perp | e_1\rangle = \sin\theta, \quad \langle e_0 | e_0\rangle = 1, \quad \langle e_1 | e_1\rangle = 1. \tag{56}$$

The reduced density matrix $\rho_s(t)$ of the two-state system is then given by

$$\rho_s(t) = Tr_e \rho(t) = \langle e_0 | \rho(t) | e_0\rangle + \langle e_0^\perp | \rho(t) | e_0^\perp\rangle, \tag{57}$$

where Tr_e denotes the trace over the environmental basis states. Substituting Eq. (55) in Eq. (57), and using Eqs. (56), one obtains

$$\rho_s(t) = |\alpha_0|^2 |0\rangle\langle 0| + \alpha_0 \alpha_1^* \cos\theta |0\rangle\langle 1|$$
(58)
$$+ \alpha_0^* \alpha_1 \cos\theta |1\rangle\langle 0| + |\alpha_1|^2 (\cos^2\theta + \sin^2\theta) |1\rangle\langle 1|.$$

If one uses the trigonometric identity $\cos^2\theta + \sin^2\theta = 1$, Eq. (58) becomes

$$\rho_s(t) = |\alpha_0|^2 |0\rangle\langle 0| + \alpha_0 \alpha_1^* \cos\theta |0\rangle\langle 1|$$
(59)
$$+ \alpha_0^* \alpha_1 \cos\theta |1\rangle\langle 0| + |\alpha_1|^2 |1\rangle\langle 1|.$$

Comparing Eq. (59) with Eq. (51), one can see that, as a result of including environmental interactions, the coherences each contain an additional factor of $\cos\theta$, the overlap between the environmental states (see Eqs. (56)). The system and its environment evolve, interacting incessantly, and because of decoherence, the overlap between the environmental states $|e_0\rangle$ and $|e_1\rangle$ can rapidly become negligible; one then has orthogonalization of the environmental basis states, namely,

(60)
$$\cos\theta \equiv \langle e_0|e_1\rangle \longrightarrow 0.$$

In this case, Eq. (59) becomes

(61)
$$\rho_s(t) \xrightarrow[\cos\theta \to 0]{} |\alpha_0|^2 |0\rangle\langle 0| + |\alpha_1|^2 |1\rangle\langle 1|.$$

Decoherence results from interactions with the environment (external and internal). As a result of the decoherence, the reduced density matrix Eq. (61) becomes, effectively, a statistical mixture. The dynamical evolution represented by Eq. (61) is, of course, nonunitary: although the evolution of the total density matrix representing a system and its environment in general evolves unitarily in accordance with the Schrödinger equation, a reduced density matrix does not. Naturally, the details of the evolution (Eq. (61)) depend on the specific structure of the total Hamiltonian of the system together with its environment, including all possible interactions. For macroscopic, mesoscopic, and many microscopic systems, the environment commonly has an enormous (or at least large) Hilbert space and a crowded energy spectrum. Heuristically, in terms of perturbation theory, close energy levels result in high sensitivity to perturbations. Two slightly different perturbations may lead to very different perturbed wave functions, which become orthogonal. Environmental wave functions have many variables, and vanishing wave-function overlap in one variable is sufficient for orthogonality. The Hilbert space of environmental states can become so enormous that two states have a small probability of not being orthogonal [37]. The resulting loss of phase correlations in the high-dimensional environmental configuration space results in orthogonalization of the environmental states that are correlated with the system states. This is the phenomenon of decoherence, resulting in orthogonalization of the environmental basis states and rapid vanishing of the off-diagonal components of the reduced density matrix [37–50,1,5]. For a complex macroscopic, or even a mesoscopic, two-state system, the orthogonalization, Eq. (61), usually occurs extremely rapidly. The qubits, representing computational degrees of freedom in quantum information processing systems, are two-state systems, and their implementation must be chosen such that interactions with noncomputational internal and external environmental degrees of freedom are small enough that decoherence of the qubits is sufficiently slow. This is, in general, difficult to achieve.

5. Quantum games

The classical theory of games [51–55] is currently being generalized to include quantum games [56–59]. These efforts will inevitably lead to a full-fledged quantum theory of games, as well as a variety of quantum game implementations. In the area of quantum communication, optimal copying of quantum states [59] and quantum eavesdropping [27,60,61] can be treated as strategic games between two or more players with the goal of extracting maximum information [57,62].

In order to gain some insight into quantum games, it is instructive to briefly consider a particular quantum game, involving coin flipping [56]. In this game, there are two players, Alice and Eve, and they must flip a penny. However, a real penny would not suffice to play the game, because the game requires a penny that can be in a superposition state of head up and tail up, and of course even if such a state could be produced, it would decohere so quickly that it would be unobservable. Cavity QED (quantum electrodynamics), ion-trap, or NMR (nuclear magnetic resonance) implementations of the game have been suggested [56]. A simple all-optical implementation might be more practical. In any case, any two quantum states forming a qubit would serve as a quantum "penny," and implementations of appropriate unitary transformations acting on the qubit would be used by Alice and Eve to do the flipping.

The object of the game for Alice is to finish with the qubit tail up; that for Eve is to finish head up. Neither player may observe the qubit in the course of the game. The game begins with Alice putting the qubit in a box in the head-up state $|H\rangle$. Next, if the game is played with classical flipping moves only, Eve either flips the qubit, using a transformation (represented by an operator F), or does not flip it (represented by the identity operator $I = 1$). The state is not observed. Next, Alice either does or does not flip the qubit, without observing it. Next, Eve again flips or does not flip the qubit. Alice and Eve finally observe the qubit, and Eve wins if the qubit ends up in the head-up state $|H\rangle$. Otherwise, Alice wins the game.

As an example, for a particular choice of strategies chosen by Alice and Eve, a possible course of the game might be as follows:

$$(62) \qquad FIF|H\rangle = FI|T\rangle = F|T\rangle = |H\rangle,$$

where $|T\rangle$ denotes the tail-up state of the qubit. Here Eve's first move is to use the flip operation to flip the qubit, which is initially in the state $|H\rangle$. In the next move, Alice does not flip it, and in the last move Eve does flip it. For this example, the qubit ends up in the state $|H\rangle$, and therefore Eve wins the game. The states $|H\rangle$ and $|T\rangle$ can be represented by

$$(63) \qquad |H\rangle = \begin{pmatrix} 1 \\ 0 \end{pmatrix}, \quad |T\rangle = \begin{pmatrix} 0 \\ 1 \end{pmatrix},$$

respectively. The flipping transformation can be represented by the unitary operator

$$(64) \qquad F = \begin{pmatrix} 0 & 1 \\ 1 & 0 \end{pmatrix},$$

and not flipping can be represented by the identity operator

$$(65) \qquad I = \begin{pmatrix} 1 & 0 \\ 0 & 1 \end{pmatrix}.$$

One notes, in passing, that the operator F in Eq. (64) corresponds to the quantum NOT gate operator, to be discussed in Sect. 6. For the above strategies, one has

$$FIF\,|H\rangle = \begin{pmatrix} 0 & 1 \\ 1 & 0 \end{pmatrix} \begin{pmatrix} 1 & 0 \\ 0 & 1 \end{pmatrix} \begin{pmatrix} 0 & 1 \\ 1 & 0 \end{pmatrix} \begin{pmatrix} 1 \\ 0 \end{pmatrix}$$

(66)
$$= \begin{pmatrix} 0 & 1 \\ 1 & 0 \end{pmatrix} \begin{pmatrix} 0 \\ 1 \end{pmatrix} = \begin{pmatrix} 1 \\ 0 \end{pmatrix} = |H\rangle.$$

In general, however, the game is conceived as a fully quantum game, in which the n-th move corresponds to a general unitary operator U_n, represented by a 2 × 2 unitary matrix for which F and I represent only special cases [56]. Thus

(67)
$$U_n = \begin{pmatrix} U_{n11} & U_{n12} \\ U_{n21} & U_{n22} \end{pmatrix},$$

where the matrix elements $U_{nij}, i,j = 1,2$ are chosen such that

(68)
$$U_n U_n^\dagger = 1.$$

An arbitrary quantum strategy might then be represented by the following sequence of unitary operators [56]:

(69)
$$U_3 U_2 U_1 |H\rangle = c_1 |H\rangle + c_2 |T\rangle,$$

in which the first move by Eve is represented by U_1, the next move by Alice is represented by U_2, the last move by Eve is represented by U_3, and c_1 and c_2 are complex numbers. The resulting state of the qubit is some superposition state of head up and tail up, as indicated. Suppose [56]

(70)
$$U_3 U_2 U_1 |H\rangle \neq |H\rangle.$$

Then Eve can improve her strategy by replacing U_3 by $U_1^{-1} U_2^{-1}$, which is also unitary, since both U_1 and U_2 are. Thus one has

(71)
$$U_3 U_2 U_1 |H\rangle = \left(U_1^{-1} U_2^{-1}\right) U_2 U_1 |H\rangle = (U_2 U_1)^{-1} (U_2 U_1) |H\rangle = |H\rangle,$$

and Eve wins. Or, suppose

(72)
$$U_3 U_2 U_1 |H\rangle \neq |T\rangle.$$

Then Alice can improve her strategy by replacing U_2 with $U_3^{-1} F U_1^{-1}$, which is unitary, since U_1, U_3, and F are. Thus

(73)
$$U_3 U_2 U_1 |H\rangle = U_3 \left(U_3^{-1} F U_1^{-1}\right) U_1 |H\rangle = F |H\rangle = |T\rangle,$$

and Alice wins. But the qubit state $U_3 U_2 U_1 |H\rangle$, when finally observed, cannot be both $|H\rangle$ and $|T\rangle$, and therefore at least one of the players can improve her strategy if the other player does not change hers. It then follows that the overall strategies represented by $U_3 U_2 U_1$ cannot be an equilibrium, since their strategies are an equilibrium only if neither player can gain by changing strategy unilaterally [56]. It has, however, also been argued that an equilibrium does exist for mixed quantum strategies in which various strategies occur with some probabilities [56]. Such a game is a manifestly stochastic quantum game.

In other work, it has been argued that in a quantum game version of the classical two-player binary choice game, *Prisoner's Dilemma* [54], there is no dilemma if quantum strategies are allowed [57]. Also, an interesting two-player protocol was conceived for quantum gambling, and it was demonstrated that neither player can increase his earning beyond some limit [58]. In other recent work, the optimal quantum copying problem is informatively described in the form of a game [59].

Although quantum decoherence is not a major issue for small quantum games such as the coin flipping game (at least if one assumes a sufficiently long-lived qubit is successfully implemented), quantum decoherence is certainly a major issue for quantum eavesdropping games, optimal quantum-state copying games, or large-scale quantum games, just as it is for any large-scale quantum information processor.

6. Quantum gates

In order to develop a multi-component qubit device, it is useful to implement various quantum gates. In the following, I present mathematical descriptions of various photonic implementations. First, consider the *quantum $\sqrt{\text{NOT}}$ gate* (square-root-of-not gate). As an example, consider the single-photon optical implementation depicted in Fig. 3, consisting of a single photon incident on a beamsplitter along two possible paths, designated by the corresponding states, $|0\rangle$ and $|1\rangle$, or more generally, in a superposition of those two paths. The exit path states are $|0'\rangle$ and $|1'\rangle$. Using the same methods as in Sect. 2, one has

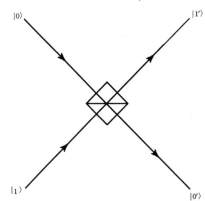

Figure 3. Quantum $\sqrt{\text{NOT}}$ gate.

(74) $$|0'\rangle = 2^{-1/2}|0\rangle + 2^{-1/2}i|1\rangle,$$

(75) $$|1'\rangle = 2^{-1/2}i|0\rangle + 2^{-1/2}|1\rangle,$$

or in matrix form,

(76) $$\begin{pmatrix} |0'\rangle \\ |1'\rangle \end{pmatrix} = 2^{-1/2} \begin{pmatrix} 1 & i \\ i & 1 \end{pmatrix} \begin{pmatrix} |0\rangle \\ |1\rangle \end{pmatrix}.$$

The matrix operator,

(77) $$\sqrt{N} \equiv 2^{-1/2} \begin{pmatrix} 1 & i \\ i & 1 \end{pmatrix},$$

is known as the "quantum square-root-of-not gate" or quantum $\sqrt{\text{NOT}}$ gate, since

(78) $$\sqrt{N}\sqrt{N} = 2^{-1} \begin{pmatrix} 1 & i \\ i & 1 \end{pmatrix} \begin{pmatrix} 1 & i \\ i & 1 \end{pmatrix} = \begin{pmatrix} 0 & i \\ i & 0 \end{pmatrix} = i \begin{pmatrix} 0 & 1 \\ 1 & 0 \end{pmatrix},$$

and the matrix operator

(79) $$N \equiv \begin{pmatrix} 0 & 1 \\ 1 & 0 \end{pmatrix}$$

is the *NOT operator*, which transforms $|0\rangle$ to $|1\rangle$ and $|1\rangle$ to $|0\rangle$, thus:

(80) $$N \begin{pmatrix} |0\rangle \\ |1\rangle \end{pmatrix} = \begin{pmatrix} 0 & 1 \\ 1 & 0 \end{pmatrix} \begin{pmatrix} |0\rangle \\ |1\rangle \end{pmatrix} = \begin{pmatrix} |1\rangle \\ |0\rangle \end{pmatrix}.$$

The overall phase factor, $i = e^{i\pi/2}$, in Eq. (78) can be ignored here, and could be physically removed with $-\pi/4$ phase shifters located at both exit ports in Fig. 3. Then, one has, effectively,

(81) $$\sqrt{N}\sqrt{N} = N.$$

This suggests that a *quantum NOT gate* can be constructed from a succession of two quantum $\sqrt{\text{NOT}}$ gates. Note also that the operator N is unitary, since

(82) $$NN^\dagger = \begin{pmatrix} 0 & 1 \\ 1 & 0 \end{pmatrix} \begin{pmatrix} 0 & 1 \\ 1 & 0 \end{pmatrix} = \begin{pmatrix} 1 & 0 \\ 0 & 1 \end{pmatrix} = 1.$$

Stacking two beamsplitters successively, as in Fig. 4, produces the succession of two quantum $\sqrt{\text{NOT}}$ gates, as suggested above. The various modes, or paths, are labeled in Fig. 4 by their path states. Using the rules for forming and combining amplitudes (see Sect. 2), one obtains

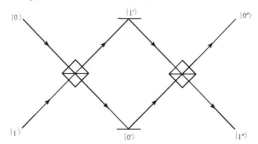

Figure 4. Quantum NOT gate.

(83) $$|0''\rangle = 2^{-1/2}|0'\rangle + 2^{-1/2}i|1'\rangle,$$

(84) $$|1''\rangle = 2^{-1/2}i|0'\rangle + 2^{-1/2}|1'\rangle.$$

Next, substituting Eqs. (74) and (75) in Eqs. (83) and (84), one obtains

(85) $|0''\rangle = 2^{-1/2}\left(2^{-1/2}|0\rangle + 2^{-1/2}i|1\rangle\right) + 2^{-1/2}i\left(2^{-1/2}i|0\rangle + 2^{-1/2}|1\rangle\right) = i|1\rangle,$

(86) $|1''\rangle = 2^{-1/2}i\left(2^{-1/2}|0\rangle + 2^{-1/2}i|1\rangle\right) + 2^{-1/2}\left(2^{-1/2}i|0\rangle + 2^{-1/2}|1\rangle\right) = i|0\rangle,$

or, equivalently, in matrix form,

(87) $$\begin{pmatrix} |0''\rangle \\ |1''\rangle \end{pmatrix} = i \begin{pmatrix} 0 & 1 \\ 1 & 0 \end{pmatrix} \begin{pmatrix} |0\rangle \\ |1\rangle \end{pmatrix},$$

which is in agreement with Eq. (78). The common phase factor, $i = e^{i\pi/2}$, can again be ignored, and could be physically removed as indicated previously, or instead by locating $-\pi/2$ phase shifters at both exit ports in Fig. 4. The unitary operator N, Eq. (79), can therefore be faithfully implemented by the device.

Consider next the one-photon device shown in Fig. 5, consisting of one beam-splitter and two $-\pi/2$ phase shifters. The possible paths for a single photon entering from the left are labeled by the corresponding path states. The entering photon may be in path state $|0\rangle$ or $|1\rangle$, or a superposition of the two. From the figure, one can see that the output states $|0'\rangle$ and $|1'\rangle$ are

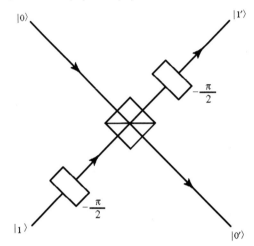

Figure 5. Hadamard gate.

(88) $$|0'\rangle = 2^{-1/2}|0\rangle + e^{-i\pi/2}2^{-1/2}i|1\rangle = 2^{-1/2}(|0\rangle + |1\rangle),$$

(89) $$|1'\rangle = 2^{-1/2}ie^{-i\pi/2}|0\rangle + e^{-i\pi/2}2^{-1/2}e^{-i\pi/2}|1\rangle = 2^{-1/2}(|0\rangle - |1\rangle).$$

Rewriting Eqs. (88) and (89) in matrix form, one has

(90) $$\begin{pmatrix} |0'\rangle \\ |1'\rangle \end{pmatrix} = 2^{-1/2} \begin{pmatrix} 1 & 1 \\ 1 & -1 \end{pmatrix} \begin{pmatrix} |0\rangle \\ |1\rangle \end{pmatrix}.$$

The unitary operator

(91) $$H = 2^{-1/2} \begin{pmatrix} 1 & 1 \\ 1 & -1 \end{pmatrix}$$

acting on the path qubit, in this example, is called the *Hadamard transform*. The *Hadamard gate*, depicted in Fig. 5, is a one-photon optical implementation of the Hadamard transform acting on a qubit. Two Hadamard gates can be combined to form a *single-photon balanced Mach-Zehnder interferometer* [34], as in the following section.

A particularly simple optical example of a single-photon device is depicted in Fig. 6: an optical implementation of the *quantum controlled-NOT gate* (also called the *quantum XOR gate*). It consists of a single polarization rotator that converts horizontal polarization into vertical, and vertical into horizontal, for one of two paths. Here, if a photon with vertical or horizontal polarization enters along path $|0\rangle$, it simply passes to the exit port $|0'\rangle$ unchanged, and if it enters along path $|1\rangle$, its polarization changes from horizontal to vertical, or vertical to horizontal. Thus,

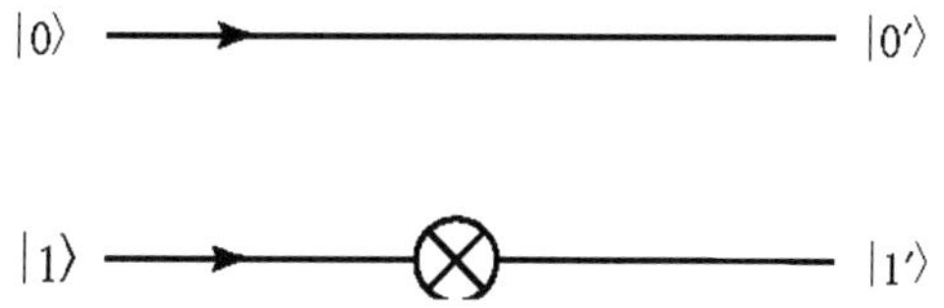

$$|0\rangle \longrightarrow \longrightarrow |0'\rangle$$

$$|1\rangle \longrightarrow \otimes \longrightarrow |1'\rangle$$

Figure 6. Quantum controlled-NOT gate.

(92) $$|0\rangle |\rightarrow\rangle \Longrightarrow |0\rangle |\rightarrow\rangle,$$

(93) $$|0\rangle |\uparrow\rangle \Longrightarrow |0\rangle |\uparrow\rangle,$$

(94) $$|1\rangle |\rightarrow\rangle \Longrightarrow |1\rangle |\uparrow\rangle,$$

(95) $$|1\rangle |\uparrow\rangle \Longrightarrow |1\rangle |\rightarrow\rangle,$$

in which tensor products are implicit. Here, the location (path) qubit is the control, and the polarization qubit is the target. Conditional dynamics occurs because the polarization of the photon flips, conditionally on its location. The quantum controlled-NOT gate is a two-qubit gate, the two qubits, in this case, being path and polarization qubits. The transformation Eqs. (92) to (95) can also be characterized as follows in terms of Boolean arithmetic (expressed in binary). If the horizontal and vertical polarization states are chosen to encode Boolean states $|0\rangle$ and $|1\rangle$, respectively, thus

(96) $$|\rightarrow\rangle = |0\rangle, \quad |\uparrow\rangle = |1\rangle,$$

and also, the path states $|0\rangle$ and $|1\rangle$ encode the binary Boolean states $|0\rangle$ and $|1\rangle$, respectively, then the transformations (92) to (95) can be written as follows:

(97) $$|0\rangle |0\rangle \Longrightarrow |0\rangle |0\rangle,$$

(98) $$|0\rangle |1\rangle \Longrightarrow |0\rangle |1\rangle,$$

(99) $$|1\rangle |0\rangle \Longrightarrow |1\rangle |1\rangle,$$

(100) $$|1\rangle |1\rangle \Longrightarrow |1\rangle |0\rangle.$$

Here, the first ket refers to a path state, the second ket refers to a polarization state, and the order must be maintained. Equations (97) to (100) can be succinctly represented as follows:

(101) $$|n\rangle |m\rangle \Longrightarrow |n\rangle |n \oplus m\rangle,$$

with $n \in \{0,1\}$, $m \in \{0,1\}$, and $\oplus$ denoting addition modulo 2. The operation in Eq. (101) is called the *controlled NOT*. In matrix notation, Eqs. (92) to (95), expressing the output states in terms of the possible input states, may be represented as follows:

(102) $$\begin{pmatrix} |0'\rangle_1 \\ |0'\rangle_2 \\ |1'\rangle_1 \\ |1'\rangle_2 \end{pmatrix} = \begin{pmatrix} 1 & 0 & 0 & 0 \\ 0 & 1 & 0 & 0 \\ 0 & 0 & 0 & 1 \\ 0 & 0 & 1 & 0 \end{pmatrix} \begin{pmatrix} |0\rangle |\rightarrow\rangle \\ |0\rangle |\uparrow\rangle \\ |1\rangle |\rightarrow\rangle \\ |1\rangle |\uparrow\rangle \end{pmatrix}.$$

The matrix operator appearing in Eq. (102) is known as the quantum controlled-NOT (or XOR) operator,

$$
(103) \qquad X \equiv \begin{pmatrix} 1 & 0 & 0 & 0 \\ 0 & 1 & 0 & 0 \\ 0 & 0 & 0 & 1 \\ 0 & 0 & 1 & 0 \end{pmatrix}.
$$

Combinations of the controlled-NOT operation and arbitrary single-qubit rotations can generate any unitary operation [4]. Although decoherence is not an issue for one-time gate operation, in the case of gate implementations for possible use in large-scale quantum computer circuits, the gate error rate must be extremely small for successful computation.

7. Single-photon balanced Mach-Zehnder interferometer

Consider the succession of two Hadamard gates, depicted in Fig. 7. Here a single photon is incident along path $|0\rangle$, path $|1\rangle$, or a superposition of both paths. One has

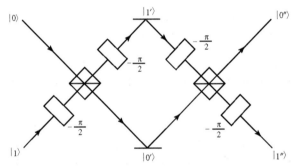

Figure 7. Single-photon Mach-Zehnder interferometer.

$$(104) \qquad |0''\rangle = e^{-i\pi/2} 2^{-1/2} i |1'\rangle + 2^{-1/2} |0'\rangle = 2^{-1/2}(|1'\rangle + |0'\rangle),$$

$$(105) \qquad |1''\rangle = e^{-i\pi/2} 2^{-1/2} e^{-i\pi/2} |1'\rangle + 2^{-1/2} i e^{-i\pi/2} |0'\rangle = 2^{-1/2}(-|1'\rangle + |0'\rangle).$$

Next, substituting Eqs. (88) and (89) in Eqs. (104) and (105), one obtains

$$(106) \qquad |0''\rangle = 2^{-1/2} \left(2^{-1/2}(|0\rangle - |1\rangle) + 2^{-1/2}(|0\rangle + |1\rangle) \right) = |0\rangle,$$

$$(107) \qquad |1''\rangle = 2^{-1/2} \left(-2^{-1/2}(|0\rangle - |1\rangle) + 2^{-1/2}(|0\rangle + |1\rangle) \right) = |1\rangle.$$

Thus

$$(108) \qquad \begin{pmatrix} |0''\rangle \\ |1''\rangle \end{pmatrix} = \begin{pmatrix} |0\rangle \\ |1\rangle \end{pmatrix}.$$

It is also clear that this must be the case, since, according to Eq. (91), one has

$$(109) \qquad HH = 2^{-1} \begin{pmatrix} 1 & 1 \\ 1 & -1 \end{pmatrix} \begin{pmatrix} 1 & 1 \\ 1 & -1 \end{pmatrix} = \begin{pmatrix} 1 & 0 \\ 0 & 1 \end{pmatrix},$$

which is the identity. Thus, the succession of two Hadamard gates produces a balanced single-photon Mach-Zehnder interferometer. If the photon enters along path $|0\rangle$, it also exits in the state $|0''\rangle = |0\rangle$. If it enters in $|1\rangle$, it exits in $|1\rangle$.

8. Qubit entanglers

By sandwiching a quantum controlled-NOT gate between two Hadamard gates, one obtains an example of a two-qubit entangler, which entangles a location qubit with a polarization qubit. The two-qubit entangler is shown in Fig. 8. If the input state is $|0\rangle\,|\rightarrow\rangle$, one can see from the figure, by taking into account the effect of the first beamsplitter and the phase shifters, that

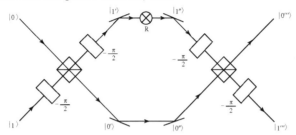

Figure 8. Two-qubit entangler.

$$|0'\rangle_1 = 2^{-1/2}\,|\rightarrow\rangle, \tag{110}$$

$$|1'\rangle_1 = 2^{-1/2}e^{-i\pi/2}i\,|\rightarrow\rangle = 2^{-1/2}\,|\rightarrow\rangle. \tag{111}$$

(The kets representing the path qubits in the right-hand sides of Eqs. (110) to (117) are suppressed for notational convenience.)

The index 1 merely distinguishes the particular choice of input state. Therefore, because of the polarization rotator R, one has

$$|0''\rangle_1 = 2^{-1/2}\,|\rightarrow\rangle, \tag{112}$$

$$|1''\rangle_1 = 2^{-1/2}\,|\uparrow\rangle. \tag{113}$$

Also, using Eqs. (88) and (89), one has

$$|0'''\rangle_1 = 2^{-1/2}\left(|0''\rangle_1 + |1''\rangle_1\right), \tag{114}$$

$$|1'''\rangle_1 = 2^{-1/2}\left(|0''\rangle_1 - |1''\rangle_1\right). \tag{115}$$

Next, substituting Eqs. (112) and (113) in (114) and (115), one obtains

$$|0'''\rangle_1 = 2^{-1}\left(|\rightarrow\rangle + |\uparrow\rangle\right), \tag{116}$$

$$|1'''\rangle_1 = 2^{-1}\left(|\rightarrow\rangle - |\uparrow\rangle\right). \tag{117}$$

The final state is then given by combining Eqs. (116) and (117), namely,

$$|\psi_1\rangle = |0'''\rangle_1 + |1'''\rangle_1 = 2^{-1}\left(|\rightarrow\rangle + |\uparrow\rangle\right)|e_{0'''}\rangle + 2^{-1}\left(|\rightarrow\rangle - |\uparrow\rangle\right)|e_{1'''}\rangle, \tag{118}$$

where we now explicitly let $|e_{0'''}\rangle$ and $|e_{1'''}\rangle$ denote unit kets corresponding to paths $|0'''\rangle$ and $|1'''\rangle$, respectively. Thus, in the case of the input state $|0\rangle\,|\rightarrow\rangle$, one has

$$|0\rangle\,|\rightarrow\rangle \Longrightarrow |\psi_1\rangle = 2^{-1}\left(|e_{0'''}\rangle\,|\rightarrow\rangle + |e_{0'''}\rangle\,|\uparrow\rangle + |e_{1'''}\rangle\,|\rightarrow\rangle - |e_{1'''}\rangle\,|\uparrow\rangle\right). \tag{119}$$

Thus, the qubit entangler converts the unentangled input state $|0\rangle\,|\rightarrow\rangle$ to a state in which the polarization qubit $(|\rightarrow\rangle, |\uparrow\rangle)$ is entangled with the path qubit $(|e_{0'''}\rangle, |e_{1'''}\rangle)$. Similarly, one obtains

$$|0\rangle\,|\uparrow\rangle \Longrightarrow |\psi_2\rangle = 2^{-1}\left(|e_{0'''}\rangle\,|\rightarrow\rangle + |e_{0'''}\rangle\,|\uparrow\rangle - |e_{1'''}\rangle\,|\rightarrow\rangle + |e_{1'''}\rangle\,|\uparrow\rangle\right), \tag{120}$$

$$(121) \quad |1\rangle|\rightarrow\rangle \implies |\psi_3\rangle = 2^{-1}\left(|e_{0'''}\rangle|\rightarrow\rangle - |e_{0'''}\rangle|\uparrow\rangle + |e_{1'''}\rangle|\rightarrow\rangle + |e_{1'''}\rangle|\uparrow\rangle\right),$$

$$(122) \quad |1\rangle|\uparrow\rangle \implies |\psi_4\rangle = 2^{-1}\left(-|e_{0'''}\rangle|\rightarrow\rangle + |e_{0'''}\rangle|\uparrow\rangle + |e_{1'''}\rangle|\rightarrow\rangle + |e_{1'''}\rangle|\uparrow\rangle\right),$$

for input states $|0\rangle|\uparrow\rangle$, $|1\rangle|\rightarrow\rangle$, and $|1\rangle|\uparrow\rangle$, respectively. The input states have become entangled states of path and polarization qubits.

In each case, Eqs. (119) to (122), the input state is separable (factorizable) into two factors, each corresponding to a state of only one qubit. Each input state is a single product of a path state with a polarization state. The input states are not entangled.

The output states, however, are not factorizable: each is a superposition of product states that is nonseparable into a single product of a path state and a polarization state. The output states are entangled. The four entangled states $|\psi_1\rangle$, $|\psi_2\rangle$, $|\psi_3\rangle$, and $|\psi_4\rangle$ can be summarized in matrix form as follows:

$$(123) \quad \begin{pmatrix} |\psi_1\rangle \\ |\psi_2\rangle \\ |\psi_3\rangle \\ |\psi_4\rangle \end{pmatrix} = 2^{-1} \begin{pmatrix} 1 & 1 & 1 & -1 \\ 1 & 1 & -1 & 1 \\ 1 & -1 & 1 & 1 \\ -1 & 1 & 1 & 1 \end{pmatrix} \begin{pmatrix} |e_{0'''}\rangle|\rightarrow\rangle \\ |e_{0'''}\rangle|\uparrow\rangle \\ |e_{1'''}\rangle|\rightarrow\rangle \\ |e_{1'''}\rangle|\uparrow\rangle \end{pmatrix}.$$

Next consider, for example, two qubits, each of which can be in two states. Qubit 1 can be in the orthonormal states $|0\rangle_1$ or $|1\rangle_1$, and qubit 2 can be in orthonormal states $|0\rangle_2$ or $|1\rangle_2$, where the subscripts on the kets distinguish the two qubits. The general combined state of the two qubits can be written as

$$(124) \quad |\psi\rangle = \alpha|0\rangle_1|0\rangle_2 + \beta|0\rangle_1|1\rangle_2 + \gamma|1\rangle_1|0\rangle_2 + \delta|1\rangle_1|1\rangle_2 = \begin{pmatrix} \alpha \\ \beta \\ \gamma \\ \delta \end{pmatrix}.$$

(Here, tensor products are implicit, and the notation does not explicitly distinguish between a ket and its representative [9,12].) Suppose that the initial combined state is given by a simple product of states, namely,

$$(125) \quad |\psi(0)\rangle = |0\rangle_1|0\rangle_2 = \begin{pmatrix} 1 \\ 0 \\ 0 \\ 0 \end{pmatrix}.$$

This state is not entangled. It is separable into two factors, each corresponding to a state of one qubit. But then, as a result of prescribed controlled interactions between the two particles (which, according to quantum mechanics, can be described by some unitary operator U), the combined state of the two qubits, after a prescribed period of time, will, in general, be

$$(126) \quad |\psi(t)\rangle = U(t)|\psi(0)\rangle = \begin{pmatrix} U_{11} & U_{12} & U_{13} & U_{14} \\ U_{21} & U_{22} & U_{23} & U_{24} \\ U_{31} & U_{32} & U_{33} & U_{34} \\ U_{41} & U_{42} & U_{43} & U_{44} \end{pmatrix} \begin{pmatrix} 1 \\ 0 \\ 0 \\ 0 \end{pmatrix},$$

expressed in terms of the matrix elements of the unitary operator U. Next, if we perform the matrix multiplication, Eq. (126) becomes

$$(127) \quad |\psi(t)\rangle = U_{11}|0\rangle_1|0\rangle_2 + U_{21}|0\rangle_1|1\rangle_2 + U_{31}|1\rangle_1|0\rangle_2 + U_{41}|1\rangle_1|1\rangle_2.$$

The state is no longer separable (if the appropriate U_{i1} are nonvanishing). It is an entangled state. Any entanglement of the two qubits can, in principle, be achieved by appropriate choice of interaction Hamiltonian between the qubits, thereby determining the appropriate unitary operator U. Thus controlled interactions between two qubits, described by a prescribed unitary operator U, produce a prescribed entangled state. A *qubit entangler* is an implementation of this process.

More generally, it is now well known that controlled qubit entanglement can be produced generically in at least three ways: (1) prescribed and controlled interactions between the qubits, (2) entangled-particle sources, such as Einstein-Podolsky-Rosen (EPR) pair sources, and (3) entanglement swapping. EPR-pair sources are discussed in Sect. 9, and entanglement swapping is discussed in Sect. 13. Concerning method (1), interactions of the qubits with the environment lead, of course, to entanglement with the environment and decoherence, while interactions between the qubits themselves in a controlled way can produce prescribed entanglements. The latter is the basis for networks of quantum gates in a quantum computer.

9. EPR-pair sources

Consider the production of entanglement, for example, in a particular *EPR-pair source* [63,64]. This is a source of two Einstein-Podolsky-Rosen (EPR) correlated particles. When an ultraviolet (UV) laser beam is incident on the nonlinear crystal beta-barium borate, one of the photons may be absorbed, with small probability, by interaction with the crystal atoms, and produce a pair of photons of lower frequency (conserving energy). The two photons are emitted into the surface of two cones whose axes (together with the path of the original UV photon) lie in the same plane, and the axes of the two cones make equal and opposite angles with the path of the UV photon (conserving momentum). This is parametric down-conversion. In type II parametric down-conversion, one of the two photons produced has vertical polarization ($\uparrow$), and the other has horizontal polarization ($\rightarrow$). The device can be configured with the two emission cones overlapping, so that the photons carry no individual polarization. Along the two directions where the cones overlap, the state of the two photons is in the entangled state

$$(128) \qquad |\psi\rangle = 2^{-1/2} \left(|\rightarrow\rangle_1 |\uparrow\rangle_2 + e^{i\alpha} |\uparrow\rangle_1 |\rightarrow\rangle_2 \right),$$

where the subscripts on the kets distinguish the two particles, and the parameter α is a relative phase resulting from crystal birefringence. This is a general EPR-pair state. Here, each of the two particles represents a qubit by its two polarization states. Thus, in the state represented by Eq. (128), the two polarization qubits are entangled. Also, the two particles will generally have two separate locations. By controlling the value of the parameter α in Eq. (128), one can produce various EPR-pair entangled states. For EPR sources to be applied to quantum communication in any practical way, the decoherence issue must be addressed: decoherence strongly limits the storage time for EPR correlated states.

10. Bell-state synthesizer

The EPR-pair source also serves as an example of a *Bell-state synthesizer*. Using a birefringent phase shifter, one can set the value of the relative phase parameter α in the EPR-pair state of Eq. (128) to the values 0 and π [63]. Also, a half-wave plate in one path can change horizontal to vertical polarization and vice versa. In this way, the following four EPR *Bell states* can be produced:

(129) $$|\psi^{\pm}\rangle = 2^{-1/2}\left(|\rightarrow\rangle_1 |\uparrow\rangle_2 \pm |\uparrow\rangle_1 |\rightarrow\rangle_2\right),$$

(130) $$|\phi^{\pm}\rangle = 2^{-1/2}\left(|\rightarrow\rangle_1 |\rightarrow\rangle_2 \pm |\uparrow\rangle_1 |\uparrow\rangle_2\right).$$

The four states given by Eqs. (129) and (130) are widely known as Bell states. The Bell states form a maximally entangled orthonormal basis for the Hilbert space of the two qubits. *Maximal entanglement* means that all information is carried by both particles jointly, and no information is stored in an individual particle. The parametric down-conversion EPR-pair source is thereby made into a *Bell-state source*.

11. Quantum dense coder

Classically, with two pennies, each having two possible states, head up or tail up, one can encode two bits of information, represented by (H, H), (H, T), (T, H), and (H, H). Here the state of one of the pennies is the first entry, and the state of the other penny is the second entry. This allows four encodings, or $\log_2 4 = 2$ bits. In *quantum dense coding* [65], two entangled two-state systems are used to encode information. It has been demonstrated experimentally [66] that, with quantum dense coding, it is presently feasible to transmit one of three messages by manipulating only one of two entangled qubits shared between a transmitter Bob and a receiver Alice. In other words, by manipulating only one of two entangled qubits, Bob can transmit $\log_2 3 = 1.58$ bits, namely a *trit*, to Alice.

In principle, however, one of four messages (two bits) could be transmitted by manipulating only one of the qubits. With two photonic qubits, each having two polarization states (vertical polarization represented by the ket $|\uparrow\rangle$, and horizontal polarization represented by the ket $|\rightarrow\rangle$), the two bits can be encoded in entangled superposition states, such as

(131) $$|\psi^{+}\rangle = 2^{-1/2}\left(|\rightarrow\rangle |\uparrow\rangle + |\uparrow\rangle |\rightarrow\rangle\right).$$

Here on the right-hand side, tensor products are implicit, and kets are taken to be ordered with the first and second kets in each term representing the first and second particles, respectively. An analogous entangled state would never be seen with two pennies, because of quantum decoherence. The state $|\psi^{+}\rangle$, along with three other states, forms a maximally entangled basis for two independent particles, the *Bell basis*. The other three states are

(132) $$|\psi^{-}\rangle = 2^{-1/2}\left(|\rightarrow\rangle |\uparrow\rangle - |\uparrow\rangle |\rightarrow\rangle\right),$$

(133) $$|\phi^{+}\rangle = 2^{-1/2}\left(|\rightarrow\rangle |\rightarrow\rangle + |\uparrow\rangle |\uparrow\rangle\right),$$

(134) $$|\phi^{-}\rangle = 2^{-1/2}\left(|\rightarrow\rangle |\rightarrow\rangle - |\uparrow\rangle |\uparrow\rangle\right).$$

Equations (131) to (134) are the four Bell states, Eqs. (129) and (130) with suppressed indices, and with ket order distinguishing the two qubits. Maximal entanglement between the two photons means that all information is carried by the photons jointly, and no information is stored in any individual photon. Note that for each of the four states, Eqs. (131) to (134), each photon is unpolarized by itself, since it is equally likely to have horizontal or vertical polarization. Also, the states are orthonormal; for example,

(135) $\langle \psi^+ | \phi^- \rangle = 2^{-1} \left(\left(\langle \rightarrow | \langle \uparrow | + \langle \uparrow | \langle \rightarrow | \right) \left(| \rightarrow \rangle | \rightarrow \rangle - | \uparrow \rangle | \uparrow \rangle \right) \right) = 0.$

The symmetries of the entanglement are made more evident when we summarize the four Bell basis states, Eqs. (131) to (134), as follows (as in Eqs. (129) and (130)):

(136) $\qquad \left| \psi^\pm \right\rangle = 2^{-1/2} \left(| \rightarrow \rangle | \uparrow \rangle \pm | \uparrow \rangle | \rightarrow \rangle \right),$

(137) $\qquad \left| \phi^\pm \right\rangle = 2^{-1/2} \left(| \rightarrow \rangle | \rightarrow \rangle \pm | \uparrow \rangle | \uparrow \rangle \right).$

With these four states, two bits of information can still be encoded. However, in the encoding with entangled states, none of the qubits carries well-defined information. For example, if the states $|\psi^+\rangle, |\psi^-\rangle, |\phi^+\rangle$, and $|\phi^-\rangle$ are chosen to encode messages 1, 2, 3, and 4, respectively, then one can see directly from Eqs. (136) and (137) that performing a local measurement of the polarization state of one of the two qubits (which may be spatially separated) is not sufficient to determine the message. The information is stored in the entanglement. It is encoded globally into relations between the two qubits, since the two particles are generally spatially separated. To obtain the information, one needs to know about the states of both qubits.

Although the four Bell states $|\psi^\pm\rangle$ and $|\phi^\pm\rangle$ can still encode only two bits of information, the encoding can be accomplished by manipulation of only one of the particles. Suppose Bob and Alice each share one of two EPR particles produced in the state $|\psi^+\rangle$, for example, by a Bell state source, as in the previous section. Bob, who wishes to transmit a message to Alice, performs one out of four possible unitary transformations on his particle alone, represented by the first ket in each term in the state $|\psi^+\rangle$. The four operators are

(1) the identity operator, for which

(138) $\qquad \left| \psi^+ \right\rangle \longrightarrow \left| \psi^+ \right\rangle,$

(2) the polarization interchange ($|\uparrow\rangle \Longleftrightarrow |\rightarrow\rangle$), for which

(139) $\qquad \left| \psi^+ \right\rangle \longrightarrow 2^{-1/2} \left(| \uparrow \rangle | \uparrow \rangle + | \rightarrow \rangle | \rightarrow \rangle \right) = \left| \phi^+ \right\rangle,$

(3) the polarization-dependent phase shift ($e^{i\alpha}$ for $|\uparrow\rangle$ and $e^{i(\alpha+\pi)}$ for $|\rightarrow\rangle$, where α is some phase), for which

(140) $\left| \psi^+ \right\rangle \longrightarrow 2^{-1/2} \left(-e^{i\alpha} | \rightarrow \rangle | \uparrow \rangle + e^{i\alpha} | \uparrow \rangle | \rightarrow \rangle \right) = -e^{i\alpha} \left| \psi^- \right\rangle$

(the factor of $-e^{i\alpha}$ is an unimportant overall phase factor), and, finally,

(4) the polarization-dependent phase shift ($e^{i\alpha}$ for $|\uparrow\rangle$ and $e^{i(\alpha+\pi)}$ for $|\rightarrow\rangle$) and polarization interchange, for which

(141) $\qquad \left| \psi^+ \right\rangle \longrightarrow 2^{-1/2} \left(-e^{i\alpha} | \uparrow \rangle | \uparrow \rangle + e^{i\alpha} | \rightarrow \rangle | \rightarrow \rangle \right) = e^{i\alpha} \left| \phi^- \right\rangle$

(again, there is an irrelevant overall phase factor).

The above four unitary operations are commonly referred to as (1) the identity operator, (2) the bit flip, (3) the phase shift, and (4) the combined phase shift/bit flip, respectively. These four manipulations by Bob of his particle result in the four orthogonal Bell states, which can represent four distinguishable messages: two bits of information. The information capacity of the transmission channel is therefore two bits, compared to the classical maximum of one bit. Alice must, of course, measure the Bell state to complete the information transfer. To do this, she uses a *Bell state analyzer*, as discussed in the following section.

A schematic diagram of a *quantum dense coder* is shown in Fig. 9. In the figure, BSS designates the Bell-state source of two entangled photons to be shared by Bob and Alice. BSE designates the Bell-state encoder with which Bob performs one of the four operations on his photon, thereby projecting the pair into a particular Bell state. He next sends the photon to Alice, who measures the state of the pair using the Bell-state analyzer BSA. Bob can thus, in principle, send two bits of information to Alice by manipulating only one of the two shared EPR photons.

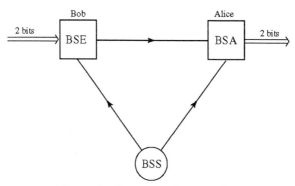

Figure 9. Quantum dense coder.

Thus, dense quantum coding can, in principle, double the transmission channel information capacity. However, to date, existing Bell state analyzers can measure only three of the four Bell states, and therefore Alice can read only three different messages (1 trit). Measurement of all four Bell states awaits the development of a gate enabling robust nonlinear interaction between two qubits [67].

Nonlinear response can be very effective for strong fields consisting of many photons [68,69]; however, at the few-photon level, the fields are far too weak for nonlinear response to be exploited. For example, combining two photons conditionally in a nonlinear crystal has an efficiency that is far too low. Several possible approaches to developing the required gate are being pursued, including cooperative two-photon interactions with pairs of atoms in a medium [70–72] and cavity-QED techniques [73–75]. The development of a robust two-qubit gate is also needed for the development of quantum computers.

Quantum decoherence is an impediment to the useful implementation of dense coding, because it limits transmission range and state storage time. However, quantum error correction methods may be implemented to change this (see Sect. 21).

12. Bell-state analyzer

The *Bell-state analyzer* addressed here exploits the quantum statistics of two qubits interacting with a beamsplitter [67,76]. Note first that the state $|\psi^-\rangle$, Eq. (132), is antisymmetric in the interchange of the two particles:

(142) $\quad |\psi^-\rangle = 2^{-1/2}(|\rightarrow\rangle|\uparrow\rangle - |\uparrow\rangle|\rightarrow\rangle) \longrightarrow 2^{-1/2}(|\uparrow\rangle|\rightarrow\rangle - |\rightarrow\rangle|\uparrow\rangle) = -|\psi^-\rangle.$

The other three Bell states, Eqs. (131), (133), and (134), are symmetric under particle interchange:

(143) $\quad |\psi^+\rangle = 2^{-1/2}(|\rightarrow\rangle|\uparrow\rangle + |\uparrow\rangle|\rightarrow\rangle) \longrightarrow 2^{-1/2}(|\uparrow\rangle|\rightarrow\rangle + |\rightarrow\rangle|\uparrow\rangle) = |\psi^+\rangle;$

and clearly,

(144) $\quad |\phi^{\pm}\rangle = 2^{-1/2}(|\rightarrow\rangle|\rightarrow\rangle \pm |\uparrow\rangle|\uparrow\rangle) \longrightarrow |\phi^{\pm}\rangle.$

The connection between spin and statistics [77,78] is that (1) half-integer spin particles are fermions and obey Fermi-Dirac statistics: that is, their total wave function is completely antisymmetric under interchange of particles; and (2) integer spin particles are bosons and obey Bose-Einstein statistics: that is, their total wave function is completely symmetric under interchange of the particles. Photons have spin-one, namely integer spin, and are therefore bosons and obey Bose-Einstein statistics. Therefore, the two-photon wave function must be totally symmetric. The total wave function has a spatial part and a spin part (polarization corresponds to spin [5]). It follows that for the two photons in the polarization Bell state $|\psi^-\rangle$, which, according to Eq. (142), is antisymmetric, the spatial part of the wave function must also be antisymmetric, so that the total wave function is completely symmetric. To see this, note that under particle interchange, one has, letting $|\Psi_A\rangle$ denote the antisymmetric spatial part of the wave function,

(145) $\quad |\Psi_A\rangle|\psi^-\rangle \longrightarrow (-|\Psi_A\rangle)(-|\psi^-\rangle) = |\Psi_A\rangle|\psi^-\rangle.$

Similarly, for the two photons in the symmetric polarization states $|\psi^+\rangle$ or $|\phi^{\pm}\rangle$, the spatial part of the wave function must also be symmetric, for under the interchange, one has, letting $|\Psi_S\rangle$ denote the symmetric spatial part of the wave function,

(146) $\quad |\Psi_S\rangle|\psi^+\rangle \longrightarrow (+|\Psi_S\rangle)(+|\psi^+\rangle) = |\Psi_S\rangle|\psi^+\rangle,$

(147) $\quad |\Psi_S\rangle|\phi^{\pm}\rangle \longrightarrow (+|\Psi_S\rangle)(+|\phi^{\pm}\rangle) = |\Psi_S\rangle|\phi^{\pm}\rangle.$

Next, consider two photons (see Fig. 10) incident from the left on a 50/50 beamsplitter, serving as a partial Bell-state analyzer. The mode $|1\rangle$ represents the upper path, and the mode $|0\rangle$ represents the lower path. Coincidence detection (designated by C in the figure) is performed at the two output ports of the beamsplitter. If one could ignore statistics, then the mode state $|1\rangle|0\rangle$ for the two particles, with one particle in each spatial mode, could be expressed as

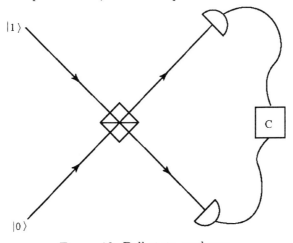

Figure 10. Bell-state analyzer.

(148) $$|1\rangle|0\rangle = 2^{-1}[|1\rangle|0\rangle + |0\rangle|1\rangle] + 2^{-1}[|1\rangle|0\rangle - |0\rangle|1\rangle].$$

I have used a trivial algebraic identity to rewrite the function in terms of a symmetric part and an antisymmetric part, corresponding to the first and second bracketed terms in Eq. (148). However, depending on the symmetry of the polarization state of the two photons, only the symmetric or the antisymmetric part of the two-photon spatial-mode state can be nonvanishing. Thus, the possible normalized spatial-mode states for the two photons incident from the left are either the normalized symmetric state,

(149) $$|\Psi_S\rangle = 2^{-1/2}(|1\rangle|0\rangle + |0\rangle|1\rangle),$$

or the normalized antisymmetric state,

(150) $$|\Psi_A\rangle = 2^{-1/2}(|1\rangle|0\rangle - |0\rangle|1\rangle).$$

Since the two-photon total wave function must be completely symmetric, the total state, including both spatial and polarization parts, must be one of the following four states:

(151) $$|\psi^+\rangle|\Psi_S\rangle, \ |\psi^-\rangle|\Psi_A\rangle, \ |\phi^+\rangle|\Psi_S\rangle, \ |\phi^-\rangle|\Psi_S\rangle.$$

Only the second state, $|\psi^-\rangle|\Psi_A\rangle$, is antisymmetric in both its polarization and spatial parts.

The beamsplitter does not affect the photon polarization state. Also, it can be shown that for all three of the states involving the spatially-symmetric part $|\Psi_S\rangle$, both photons emerge together in only one of the exit ports of the beamsplitter [67,79]; therefore, the coincidence detectors will not respond to those three states. It is also true that the antisymmetric state $|\Psi_A\rangle$ is an eigenstate of the beamsplitter [67,80], which I demonstrate explicitly below. It follows that the beamsplitter can discriminate the state $|\psi^-\rangle|\Psi_A\rangle$ from all the other states, since only that state will register coincidences [67]. Furthermore, for $|\psi^+\rangle$, the two photons have different polarizations, while for both $|\phi^+\rangle$ and $|\phi^-\rangle$, both photons have the same polarization state. Therefore, by performing polarization measurements, one can decide whether the photons are in the state $|\psi^+\rangle$ or one of the remaining states $|\phi^+\rangle$ or $|\phi^-\rangle$. To complete the Bell-state analysis, a robust optical element involving nonlinear interactions is needed (see the previous section) [70–73,80].

It is appropriate to independently demonstrate that the spatially antisymmetric state $|\Psi_A\rangle$ is, in fact, an eigenstate of the beamsplitter. The beamsplitter does not change this state, except for an irrelevant overall phase factor. To see this, one can use the following vector representation of $|\Psi_A\rangle$, Eq. (150), in terms of vectors $\binom{1}{0}$ and $\binom{0}{1}$, representing spatial modes $|1\rangle$ and $|0\rangle$, respectively; thus

(152) $$|\Psi_A\rangle = 2^{-1/2}\left[\binom{1}{0}\binom{0}{1} - \binom{0}{1}\binom{1}{0}\right].$$

Tensor products are again implicit here. A general state

(153) $$\binom{a}{b} = a\binom{1}{0} + b\binom{0}{1},$$

incident on the beamsplitter, can be depicted as in Fig. 11. Since the probability amplitude [9–12] that a photon enters the beamsplitter on the upper left is a, and that it enters on the lower left is b, then the amplitude that it exits on the upper right is $2^{-1/2}ia + 2^{-1/2}b$ (the factor of $2^{-1/2}$ is due to the 50/50 beamsplitter, and the factor of i is due to the reflection [13–15]). Analogously, the amplitude that the photon exits on the lower right is $2^{-1/2}a + 2^{-1/2}ib$. The exit state can therefore be represented by

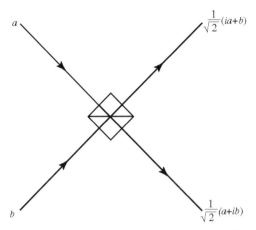

Figure 11. Beamsplitter input-output modes.

$$(154) \qquad 2^{-1/2}\begin{pmatrix} ia+b \\ a+ib \end{pmatrix} = 2^{-1/2}\begin{pmatrix} i & 1 \\ 1 & i \end{pmatrix}\begin{pmatrix} a \\ b \end{pmatrix}.$$

Thus, the effect of the beamsplitter is represented by the operator

$$(155) \qquad M = 2^{-1/2}\begin{pmatrix} i & 1 \\ 1 & i \end{pmatrix}.$$

The operator M is the quantum $\sqrt{\text{NOT}}$ gate, Eq. (77), and it is unitary:

$$(156) \qquad MM^\dagger = 2^{-1}\begin{pmatrix} i & 1 \\ 1 & i \end{pmatrix}\begin{pmatrix} -i & 1 \\ 1 & -i \end{pmatrix} = \begin{pmatrix} 1 & 0 \\ 0 & 1 \end{pmatrix} = 1.$$

It then follows that the effect of the beamsplitter on the spatially antisymmetric state of two photons, Eq. (152), is faithfully represented by

$$(157) \qquad |\Psi'_A\rangle = MM\,|\Psi_A\rangle,$$

in which $|\Psi'_A\rangle$ is the output state, the first operator M represents the beamsplitter acting on the first photon appearing in the expression for the Bell state, and the second operator M represents the same beamsplitter acting on the second photon (tensor products are again implicit). Thus, using Eqs. (155), (152), and (157), one has

$$|\Psi'_A\rangle = 2^{-1/2}\begin{pmatrix} i & 1 \\ 1 & i \end{pmatrix} 2^{-1/2} \begin{pmatrix} i & 1 \\ 1 & i \end{pmatrix} 2^{-1/2}\left[\begin{pmatrix}1\\0\end{pmatrix}\begin{pmatrix}0\\1\end{pmatrix} - \begin{pmatrix}0\\1\end{pmatrix}\begin{pmatrix}1\\0\end{pmatrix}\right]$$

$$= 2^{-3/2}\left[\begin{pmatrix}i\\1\end{pmatrix}\begin{pmatrix}1\\i\end{pmatrix} - \begin{pmatrix}1\\i\end{pmatrix}\begin{pmatrix}i\\1\end{pmatrix}\right]$$

$$= 2^{-3/2}\left\{\left[i\begin{pmatrix}1\\0\end{pmatrix} + \begin{pmatrix}0\\1\end{pmatrix}\right]\left[\begin{pmatrix}1\\0\end{pmatrix} + i\begin{pmatrix}0\\1\end{pmatrix}\right]\right.$$

$$\left. - \left[\begin{pmatrix}1\\0\end{pmatrix} + i\begin{pmatrix}0\\1\end{pmatrix}\right]\left[i\begin{pmatrix}1\\0\end{pmatrix} + \begin{pmatrix}0\\1\end{pmatrix}\right]\right\}$$

$$= -2^{-1/2}\left[\begin{pmatrix}1\\0\end{pmatrix}\begin{pmatrix}0\\1\end{pmatrix} - \begin{pmatrix}0\\1\end{pmatrix}\begin{pmatrix}1\\0\end{pmatrix}\right]$$

(158)
$$= -|\Psi_A\rangle.$$

One concludes that, in fact, the spatially antisymmetric state $|\Psi_A\rangle$ is an eigenstate of the beamsplitter, as already stated. Bell state analyzers can be particularly useful in *entanglement swappers*.

13. Entanglement swappers

With an entanglement swapper, two well-separated particles that have never interacted can become entangled. This *entanglement swapping* [81–85,67], has been accomplished experimentally [86]. Two separate EPR-pair down-conversion sources were used to produce two separate pairs of entangled photons, and a Bell-state measurement was then performed on two of the photons, one from each of the separate entangled pairs. The Bell-state measurement results in projecting the remaining two photons (one from each pair), which have never interacted, into an entangled state. The device is depicted schematically in Fig. 12. EPR_1 and EPR_2 designate the two separate EPR-pair down-conversion sources that produce two separate EPR pairs of photons, (1, 2) and (3, 4). BSM designates the Bell-state measurement performed on photons 2 and 3. The Bell-state measurement of photons 2 and 3 results in the entanglement of photons 1 and 4, which have never interacted. Thus, entanglement can occur by entanglement swapping, and not only by a common source, or by interaction in the past.

Proceeding then to examine entanglement swapping in more detail, assume the two EPR-pair sources EPR_1 and EPR_2 each produce a pair of particles in the antisymmetric polarization state given by the Bell state $|\psi^-\rangle$, Eq. (132):

(159)
$$|\psi^-\rangle_{12} = 2^{-1/2}(|\rightarrow\rangle_1|\uparrow\rangle_2 - |\uparrow\rangle_1|\rightarrow\rangle_2),$$

(160)
$$|\psi^-\rangle_{34} = 2^{-1/2}(|\rightarrow\rangle_3|\uparrow\rangle_4 - |\uparrow\rangle_3|\rightarrow\rangle_4).$$

Here the states are labeled explicitly with subscripts explicitly designating the particles, so that one can ignore the order of the kets, for ease of algebraic analysis. The total combined state of all four photons is then given by [86]

(161)
$$|\psi\rangle_{1234} = |\psi^-\rangle_{12}|\psi^-\rangle_{34}.$$

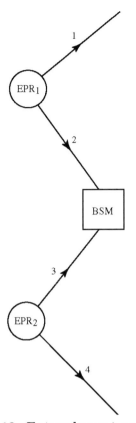

Figure 12. Entanglement swapper.

Although each separate pair is separately entangled, neither pair is entangled with the other pair. However, Eq. (161) can also be rewritten as follows:

$$|\psi\rangle_{1234} = 2^{-1}\left(|\psi^+\rangle_{14}|\psi^+\rangle_{23} - |\psi^-\rangle_{14}|\psi^-\rangle_{23} - |\phi^+\rangle_{14}|\phi^+\rangle_{23} + |\phi^-\rangle_{14}|\phi^-\rangle_{23}\right) \quad (162)$$

in terms of the Bell states $|\psi^\pm\rangle_{14}$, $|\psi^\pm\rangle_{23}$, $|\phi^\pm\rangle_{14}$, and $|\phi^\pm\rangle_{23}$ (see Eqs. (131) to (134)), where

$$|\psi^\pm\rangle_{14} = 2^{-1/2}\left(|\rightarrow\rangle_1|\uparrow\rangle_4 \pm |\uparrow\rangle_1|\rightarrow\rangle_4\right), \quad (163)$$

$$|\psi^\pm\rangle_{23} = 2^{-1/2}\left(|\rightarrow\rangle_2|\uparrow\rangle_3 \pm |\uparrow\rangle_2|\rightarrow\rangle_3\right), \quad (164)$$

$$|\phi^\pm\rangle_{14} = 2^{-1/2}\left(|\rightarrow\rangle_1|\rightarrow\rangle_4 \pm |\uparrow\rangle_1|\uparrow\rangle_4\right), \quad (165)$$

$$|\phi^\pm\rangle_{23} = 2^{-1/2}\left(|\rightarrow\rangle_2|\rightarrow\rangle_3 \pm |\uparrow\rangle_2|\uparrow\rangle_3\right). \quad (166)$$

(Equation (162) corrects the sign errors in Eq. (3) in Ref. 86.) To see that Eq. (162) is equivalent to Eq. (161) above, first note, using Eqs. (163) to (166), that

$$\begin{aligned}|\psi^\pm\rangle_{14}|\psi^\pm\rangle_{23} = 2^{-1}(&|\rightarrow\rangle_1|\uparrow\rangle_4|\rightarrow\rangle_2|\uparrow\rangle_3 \pm |\rightarrow\rangle_1|\uparrow\rangle_4|\uparrow\rangle_2|\rightarrow\rangle_3 \\ &\pm |\uparrow\rangle_1|\rightarrow\rangle_4|\rightarrow\rangle_2|\uparrow\rangle_3 + |\uparrow\rangle_1|\rightarrow\rangle_4|\uparrow\rangle_2|\rightarrow\rangle_3)\end{aligned} \quad (167)$$

and

$$|\phi^\pm\rangle_{14}|\phi^\pm\rangle_{23} = 2^{-1}(|\rightarrow\rangle_1|\rightarrow\rangle_4|\rightarrow\rangle_2|\rightarrow\rangle_3 \pm |\rightarrow\rangle_1|\rightarrow\rangle_4|\uparrow\rangle_2|\uparrow\rangle_3$$
(168)
$$\pm|\uparrow\rangle_1|\uparrow\rangle_4|\rightarrow\rangle_2|\rightarrow\rangle_3 + |\uparrow\rangle_1|\uparrow\rangle_4|\uparrow\rangle_2|\uparrow\rangle_3).$$

Next, substituting Eqs. (167) and (168) in Eq. (162), one obtains

$$|\psi\rangle_{1234} = 2^{-1}(|\rightarrow\rangle_1|\uparrow\rangle_4|\uparrow\rangle_2|\rightarrow\rangle_3 + |\uparrow\rangle_1|\rightarrow\rangle_4|\rightarrow\rangle_2|\uparrow\rangle_3$$
(169)
$$-|\rightarrow\rangle_1|\rightarrow\rangle_4|\uparrow\rangle_2|\uparrow\rangle_3 - |\uparrow\rangle_1|\uparrow\rangle_4|\rightarrow\rangle_2|\rightarrow\rangle_3).$$

By reordering terms and factors within each term, Eq. (169) becomes

$$|\psi\rangle_{1234} = 2^{-1}(|\rightarrow\rangle_1|\uparrow\rangle_2|\rightarrow\rangle_3|\uparrow\rangle_4 - |\rightarrow\rangle_1|\uparrow\rangle_2|\uparrow\rangle_3|\rightarrow\rangle_4$$
(170)
$$-|\uparrow\rangle_1|\rightarrow\rangle_2|\rightarrow\rangle_3|\uparrow\rangle_4 + |\uparrow\rangle_1|\rightarrow\rangle_2|\uparrow\rangle_3|\rightarrow\rangle_4).$$

However, Eq. (170) factors directly into the following form:

(171) $$|\psi\rangle_{1234} = 2^{-1}(|\rightarrow\rangle_1|\uparrow\rangle_2 - |\uparrow\rangle_1|\rightarrow\rangle_2)(|\rightarrow\rangle_3|\uparrow\rangle_4 - |\uparrow\rangle_3|\rightarrow\rangle_4).$$

Finally, substituting Eqs. (159) and (160) in Eq. (171), one obtains Eq. (161). Thus, in fact, Eq. (162) is true.

Next, from Eq. (162), it is evident that a Bell-state projective measurement of photons 2 and 3 also projects photons 1 and 4 onto a Bell state. Thus, for example, if the Bell-state projective measurement operator $|\psi^+\rangle_{23\ 23}\langle\psi^+|$ for the Bell state $|\psi^+\rangle_{23}$ acts on $|\psi\rangle_{1234}$ in Eq. (162), one obtains

(172) $$\left(|\psi^+\rangle_{23\ 23}\langle\psi^+|\right)|\psi\rangle_{1234} = 2^{-1}|\psi^+\rangle_{14}|\psi^+\rangle_{23},$$

because of the orthonormality of the Bell states (see Eq. (135), for example). Analogously, one obtains

(173) $$\left(|\psi^-\rangle_{23\ 23}\langle\psi^-|\right)|\psi\rangle_{1234} = -2^{-1}|\psi^-\rangle_{14}|\psi^-\rangle_{23},$$

(174) $$\left(|\phi^\pm\rangle_{23\ 23}\langle\phi^\pm|\right)|\psi\rangle_{1234} = \mp 2^{-1}|\phi^\pm\rangle_{14}|\phi^\pm\rangle_{23}.$$

Note that in each case both projected Bell states are the same as the measured Bell state. That is, if the result of the Bell state measurement of particles 2 and 3 is $|\psi^\pm\rangle_{23}$ or $|\phi^\pm\rangle_{23}$, then the state of particles 1 and 4 is projected onto the entangled states $|\psi^\pm\rangle_{14}$ or $|\phi^\pm\rangle_{14}$, respectively. Also, in every case photons 1 and 4 become entangled, even though they never interacted.

In the experimental demonstration [86], an ultraviolet pulse first passed through a nonlinear crystal to produce the two separate entangled pairs of photons (1, 2) and (3, 4). A beamsplitter, phase plates, and coincidence detectors were used to perform the Bell-state measurement of photons 2 and 3. The Bell-state analyzer, for analysis of photons 1 and 4, consisted of a polarizing beamsplitter, phase plates, polarizer, narrow bandwidth filters, and coincidence detectors.

Entanglement swapping has also been generalized to manipulate entangled multiparticle systems [84,85]. Entangled states of many particles can, in principle, be generated in a controlled way. Potential applications include cryptographic conferencing, multiparticle generalizations of superdense coders, message reading from more than one source by making only one measurement, the construction of a quantum telephone exchange, the speeding up of the distribution of entangled particles, and series purification [84,85]. Decoherence amelioration will also be essential to further practical development of entanglement swappers.

14. Quantum teleporter

Another exciting application of entanglement is the quantum teleporter. In *quantum teleportation* [82,83], an unknown quantum state can be disassembled by a sender (Alice) into ordinary classical information, which can then be used together with an EPR pair of particles, shared by both Alice and a receiver (Bob), to enable Bob to reconstruct the initial unknown state. To accomplish this, Alice must make a joint measurement on her EPR particle, together with the unknown quantum system, which of course destroys the unknown quantum state. She next sends Bob two bits of information, depending on the result of her measurement, over a classical communication channel. With this information, Bob can then convert the state of his EPR particle into an exact replica of the initial unknown state.

A quantum teleporter is depicted schematically in Fig. 13. EPR designates a down-conversion source producing an EPR pair of photons, 2 and 3, shared by Alice and Bob, respectively. Alice first performs a Bell-state measurement, designated by BSM, on particle 1, whose state $|\chi\rangle_1$ is unknown, together with her EPR photon 2. The measurement randomly projects the two-particle state onto any one of four Bell states. She gains no information about any of the particles, but she does obtain one of four possible joint states. She next transmits the corresponding two classical bits of information to Bob, informing him which of four unitary operations (designated U in the figure) Bob must perform to transform the state of his EPR particle 3 into the original unknown state $|\chi\rangle_3$. Bob thereby has a teleported reappearance of the original state $|\chi\rangle_1$. Quantum teleportation has been experimentally demonstrated by several groups [87–90].

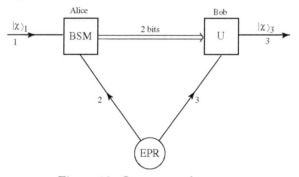

Figure 13. Quantum teleporter.

Let us examine quantum teleportation in more detail. Assume the EPR pair of photons 2 and 3 is in the antisymmetric two-particle Bell state $|\psi^-\rangle_{23}$, Eq. (164), namely,

$$(175) \qquad |\psi^-\rangle_{23} = 2^{-1/2} \left(|\rightarrow\rangle_2 |\uparrow\rangle_3 - |\uparrow\rangle_2 |\rightarrow\rangle_3 \right).$$

(The states are labeled here explicitly with subscripts, designating the particles.) Denote the unknown state of Alice's particle 1 (also a photon) by

$$(176) \qquad |\chi\rangle_1 = a |\rightarrow\rangle_1 + b |\uparrow\rangle_1,$$

with unknown complex coefficients a and b. The combined state of Alice's unknown state and the EPR pair is then given by

$$(177) \quad |\psi\rangle_{123} = |\psi^-\rangle_{23} |\chi\rangle_1 = 2^{-1/2} \left(|\rightarrow\rangle_2 |\uparrow\rangle_3 - |\uparrow\rangle_2 |\rightarrow\rangle_3 \right) \left(a |\rightarrow\rangle_1 + b |\uparrow\rangle_1 \right).$$

Next, expanding the state $|\psi\rangle_{123}$ in terms of the Bell-state basis for particles 1 and 2, namely, $|\psi^\pm\rangle_{12}$ and $|\phi^\pm\rangle_{12}$ (see Eqs. (129) and (130)), one obtains

$$|\psi\rangle_{123} = |\psi^-\rangle_{12}\,{}_{12}\langle\psi^-|\psi\rangle_{123} + |\psi^+\rangle_{12}\,{}_{12}\langle\psi^+|\psi\rangle_{123}$$
(178)
$$+ |\phi^-\rangle_{12}\,{}_{12}\langle\phi^-|\psi\rangle_{123} + |\phi^+\rangle_{12}\,{}_{12}\langle\phi^+|\psi\rangle_{123}.$$

Next, using Eqs. (129), (130), and (177), one obtains

$$\begin{aligned}
{}_{12}\langle\psi^-|\psi\rangle_{123} &= 2^{-1}\left({}_1\langle\rightarrow|\,{}_2\langle\uparrow| - {}_1\langle\uparrow|\,{}_2\langle\rightarrow|\right) \\
&\quad \times (|\rightarrow\rangle_2|\uparrow\rangle_3 - |\uparrow\rangle_2|\rightarrow\rangle_3)(a|\rightarrow\rangle_1 + b|\uparrow\rangle_1) \\
&= 2^{-1}(-a|\rightarrow\rangle_3 - b|\uparrow\rangle_3),
\end{aligned}$$
(179)

$$\begin{aligned}
{}_{12}\langle\psi^+|\psi\rangle_{123} &= 2^{-1}\left({}_1\langle\rightarrow|\,{}_2\langle\uparrow| + {}_1\langle\uparrow|\,{}_2\langle\rightarrow|\right) \\
&\quad \times (|\rightarrow\rangle_2|\uparrow\rangle_3 - |\uparrow\rangle_2|\rightarrow\rangle_3)(a|\rightarrow\rangle_1 + b|\uparrow\rangle_1) \\
&= 2^{-1}(-a|\rightarrow\rangle_3 + b|\uparrow\rangle_3),
\end{aligned}$$
(180)

$$\begin{aligned}
{}_{12}\langle\phi^-|\psi\rangle_{123} &= 2^{-1}\left({}_1\langle\rightarrow|\,{}_2\langle\rightarrow| - {}_1\langle\uparrow|\,{}_2\langle\uparrow|\right) \\
&\quad \times (|\rightarrow\rangle_2|\uparrow\rangle_3 - |\uparrow\rangle_2|\rightarrow\rangle_3)(a|\rightarrow\rangle_1 + b|\uparrow\rangle_1) \\
&= 2^{-1}(a|\uparrow\rangle_3 + b|\rightarrow\rangle_3),
\end{aligned}$$
(181)

$$\begin{aligned}
{}_{12}\langle\phi^+|\psi\rangle_{123} &= 2^{-1}\left({}_1\langle\rightarrow|\,{}_2\langle\rightarrow| + {}_1\langle\uparrow|\,{}_2\langle\uparrow|\right) \\
&\quad \times (|\rightarrow\rangle_2|\uparrow\rangle_3 - |\uparrow\rangle_2|\rightarrow\rangle_3)(a|\rightarrow\rangle_1 + b|\uparrow\rangle_1) \\
&= 2^{-1}(a|\uparrow\rangle_3 - b|\rightarrow\rangle_3).
\end{aligned}$$
(182)

Then, substituting Eqs. (179) to (182) in Eq. (178), one obtains

$$|\psi\rangle_{123} = 2^{-1}\left[|\psi^-\rangle_{12}(-a|\rightarrow\rangle_3 - b|\uparrow\rangle_3) + |\psi^+\rangle_{12}(-a|\rightarrow\rangle_3 + b|\uparrow\rangle_3)\right.$$
(183)
$$\left. + |\phi^-\rangle_{12}(b|\rightarrow\rangle_3 + a|\uparrow\rangle_3) + |\phi^+\rangle_{12}(-b|\rightarrow\rangle_3 + a|\uparrow\rangle_3)\right].$$

The four Bell-state measurement outcomes are equally likely. This follows, since the probabilities of Bell states $|\psi^\pm\rangle_{12}$ are

(184) $\quad 4^{-1}(-a^*\,{}_3\langle\rightarrow|\pm b^*\,{}_3\langle\uparrow|)(-a|\rightarrow\rangle_3 \pm b|\uparrow\rangle_3) = 4^{-1}\left(|a|^2 + |b|^2\right) = 1/4,$

assuming that state $|\chi\rangle_1$, Eq. (176), is normalized to unity. Similarly, the probabilities of Bell states $|\phi^\pm\rangle_{12}$ are

(185) $\quad 4^{-1}(\mp b^*\,{}_3\langle\rightarrow|+a^*\,{}_3\langle\uparrow|)(\mp b|\rightarrow\rangle_3 + a|\uparrow\rangle_3) = 4^{-1}\left(|b|^2 + |a|^2\right) = 1/4.$

Each has a probability of 1/4. After Alice's measurement, Bob's EPR particle 3 will have been projected into one of the four pure-state superpositions in Eq. (183), according to the Bell-state measurement outcome. Each of the possible resulting states of Bob's EPR particle 3 is simply related to the initial state $|\chi\rangle_1$. The state $|\chi\rangle_1$, given by Eq. (176), can be represented by the vector

(186) $$|\chi\rangle_1 = \begin{pmatrix} a \\ b \end{pmatrix}$$

in a basis in which vertical and horizontal polarizations are represented by the basis vectors

(187) $$|\uparrow\rangle = \begin{pmatrix} 0 \\ 1 \end{pmatrix}, \quad |\rightarrow\rangle = \begin{pmatrix} 1 \\ 0 \end{pmatrix},$$

respectively. If Bob's measurement outcome is $|\psi^-\rangle_{12}$, then the state of Bob's EPR particle 3, according to Eq. (183), is given by

$$（188）\qquad |1\rangle_3 = -a|\rightarrow\rangle_3 - b|\uparrow\rangle_3 = -\begin{pmatrix} a \\ b \end{pmatrix},$$

which is the same as Alice's initial state $|\chi\rangle_1$, except for an irrelevant phase factor ($e^{i\pi} = -1$). If Bob's measurement outcome is $|\psi^+\rangle_{12}$, then the state of Bob's EPR particle 3, according to Eq. (183), is

$$（189）\qquad |2\rangle_3 = -a|\rightarrow\rangle_3 + b|\uparrow\rangle_3 = \begin{pmatrix} -1 & 0 \\ 0 & 1 \end{pmatrix} \begin{pmatrix} a \\ b \end{pmatrix}.$$

Bob can convert this state to a replica of Alice's initial state $|\chi\rangle_1$ by applying an implementation of the unitary operator

$$（190）\qquad U_2 = \begin{pmatrix} -1 & 0 \\ 0 & 1 \end{pmatrix}^{-1} = \begin{pmatrix} -1 & 0 \\ 0 & 1 \end{pmatrix},$$

since $U_2|2\rangle_3 = \begin{pmatrix} a \\ b \end{pmatrix}$. I note in passing that the operator Eq. (190) is proportional to the operator HNH, since, according to Eqs. (79), (91), and (190), one has

$$（191）\qquad HNH = 2^{-1/2}\begin{pmatrix} 1 & 1 \\ 1 & -1 \end{pmatrix}\begin{pmatrix} 0 & 1 \\ 1 & 0 \end{pmatrix}2^{-1/2}\begin{pmatrix} 1 & 1 \\ 1 & -1 \end{pmatrix} = \begin{pmatrix} 1 & 0 \\ 0 & -1 \end{pmatrix} = -U_2.$$

If Bob's measurement outcome is $|\phi^-\rangle_{12}$, then the state of Bob's EPR particle 3, according to Eq. (183), is given by

$$（192）\qquad |3\rangle_3 = b|\rightarrow\rangle_3 + a|\uparrow\rangle_3 = \begin{pmatrix} 0 & 1 \\ 1 & 0 \end{pmatrix}\begin{pmatrix} a \\ b \end{pmatrix}.$$

This state can be converted to a replica of Alice's initial state $|\chi\rangle_1$ by the application of the unitary operator

$$（193）\qquad U_3 = \begin{pmatrix} 0 & 1 \\ 1 & 0 \end{pmatrix}^{-1} = \begin{pmatrix} 0 & 1 \\ 1 & 0 \end{pmatrix},$$

since $U_3|3\rangle_3 = \begin{pmatrix} a \\ b \end{pmatrix}$. Note that the operator (193) is the NOT operator, Eq. (79). Finally, if Bob's measurement outcome is $|\phi^+\rangle_{12}$, then the state of Bob's EPR particle 3, according to Eq. (183), is given by

$$（194）\qquad |4\rangle_3 = -b|\rightarrow\rangle_3 + a|\uparrow\rangle_3 = \begin{pmatrix} 0 & -1 \\ 1 & 0 \end{pmatrix}\begin{pmatrix} a \\ b \end{pmatrix},$$

which Bob can convert to a replica of Alice's initial state $|\chi\rangle_1$ by applying the unitary operator

$$（195）\qquad U_4 = \begin{pmatrix} 0 & -1 \\ 1 & 0 \end{pmatrix}^{-1} = \begin{pmatrix} 0 & 1 \\ -1 & 0 \end{pmatrix},$$

since $U_4|4\rangle_3 = \begin{pmatrix} a \\ b \end{pmatrix}$. Bob can use a suitable combination of half-wave plates to perform the unitary operations U_2, U_3, and U_4. In each case an accurate teleportation is obtained if Alice communicates (classically) to Bob the classical outcome of her measurement, after which Bob applies the required operation with his wave plates to transform the state of his EPR photon into a replica of Alice's original state $|\chi\rangle_1$. The state of qubit $|\chi\rangle_1$, Eq. (176), has been transferred to the state $|\chi\rangle_3 = a|\rightarrow\rangle_3 + b|\uparrow\rangle_3$. Alice is left with particles 1 and 2 in one Bell state $|\psi^\pm\rangle_{12}$

or $|\phi^{\pm}\rangle_{12}$, and she acquires no information on the original state $|\chi\rangle_1$. Although a perfect copy of the original qubit can be created in quantum teleportation, the original qubit is completely destroyed. This is compatible with the fact that arbitrary quantum states cannot be cloned [32,33]. In recent teleportation experiments, the entire quadrature phase amplitude of a light beam was teleported [89], instead of just discrete polarization states.

Quantum teleportation is in principle possible even if Alice and Bob lose track of each other's location after sharing their EPR pair, provided Alice can broadcast the requisite classical information, and provided the EPR particle states are sufficiently long lived. However, decoherence severely limits the storage time of any qubits. Thus, decoherence is a critical issue for the development of practical quantum teleporters. It is presently not possible to store separated EPR particles for a sufficiently long period of time, because of decoherence. For example, the effects of attenuation and noise on a single photon, sent through an optical fiber, lead to coherence lengths $\sim 10 km$ and decoherence times of order $(10 km)/(3 \times 10^8 m/s)$ $\sim 10^{-5}$ s. The implementation of error-correction methods (see Sect. 21) may, however, lead to significant improvements.

15. Quantum copiers

According to the no-cloning theorem, arbitrary quantum states cannot be cloned because of the linearity of quantum mechanics [32,33]. The implication is that one cannot produce an exact copy of an arbitrary qubit. This does not mean, however, that an approximate copy cannot be made. The function of one type of *quantum copier* is to produce an approximate copy of a qubit that is as close to being an exact copy as possible, and with the original qubit changed as little as possible in the process. A number of types of quantum copiers have been considered in the literature [91–104,59].

A universal copier is one that produces two identical copies whose quality is independent of the input state [91,92]. The universal quantum copier must copy an arbitrary pure state, which can be written in a chosen basis, $\{|0\rangle_{a_0}, |1\rangle_{a_0}\}$ as [96]

$$|\psi\rangle_{a_0} = \alpha |0\rangle_{a_0} + \beta |1\rangle_{a_0}, \tag{196}$$

for which a general parameterization of the coefficients is

$$\alpha = \sin\theta \, e^{i\phi}, \qquad \beta = \cos\theta. \tag{197}$$

A universal quantum copier satisfies three basic requirements [92]:

(1) If the state of the original qubit at the output of the quantum copier is denoted by the density operator $\rho_{a_0}^{(\text{out})}$, and that of the quantum copy is $\rho_{a_1}^{(\text{out})}$, one requires that

$$\rho_{a_0}^{(\text{out})} = \rho_{a_1}^{(\text{out})}. \tag{198}$$

(2) If the measure of distance $d(\rho_1, \rho_2)$ between two states with density operators ρ_1 and ρ_2 is taken to be the square of the Hilbert-Schmidt norm, that is,

$$d(\rho_1, \rho_2) \equiv Tr\left[(\rho_1 - \rho_2)^2\right], \tag{199}$$

then the requirement that pure states should be copied equally well can be expressed by

$$d(\rho_{a_i}^{(\text{out})}, \rho_{a_i}^{(\text{id})}) = C, \qquad i = 0, 1, \tag{200}$$

where $\rho_{a_i}^{(id)}$ is the ideal density operator describing the input state, and C is a constant, independent of the input state.

(3) Equation (200) should be minimized with respect to all unitary transformations within the Hilbert space of the two qubits and the quantum copier.

It can be shown that the unitary transformation that implements the universal quantum copier by satisfying requirements (1) to (3) is given by [92,96]

$$(201) \quad |0\rangle_{a_0} |Q\rangle_x \Rightarrow (3/2)^{-1/2} |0\rangle_{a_0} |0\rangle_{a_1} |\uparrow\rangle_x + 3^{-1/2} |+\rangle_{a_0 a_1} |\downarrow\rangle_x ,$$

and

$$(202) \quad |1\rangle_{a_0} |Q\rangle_x \Rightarrow (3/2)^{-1/2} |1\rangle_{a_0} |1\rangle_{a_1} |\downarrow\rangle_x + 3^{-1/2} |+\rangle_{a_0 a_1} |\uparrow\rangle_x ,$$

where

$$(203) \quad |+\rangle_{a_0 a_1} = 2^{-1/2} \left(|1\rangle_{a_0} |0\rangle_{a1} + |0\rangle_{a_0} |1\rangle_{a1} \right).$$

Here, indices a_0, a_1, and x designate the original qubit, the copy, and the copier, respectively. The copier has a two-dimensional state space with basis vectors $|\uparrow\rangle_x$ and $|\downarrow\rangle_x$, and $|Q\rangle_x$ denotes the initial state of the copier. The implication of Eqs. (201) and (202) is that the copy contains 5/6 of the desired state and 1/6 undesired.

The universal quantum copier can be implemented with a network of simple quantum logic gates [92,96]. Also, a quantum copier has been designed that produces multiple identical copies from one or more original qubits [96–99,59].

In other work, it has been shown that, owing to residual correlations between the copy and the quantum copier, quantum copying degrades entanglement [100]. A quantum copier can also involve quantum teleportation schemes [101,102]. Quantum copiers implementing a quality measure based on state distinguishability have also been investigated [103,104]. Decoherence is an obstacle to useful implementations of quantum copying, because it limits the state storage time.

16. All-optical quantum information processors

It is next appropriate to begin to address various possible qubit devices for quantum computation. Consider first *all-optical quantum information processors*. All-optical quantum computers are of two general types, depending on whether nonlinear optical elements are employed. In a quantum computer based on linear optics, the basic qubits consist of two path states and/or polarization states of a single photon [105,106]. A photon leaving an optical element, such as a beamsplitter with two exit ports, has a propensity to exit along either path, so the photon becomes a two-path-state system. The linear optical elements composing the device include beamsplitters, mirrors, polarizers, wave plates, etc. The initial state of the device need only consist of a single photon entering the device at a beamsplitter. All the necessary quantum gates can be implemented (see Sect. 6). Even two-qubit gates such as the controlled-NOT gate can be implemented. By cascading the number of beamsplitters, locating one at each alternative path in a network of optical elements, we make the device a multiple-qubit system. However, because of the resulting exponential proliferation of optical elements needed to form a large number of path qubits, such a linear-optical quantum computer is limited to a relatively small number of qubits. The device is therefore limited to implementing small-scale quantum networks for performing small quantum algorithms, such as those involved in simple quantum error correction and quantum teleportation [105].

The second type of all-optical quantum computer is a multi-photon device employing nonlinear optical elements [107–109]. Nonlinear optical elements are needed so that the state of one photonic qubit can control the state of another at certain nodes in the network. The use of many photons circumvents the exponential cascading required by large linear optical quantum computers. The problem with the use of traditional nonlinear optical elements for implementing conditional dynamics, in which the state of one photon conditionally modulates the state of another, is the huge nonlinear susceptibility required to produce the necessary phase shifts at the two-photon level of intensity (see Sect. 11). Practical nonlinear photon gates operating at the two-photon level of intensity are not presently available; however, innovative approaches are currently under investigation. These involve mutual interactions between two gate photons and atoms in a medium [70–72,110,111] or in a QED cavity [73].

Although decoherence is not an obstacle to the development of a reasonably small, special-purpose, all-optical quantum information processor, it would become an issue in any attempt to scale up the device to include large numbers of optical elements. Before considering other physical quantum computer implementations, it is well to review the concept of a *universal quantum computer*.

17. Universal quantum computer

A universal quantum computer [112–116] ideally consists of a set of n qubits on which the following operations can be performed: (1) each qubit can be initialized in some state $|0\rangle$; (2) a universal quantum gate, or set of gates, can be applied to any subset of qubits; (3) each qubit can be measured in the basis $\{|0\rangle, |1\rangle\}$; and (4) the qubits evolve only as a result of these transformations.

In the network model [115], logic gates are applied sequentially. The logic gates are typically implemented by interactions that can be turned on or off at appropriate times.

A universal quantum gate is one that, when applied to different combinations of qubits, can produce the same result as any other gate. Since all quantum systems evolve according to the Schrödinger equation, and quantum evolution is unitary, the set of all possible quantum gates must be such that all possible unitary transformations on the n qubits of the computer can be generated. The controlled-NOT gate and single-qubit rotations are universal, since any $n \times n$ unitary matrix can be formed by combining two-qubit controlled-NOT operators and single-qubit rotations. The controlled-NOT gate and rotation can be combined into a single universal quantum gate, which is a conditional rotation [4,117–120]. The universal quantum computer can, in principle, simulate, by finite means, and arbitrarily closely, any finitely realizable physical system. This is the quantum extension of the Church-Turing principle [114–116]. The extended principle holds, since (1) the state of any finite quantum system can be represented by a vector in Hilbert space, and can therefore be precisely represented by a finite number of qubits; and (2) the evolution of the state of any finite quantum system can be represented by a unitary transformation, and can therefore be simulated on a quantum computer capable of generating any unitary transformation with arbitrary precision.

It must be assumed that the tasks performed by the quantum computer are such that the number of steps is predictable, or that the quantum computer can indicate completion of a computation by means of a dedicated qubit not otherwise involved in the computation. Otherwise, there are fundamental unavoidable obstacles to the construction of a halt qubit to control task completion [121–123]. Also, possible contradictions have recently been suggested between the Church-Turing principle and the inclusion of unrestricted classes of quantum observables and unitary operators [124].

It must be emphasized that, as described above, the universal quantum computer is an ideal theoretical construct that does not take into account imprecise gate operations, quantum decoherence, imprecise measurements, and quantum error correction. Only with qualifications to the stated requirements might the prescription for a universal quantum computer be implementable.

A quantum computer is not a closed system, and therefore the computational degrees of freedom cannot evolve unitarily. Interactions between the computational degrees of freedom and noncomputational degrees of freedom, both outside and inside the computer, will result in qubit decoherence, and consequent deterioration of the qubit entanglements necessary for successful computer operation. Inexact tuning of quantum gates and structural inaccuracies will also result in erroneous evolution of quantum states. Quantum error correction must therefore be implemented (see Sect. 21).

18. Quantum simulators

A quantum computer can be used to simulate other quantum systems [125–138]. Such a quantum computer is a *quantum simulator*. A quantum computer needs at least n qubits to simulate a state vector in a 2^n-dimensional Hilbert space. It must implement unitary transformations in the 2^n-dimensional Hilbert space, and this will typically require an exponential number of elementary quantum logic gates. The implication is that a quantum computer cannot efficiently simulate every physical system. However, many physical systems can be simulated with a quantum computer that could not, in practice, be simulated with a classical computer because of intractability [125–134]. A computation is only tractable if its complexity is such that the resources required to perform it do not increase exponentially with the number of digits in the input. The time evolution of the wave function of a quantum many-body system could be faithfully simulated on a quantum computer [131–137]. Quantum chaos may also be efficiently calculated on a quantum computer [138]. Ultimately, the simulation of quantum field theory may also be possible on a large quantum computer [134]. However, presently, no quantum simulator exists of sufficient scale to perform any significant simulations of quantum systems.

19. Quantum factorizer

A strong incentive for attempts to develop practical quantum computers arises from their possible use in speedily factoring very large numbers for cryptographic applications. A quantum computer could be used to factor large L-digit numbers in $\sim L^3$ time compared to $\sim e^{(\ln L)^{2/3} L^{1/3}}$ time for a classical computer [139–141]. This would exploit the coherence of the quantum wave function of a quantum register implementing an array of qubits. To factor a number N, choose a number x at random that is coprime with N. Use the quantum computer to calculate the order

r of x mod N; that is, to find r such that

(204) $$x^r = 1 \text{ mod } N.$$

If r is even, then the greatest common divisor of $\left(x^{r/2} \pm 1\right)$ and N is a factor of N, and can be determined with Euclid's algorithm. For example, if $N = 1295$ and $x = 6$, one has

(205) $$6^r = 1 \text{ mod } 1295 \Rightarrow r = 4,$$

(206) $$6^{4/2} \pm 1 = 36 \pm 1 = \{35, 37\},$$

(207) $$1295 = 5 \cdot 7 \cdot 37.$$

A *quantum factorizer* would implement Shor's quantum factoring algorithm to calculate the order r [139–141]. A number q having small prime factors is first chosen, such that

(208) $$N^2 < q < 2N^2.$$

By means of appropriate quantum gates, the qubits constituting the quantum register (see Sect. 20) can be manipulated to produce the state

(209) $$|\psi_1\rangle = q^{-1/2} \sum_{a=0}^{q-1} |a, 0\rangle,$$

where $|a, 0\rangle$ denotes the tensor-product state $|a\rangle |0\rangle$. Next, an additional set of quantum gates must be used to implement a unitary transformation of the state $|\psi_1\rangle$ to produce the state

(210) $$|\psi_2\rangle = q^{-1/2} \sum_{a=0}^{q-1} |a, x^a \text{ mod } N\rangle.$$

Here $|a, x^a \text{ mod } N\rangle$ denotes $|a\rangle |x^a \text{ mod } N\rangle$. Next, the state $|\psi_2\rangle$ must be Fourier transformed with the quantum computer to produce the state

(211) $$|\psi_3\rangle = q^{-1/2} \sum_{m=0}^{q-1} \sum_{a=0}^{q-1} e^{i2\pi am/q} |m, x^a \text{ mod } N\rangle.$$

Both arguments of the superposition must be measured, resulting in $\{c, x^k\}$ for

(212) $$m = c, \quad x^a = x^k, \quad 0 < k < r.$$

Then, the probability of the result $\{c, x^k\}$ is

(213) $$P(c, x^k) = \left| q^{-1/2} \sum_{a=0, x^a = x^k \text{ mod } N}^{q-1} e^{i2\pi ac/q} \right|^2.$$

The probability is periodic in c with period q/r and is sharply peaked at $c = pq/r$ for integer p. Therefore, the period yields r after a few trial runs.

Note that the state $|\psi_2\rangle$ in Eq. (210) is a superposition of product states, and cannot be expressed as a single-product state. It is entangled. Entanglement, in addition to superposition, is often an essential feature of quantum computation. The performance of the operations leading from $|\psi_1\rangle$ in Eq. (209) to $|\psi_2\rangle$ in Eq. (210), for example, would exploit the massive quantum parallelism characteristic of a quantum computer. Because the input $|\psi_1\rangle$ for large N is set up in a large superposition of states, the quantum computer would carry out all the computations for each value of a simultaneously. Although there are $q > N^2$ inputs in Eq. (209), the only outputs, correspond to the measured values of Eq. (212). The computation must be repeated enough times to determine the peaks of the probability distribution in c, Eq. (213).

It is necessary that the rate of decoherence be sufficiently low for the computation to be completed [140,142,143]. Although a quantum factorizer capable of factoring a 250-digit number does not presently exist, if and when one does, the widespread cryptosystems relying on the difficulty of factoring large numbers will be rendered insecure and obsolete. However, to date, not even a 5-digit number has been factored with a quantum computer.

20. Quantum register

One of the main ingredients of any sizable quantum computer would be the *quantum register*. A quantum register may be thought of as a row of N qubits. A binary number,

$$(214) \qquad n = \sum_{k=0}^{N-1} n_k 2^k, \qquad n_k = 0 \text{ or } 1$$

can be stored in the quantum register, and is represented by a product state of the N qubits, namely,

$$(215) \qquad |n\rangle = |n_{N-1}\rangle |n_{N-2}\rangle \cdots |n_1\rangle |n_0\rangle .$$

Here, tensor products are implicit, and the order of the kets corresponds to the order of the qubits in the register. Each ket in the product corresponds to a single qubit. A general state $|\psi\rangle$ of the free quantum register is given by the N-qubit entangled state,

$$(216) \qquad |\psi\rangle = \sum_{n=0}^{2^N-1} \alpha_n |n\rangle ,$$

where the α_n are complex numbers, and the sum is over all 2^N possible product Boolean states, Eq. (215), thereby forming a 2^N-dimensional Hilbert space.

For example, a three-bit classical register can store only one of eight different numbers, $\{000,001,010,011,100,101,110,111\}$, using a binary representation of numbers between 0 and 7. But a quantum register consisting of three qubits can store up to eight numbers at the same time in a quantum superposition. The state of the three-qubit quantum register is, in general, given by

$$(217) \qquad \begin{aligned} |\psi\rangle &= \alpha_{000} |0\rangle |0\rangle |0\rangle + \alpha_{001} |0\rangle |0\rangle |1\rangle + \alpha_{010} |0\rangle |1\rangle |0\rangle + \alpha_{011} |0\rangle |1\rangle |1\rangle \\ &+ \alpha_{100} |1\rangle |0\rangle |0\rangle + \alpha_{101} |1\rangle |0\rangle |1\rangle + \alpha_{110} |1\rangle |1\rangle |0\rangle + \alpha_{111} |1\rangle |1\rangle |1\rangle . \end{aligned}$$

This is a coherent superposition of the numbers from 0 to 7.

For the N-qubit quantum register, although measuring the register's contents will yield only one number, a quantum computation can effectively manipulate all 2^N numbers at once. For example, if each qubit consists of two states of an atom, tuned laser pulses can switch the electronic states so that an initial superposition of 2^N numbers will evolve into a different superposition. This evolution results in massive parallelism, since each number in the superposition is affected.

To prepare a specific number in a quantum register, N elementary operations must be performed. Each of the N qubits must be put in one of two states. N elementary operations, which can be represented by unitary transformations on each qubit, can prepare the register in a coherent superposition of 2^N numbers, to be stored in the register. This process can be seen as follows. One can represent the Boolean states $|0\rangle$ and $|1\rangle$ by the vectors

$$|0\rangle = \begin{pmatrix} 1 \\ 0 \end{pmatrix}, \quad |1\rangle = \begin{pmatrix} 0 \\ 1 \end{pmatrix}. \tag{218}$$

If, for example, the first qubit of a three-qubit register is in the state $|0\rangle$, then applying the Hadamard operator H, Eq. (91), causes the state of the qubit to become

$$H|0\rangle = 2^{-1/2} \begin{pmatrix} 1 & 1 \\ 1 & -1 \end{pmatrix} \begin{pmatrix} 1 \\ 0 \end{pmatrix} = 2^{-1/2} \begin{pmatrix} 1 \\ 1 \end{pmatrix} = 2^{-1/2}(|0\rangle + |1\rangle), \tag{219}$$

which is an equally weighted superposition of Boolean states $|0\rangle$ and $|1\rangle$. The Hadamard transform applied to each qubit of a three-qubit quantum register, each initially in state $|0\rangle$, yields the state,

$$|\psi\rangle = \prod_{i=1}^{3}(H|0\rangle) = H|0\rangle H|0\rangle H|0\rangle = 2^{-3/2}(|0\rangle + |1\rangle)(|0\rangle + |1\rangle)(|0\rangle + |1\rangle)$$

$$= 2^{-3/2}(|0\rangle + |1\rangle)(|0\rangle|0\rangle + |0\rangle|1\rangle + |1\rangle|0\rangle + |1\rangle|1\rangle)$$

$$= 2^{-3/2}(|0\rangle|0\rangle|0\rangle + |0\rangle|0\rangle|1\rangle + |0\rangle|1\rangle|0\rangle + |0\rangle|1\rangle|1\rangle$$
$$+ |1\rangle|0\rangle|0\rangle + |1\rangle|0\rangle|1\rangle + |1\rangle|1\rangle|0\rangle + |1\rangle|1\rangle|1\rangle), \tag{220}$$

in which the order of the kets is preserved and corresponds to the order of the qubits in the register. Next, using a notation for states representing numbers to the base 10, that is,

$$\begin{aligned}|0\rangle &\equiv |0\rangle|0\rangle|0\rangle, \quad |1\rangle \equiv |0\rangle|0\rangle|1\rangle, \quad |2\rangle \equiv |0\rangle|1\rangle|0\rangle, \quad |3\rangle \equiv |0\rangle|1\rangle|1\rangle, \\ |4\rangle &\equiv |1\rangle|0\rangle|0\rangle, \quad |5\rangle \equiv |1\rangle|0\rangle|1\rangle, \quad |6\rangle \equiv |1\rangle|1\rangle|0\rangle, \quad |7\rangle \equiv |1\rangle|1\rangle|1\rangle,\end{aligned} \tag{221}$$

one can rewrite Eq. (220) as

$$|\psi\rangle = 2^{-3/2}(|0\rangle + |1\rangle + |2\rangle + |3\rangle + |4\rangle + |5\rangle + |6\rangle + |7\rangle) = 2^{-3/2}\sum_{n=0}^{2^3-1}|n\rangle. \tag{222}$$

Thus, more generally, if an N-qubit register is initially in the state $|0\rangle|0\rangle|0\rangle...|0\rangle$, one can apply the Hadamard operator H, Eq. (219), to each qubit, and the resulting state of the register is an equally weighted superposition of all 2^N numbers, namely,

$$|\psi\rangle = \prod_{i=1}^{N}(H|0\rangle) = H|0\rangle H|0\rangle ... H|0\rangle = 2^{-N/2}\sum_{n=0}^{2^N-1}|n\rangle. \tag{223}$$

The N elementary operations generate a state containing all 2^N possible numerical values of the register. This provides a method for generating the state, Eq. (209), in Shor's quantum factoring algorithm. If the quantum register is prepared in such a coherent superposition of numbers, and all subsequent computational transformations are unitary (preserving the superposition of states), then in each step the computation is performed on each of the numbers in the superposition simultaneously.

21. Quantum error correctors

Quantum error-correction methods may provide the means to successfully combat decoherence in quantum computers and other qubit devices [144–176,83,116]. *Quantum error correctors* are implementations of these methods that involve quantum circuits consisting of networks of quantum gates. The interaction of a qubit in a general state,

$$|\phi\rangle = a|0\rangle + b|1\rangle, \tag{224}$$

with its environment in state $|e\rangle$, results, in general, in the following entanglement between the qubit and its environment [116,170]:

$$|e\rangle|\phi\rangle = |e\rangle(a|0\rangle + b|1\rangle) \Rightarrow a(c_{00}|e_{00}\rangle|0\rangle + c_{01}|e_{01}\rangle|1\rangle) \\ + b(c_{10}|e_{10}\rangle|1\rangle + c_{11}|e_{11}\rangle|0\rangle), \tag{225}$$

where $|e_{ij}\rangle$ denote states of the environment, and c_{ij} are complex coefficients that depend on the environmental interactions with the qubit. Equivalently, Eq. (225) can be rewritten as follows [116]:

$$|e\rangle|\phi\rangle \Rightarrow (|e_I\rangle I + |e_X\rangle X + |e_Y\rangle Y + |e_Z\rangle Z)|\phi\rangle, \tag{226}$$

where the operators $\{I, X, Y, Z\}$ are defined by

$$I \equiv |0\rangle\langle 0| + |1\rangle\langle 1| = \begin{pmatrix} 1 & 0 \\ 0 & 1 \end{pmatrix}, \tag{227}$$

$$X \equiv |0\rangle\langle 1| + |1\rangle\langle 0| = \begin{pmatrix} 0 & 1 \\ 1 & 0 \end{pmatrix}, \tag{228}$$

$$Z \equiv |0\rangle\langle 0| - |1\rangle\langle 1| = \begin{pmatrix} 1 & 0 \\ 0 & -1 \end{pmatrix}, \tag{229}$$

$$Y \equiv XZ = |1\rangle\langle 0| - |0\rangle\langle 1| = \begin{pmatrix} 0 & -1 \\ 1 & 0 \end{pmatrix}. \tag{230}$$

The corresponding matrix representatives are also given in Eqs. (227) to (230) (these matrices can also be simply related to the Pauli spin matrices [36]). Note that by completeness, I, in Eq. (227), is the identity operator, which corresponds to the unit matrix. Also, the operator X is the NOT operator, Eq. (79), since in matrix form one has

$$X = [\langle m|X|n\rangle] = [\langle m|(|0\rangle\langle 1| + |1\rangle\langle 0|)|n\rangle] = \begin{pmatrix} 0 & 1 \\ 1 & 0 \end{pmatrix} = N. \tag{231}$$

Analogously, one obtains the matrix representatives shown for Z and Y in Eqs. (229) and (230), respectively. Also, in Eq. (226), the states $|e_I\rangle$, $|e_X\rangle$, $|e_Y\rangle$, and $|e_Z\rangle$ are given by

$$|e_I\rangle = 2^{-1}(c_{00}|e_{00}\rangle + c_{10}|e_{10}\rangle), \tag{232}$$

$$|e_X\rangle = 2^{-1}(c_{01}|e_{01}\rangle + c_{11}|e_{11}\rangle), \tag{233}$$

$$|e_Y\rangle = 2^{-1}(c_{01}|e_{01}\rangle - c_{11}|e_{11}\rangle), \tag{234}$$

$$|e_Z\rangle = 2^{-1}(c_{00}|e_{00}\rangle - c_{10}|e_{10}\rangle). \tag{235}$$

Equation (226) represents three types of errors, corresponding to the operators X, Y, and Z. The operator X represents a bit flip, since it interchanges the basis states, thus

$$X \begin{pmatrix} |0\rangle \\ |1\rangle \end{pmatrix} = \begin{pmatrix} |1\rangle \\ |0\rangle \end{pmatrix}. \tag{236}$$

The operator Z represents a phase error, since it introduces a relative phase $e^{i\pi} = -1$:

$$Z \begin{pmatrix} |0\rangle \\ |1\rangle \end{pmatrix} = \begin{pmatrix} 1 & 0 \\ 0 & -1 \end{pmatrix} \begin{pmatrix} |0\rangle \\ |1\rangle \end{pmatrix} = \begin{pmatrix} |0\rangle \\ -|1\rangle \end{pmatrix}. \tag{237}$$

The operator $Y = XZ$ represents a phase change together with a bit flip, since

$$Y \begin{pmatrix} |0\rangle \\ |1\rangle \end{pmatrix} = \begin{pmatrix} 0 & -1 \\ 1 & 0 \end{pmatrix} \begin{pmatrix} |0\rangle \\ |1\rangle \end{pmatrix} = \begin{pmatrix} -|1\rangle \\ |0\rangle \end{pmatrix}. \tag{238}$$

To see that Eq. (226) is equivalent to Eq. (225), note that if we use Eqs. (227) to (230), (232) to (235), and Eq. (224),

$$|e\rangle |\phi\rangle \Rightarrow (|e_I\rangle I + |e_X\rangle X + |e_Y\rangle Y + |e_Z\rangle Z) |\phi\rangle$$
$$= 2^{-1} (c_{00} |e_{00}\rangle + c_{10} |e_{10}\rangle) (|0\rangle \langle 0| + |1\rangle \langle 1|) (a |0\rangle + b |1\rangle)$$
$$+ 2^{-1} (c_{01} |e_{01}\rangle + c_{11} |e_{11}\rangle) (|0\rangle \langle 1| + |1\rangle \langle 0|) (a |0\rangle + b |1\rangle)$$
$$+ 2^{-1} (c_{01} |e_{01}\rangle - c_{11} |e_{11}\rangle) (|1\rangle \langle 0| - |0\rangle \langle 1|) (a |0\rangle + b |1\rangle)$$
$$+ 2^{-1} (c_{00} |e_{00}\rangle - c_{10} |e_{10}\rangle) (|0\rangle \langle 0| - |1\rangle \langle 1|) (a |0\rangle + b |1\rangle)$$

$$= 2^{-1} (c_{00} |e_{00}\rangle + c_{10} |e_{10}\rangle) (a |0\rangle + b |1\rangle)$$
$$+ 2^{-1} (c_{01} |e_{01}\rangle + c_{11} |e_{11}\rangle) (b |0\rangle + a |1\rangle)$$
$$+ 2^{-1} (c_{01} |e_{01}\rangle - c_{11} |e_{11}\rangle) (-b |0\rangle + a |1\rangle)$$
$$+ 2^{-1} (c_{00} |e_{00}\rangle - c_{10} |e_{10}\rangle) (a |0\rangle - b |1\rangle)$$

$$= a (c_{00} |e_{00}\rangle |0\rangle + c_{01} |e_{01}\rangle |1\rangle) + b (c_{10} |e_{10}\rangle |1\rangle + c_{11} |e_{11}\rangle |0\rangle). \tag{239}$$

Thus Eq. (226) is equivalent to Eq. (225).

Suppose that a quantum computer manipulates k qubits in the general state $|\phi_k\rangle$. Then add $n - k$ qubits in the state $|0\rangle$ to the computer, so that there are n qubits. Next, perform the encoding operation

$$E(|\phi_k\rangle |0\rangle) = |\phi_E\rangle, \tag{240}$$

which produces, in general, some entangled state $|\phi_E\rangle$ of all n qubits. Then let noise affect all n qubits. The noise can be represented as a sum of error operators M, where each M is a tensor product of n operators I, X, Y, Z, one acting on each qubit. For example, if I operates on qubit 1, X on 2, Z on 3, X on 4, Y on 5, X on 6, Y on 7, and I on 8, this can be represented by the operator

$$M = I_1 X_2 Z_3 X_4 Y_5 X_6 Y_7 I_8. \tag{241}$$

Then, general interactions between the n qubits and the environment produce the general noisy state

$$|\psi\rangle_N = \sum_s |e_s\rangle M_s |\phi_E\rangle, \tag{242}$$

where each M_s is an operator involving products of the four operators I,X,Y,Z, such that each of the n qubits is acted on by one of them. Next, add another $n-k$ ancilla qubits, prepared in the state $|0\rangle_a$. For any encoding E, there is some operator A, called the syndrome extraction operator, which identifies the type of corrected error, namely [116],

$$A\left(M_s |\phi_E\rangle |0\rangle_a\right) = \left(M_s |\phi_E\rangle\right) |s\rangle_a \ \forall M_s \in S, \tag{243}$$

where S is the set of correctable errors and depends on the encoding. Here, the symbol s in $|s\rangle_a$ is a binary number that identifies the error operator M_s considered, and the states $|s\rangle_a$ are mutually orthogonal. For the simple case that the general noise state $|\psi\rangle_N$, Eq. (242), contains only $M_s \in S$, the joint state of the n non-ancilla qubits, environment, and ancilla (following the syndrome extraction) is given by

$$|\psi\rangle_{Na} = A |\psi\rangle_N |0\rangle_a = A\left[\left(\sum_s |e_s\rangle M_s |\phi_E\rangle\right) |0\rangle_a\right] = \sum_s |e_s\rangle \left(M_s |\phi_E\rangle\right) |s\rangle_a. \tag{244}$$

If the ancilla state is measured with the measurement operator $|s\rangle_a{}_a\langle s|$, then

$$\left(|s\rangle_a{}_a\langle s|\right) |\psi\rangle_{Na} = |s\rangle_a{}_a\langle s| \sum_{s'} |e_{s'}\rangle \left(M_{s'} |\phi_E\rangle\right) |s'\rangle_a$$

$$= |s\rangle_a \sum_{s'} |e_{s'}\rangle \left(M_{s'} |\phi_E\rangle\right) \delta_{ss'}$$

$$= |e_s\rangle \left(M_s |\phi_E\rangle\right) |s\rangle_a ; \tag{245}$$

that is, the entire state collapses into $|e_s\rangle \left(M_s |\phi_E\rangle\right) |s\rangle_a$ for a particular s. Thus, the measurement reveals the value s, thereby determining the error operator M_s (s is the error syndrome).

Next, if the operator M_s^{-1} is applied to the measured state by means of various quantum gates X,Y, or Z, there results

$$M_s^{-1}\left(|s\rangle_a{}_a\langle s| |\psi\rangle_{Na}\right) = M_s^{-1}\left[|e_s\rangle \left(M_s |\phi_E\rangle\right) |s\rangle_a\right] = |e_s\rangle |\phi_E\rangle |s\rangle_a, \tag{246}$$

resulting in the noise-free state $|\phi_E\rangle$. The state $|e_s\rangle$ of the environment, appearing here, is of no interest, and the ancilla state $|s\rangle_a$ can be put back in state $|0\rangle_a$ and used again. If the noise in $|\psi\rangle_N$, Eq. (242), contains errors M_s that are not in the correctable set S, then the probability must be large that when the syndrome is extracted, the state collapses onto a correctable state.

The error-correction procedure must be such that the encoding operation E and the extraction operation A are such that the set S of correctable errors includes all likely errors. If uncorrelated stochastic noise is assumed, for which the effect on a qubit at different times is uncorrelated (and the effect on different qubits is uncorrelated also), then all possible error operators can be categorized in terms of their likelihood. Those affecting fewer qubits are more likely. If a quantum error-correcting code is such that all errors affecting up to t qubits are correctable, then the code is a t-error correcting code [147,148,151,154,83].

Errors can also occur in the ancilla, quantum gates, and measurements. Methods have been discovered by which the error correction suppresses more noise than it produces [156,158,168,170,175]. Error-correcting codes may require an extremely large overhead in terms of the numbers of qubits (required for sufficient redundancy to recover from errors) and gates (required to process the redundantly encoded data and to diagnose and reverse errors). The error probability per qubit per gate must be very small (below the *accuracy threshold*) if the error correction is to succeed for arbitrarily long computations [161–163,165]. The requirements are formidable for reliable quantum computing using such fault-tolerant quantum error-correcting codes [160,171]. Furthermore, quantum error-correcting codes usually assume that errors in distant qubits are, at most, weakly correlated, and the codes are inadequate to deal with strongly correlated errors involving many qubits [171].

Note that much simpler quantum error-correction methods can be used if enough is known about the sources of noise [177,172,173]. Several passive error-prevention schemes have been proposed, in which the encoding occurs within subspaces that do not decohere because of symmetry properties [178–181]. It has been argued [181], on the basis of a semigroup description of quantum decoherence [182,183], that error-free quantum computation is possible in decoherence-free subspaces. The evolution of the computational degrees of freedom, which form a subspace of the total Hilbert space describing the quantum dynamics of the qubit device and its environment, is nonunitary, and is described by a semigroup. The decoherence-free subspaces are spaces spanned by states annihilated by all error generators (the operators X, Y, and Z in Eq. (226) are error generators). Also, various methods of decoherence control are currently under investigation. These include the application of feedback [184,185] and of external controllable interactions [186–189]. It is argued that the effects of qubit-environment interactions can be removed by suitable decoupling perturbations acting on the qubit device over time scales comparable to the correlation time of the environment [187–189].

22. Quantum computers

I proceed to discuss possible quantum computer implementations that are currently under development [114–116,190–195]. A quantum computer is a qubit device that provides a physical implementation of a quantum mechanical unitary transformation that acts on an array of qubits containing computational inputs in the form of binary-labeled quantum states. The device transforms the corresponding initial quantum superposed state into a final state for which quantum mechanical interference results in a very high probability that when the computer output is read, it will yield the correct answer.

In the ion-trap quantum computer [196], a one-dimensional lattice of identical ions is stored and laser cooled in a linear Paul trap (radio-frequency (rf) quadrupole trap) [197–200,75]. The linear array of ions acts as a quantum register (see Section 20). The rf trap potential strongly confines the ions radially about the trap axis, and an electrostatic potential causes the ions to oscillate along the trap axis in an effective harmonic potential. Laser cooling results in localization of the ions along the trap axis, with spacing determined by Coulomb repulsion and the confining axial potential. The lowest frequency mode of collective oscillation of the ions is the axial center-of-mass mode, in which all the ions oscillate with the same phase together. Each of the trapped ions acts as a qubit, in which the two pertinent states are the electronic ground state and a long-lived excited state. By means of coherent interaction of a precisely controlled laser pulse with any one of the ions in a standing-wave configuration, one can manipulate the ion's electronic state and the quantum state of the collective center of mass mode of the oscillator. The center of mass mode can then be used as a bus, quantum dynamically connecting the qubits, to implement the necessary quantum logic gates. The general state of the line of ions that the quantum register comprises is an entangled linear superposition of their states. A completed computation can be read out by a quantum jump measurement technique [196]. Experimental demonstration of the ion-trap approach began with state preparation, quantum gates, and measurement for a single trapped ion [201]. Since then, a number of experimental and theoretical issues regarding the ion-trap approach to quantum computation have been explored [202–229,116,85]. Presently, the main experimental difficulty in implementing this approach is cooling the ions to the ground state in the trap. The primary source of decoherence is apparently the heating due to coupling between the ions and noise voltages in the trap electrodes [204,205]. In the near term, it is contemplated that 100 quantum gate operations could be applied to a few ions [116]. It is, however, very questionable that sufficient storage capacity and coherence will ever be achieved to enable factorization of hundred-digit numbers by the trapped-ion approach [85,227–229]. Also, the speed of an ion-trap quantum computer would apparently be limited by the frequencies of vibrational modes in the trap.

Another popular approach to the development of a quantum computer is provided by cavity quantum electrodynamics (QED). In the cavity QED [230–233] approach, a number of neutral atoms are trapped inside a high-finesse optical cavity [234]. Electronic states of the atom act as qubits to store information. The atoms in the cavity interact with a quantized mode of the cavity. The separations between the atoms are much greater than the wavelength of the cavity mode, and the atoms can interact individually with laser pulses. This permits sequences of operations between two qubits and the implementation, in principle, of an entire quantum network. The qubits consist of ground-state levels of the trapped atoms. Quantum gates can be implemented by the atoms being coupled to individual laser pulses and entangled by exchange of a cavity photon. Pulsed lasers can be used to drive transitions in one atom conditionally on the internal states of another atom. Also, the polarization states of a photon can serve as a qubit. An atom trapped in the cavity can serve as an effective nonlinear medium to mediate interactions between two photons, and thereby implement a two-photon quantum gate, in which the polarization state of one photon alters the phase of the other photon [73,235]. Letting $|l\rangle_i$ and $|r\rangle_i$ denote left and right circular polarization states of photon i $(i = 1,2)$, one has, effectively,

(247) $$|l\rangle_1 |l\rangle_2 \implies |l\rangle_1 |l\rangle_2,$$

(248) $$|l\rangle_1 |r\rangle_2 \implies e^{i\phi_2} |l\rangle_1 |r\rangle_2,$$

(249) $$|r\rangle_1 |l\rangle_2 \implies e^{i\phi_1} |r\rangle_1 |l\rangle_2,$$

(250) $$|r\rangle_1 |r\rangle_2 \implies e^{i(\phi_1+\phi_2+\Delta)} |r\rangle_1 |r\rangle_2,$$

where ϕ_1 and ϕ_2 are differential phases between the two polarization states, and Δ is the conditional phase shift. The transformations, Eqs. (247) to (250), are accomplished first by one photon being stored in the cavity, in which the right circular polarization state couples strongly to the atom, but the left circular polarization state does not. Next, another photon traverses the cavity, also interacting preferentially in one polarization state with the atom, and acquiring the conditional phase shift only if the photons are in the right circular polarization state. Thus, the phase shift is conditional on the polarization state of both photons; the result is a two-qubit quantum logic gate. The gate exploits the extremely large optical nonlinearities that are achievable in cavity QED. The cavity operates in a pioneering parameter regime, in which [73]

$$\kappa > \left(g^2/\kappa\right) > \gamma, \tag{251}$$

where κ is the cavity-field damping rate, g is the dipole coupling rate of the atom to the cavity, and γ is the transverse atomic decay rate to noncavity modes. Under these conditions, the coherent coupling of the atom to the cavity mode (at rate g^2/κ) dominates incoherent emission into free space (at rate γ). This enables strong coupling of a single atom to the cavity mode, allowing efficient transfer of electromagnetic fields from input to output channels (at rate κ). Conditional dynamics at the single quantum level has also been achieved with single atoms interacting with very weak microwave fields in superconducting cavities [74]. Atomic wave function phase shifts were produced by microwave fields with, on average, much less than one photon in the cavity. In related work, a quantum memory was implemented, in which the quantum information carried by a two-level atom was transferred to a cavity and subsequently to another atom [236]. Within the same framework, a methodology was developed for the construction of arbitrary quantum computational networks with all the necessary quantum gates to perform all quantum logic operations [237,238]. In cavity QED, sources of decoherence include spontaneous emission from excited states of atoms and cavity decay during gate operation. Maintaining coherence between multiple cavities is problematic. Also, the trapping and localization of atoms inside cavities present formidable difficulties. Possible scaling up of the cavity-QED approaches to more than several qubits remains to be accomplished and poses serious problems that may limit the practical utility of *cavity-QED quantum computers* to special-purpose, small-scale quantum computations (for use in quantum communication, for example).

Cavity QED is also being implemented in the development of possible *quantum computer communication networks*. In the cavity-QED approach to quantum information processing, both the states of atoms confined in cavities and the states of photons interacting with the atoms may serve as qubits to store and transfer quantum information. Although the difficulties in the successful trapping and localization of atoms inside high-finesse optical cavities are considerable (possibly making the development of large-scale universal quantum computers based on the cavity-QED concept an unattainable goal), the development of small-scale, special-purpose quantum information processors involving limited numbers of trapped atoms will likely be possible. For example, the cavity-QED paradigm may provide a practical approach to the development of controlled single-photon sources [239], the synthesis of entangled states [240], and quantum teleportation between cavities [241]. Both photonic and atomic qubits may be exploited with the cavity-QED paradigm, in the form of quantum information networks that enable the implementation of quantum communication protocols and distributed quantum computation [242–246]. Multiple atom-cavity systems located at distant network nodes may be interconnected with optical fibers, or perhaps even use free-space transmission. Analysis has been performed of basic network operations, including local quantum information processing , quantum-state transmission between network nodes, and quantum entanglement distribution [242–248]. Ideal transmission may be permitted after a finite number of trials, without disturbing the quantum information. Possible sources of error include absorption of photons in the optical fibers and cavity mirrors, cavity and laser design errors, and spontaneous emission from excited states.

A recent proof of principle of quantum computation has been accomplished by another approach, which makes innovative use of established nuclear magnetic resonance (NMR) technology. NMR [249,250] can be used as the basis for quantum computation when certain liquids are used along with available NMR instrumentation [251–256]. The qubits are the spins of atomic nuclei in the molecules constituting the liquid. These qubits are extremely well isolated from their environment. The nuclear spin orientations in a single molecule form a quantum data register. The liquid contains about 10^{23} molecules at room temperature and undergoes strong random thermal fluctuations. The liquid is located in a large magnetic field, and each spin can be oriented either in the direction of the magnetic field ($|\uparrow\rangle = |0\rangle$) or opposite ($|\downarrow\rangle = |1\rangle$). An *NMR quantum computer* operating on N qubits uses molecules having N atoms with distinguishable spins in the frequency domain. The input to the computer is an ensemble of nuclear spins initially in thermal equilibrium. Each spin can be manipulated with resonant rf pulses, and the coupling between neighboring nuclear spins can be exploited to produce quantum logic gates. The spins have scalar coupling, and a driving pulse in resonance can tip a spin conditional on the state of another spin, thus providing a quantum bus channel. A sequence of rf pulses and delays produces a series of quantum logic gates connecting the initial state to a desired final state. By suitable timing of each pulse, a desired unitary transformation can be resonantly performed on a single spin of a molecule, even though all the spins in the molecule are exposed, since they all have slightly different resonant frequencies. The decoherence times of the spins are long enough that the qubits can be stored for a sufficiently long time. The average magnetic moment of all the nuclei together is big enough to produce a detectable magnetic field for measurement purposes. The liquid consists effectively of a statistical ensemble of single-molecule quantum computers, which can be described by a density matrix. The method exploits the structure present in thermal equilibrium to produce a perturbation in the system's large density matrix that is effectively equivalent to a pure state of much smaller dimension, a pseudo-pure state. The system of molecules, each having N nuclear spins, can be described by a density matrix [251–253],

(252)
$$\rho = 2^{-N} I + \rho_\Delta,$$

in which the first term describes an equilibrium part that is proportional to the identity I, and the second term ρ_Δ is a traceless matrix representing the deviation from equilibrium. For an appropriate pulsed field sequence, the deviation transforms as a density matrix, the *deviation density matrix*, and represents the statistical ensemble of single-molecule quantum computers in the form of a bulk effective quantum computer. An effective pure state can be distilled out of ρ_Δ by means of a data compression pulse sequence. An appropriate computational procedure yields a deterministic result in which measuring the ensemble yields a nonvanishing average. Readout is performed by measurement of the magnetization of the bulk sample. This is bulk quantum computation employing large ensembles of quantum systems instead of single systems. Such a bulk quantum computer acts as an ensemble of many small quantum computers carrying out computations independently in parallel. The initial state of each is random, and only ensemble averages of each computer register can be measured. The ensemble can effectively behave like a pure state, since even if, for example, only a small fraction of the systems are in their ground states, the ones that are not can be arranged so that their signals cancel each other, and only the fraction in the ground state produces a nonvanishing signal, making the ensemble appear to be pure. Generally, if a chosen fraction of the states can be labeled, and the rest caused to average away, then an effective pure state can be produced [251–253]. Pioneering experimental implementations included (1) synthesis of high-resolution samples of several two- and three-qubit molecules [257], (2) implementation of Grover's fast quantum search algorithm for a system with only four states [257–259], (3) a proof of principle for three-qubit quantum computation [260,261], (4) proof-of-principle quantum error correction [262], (5) implementation of a quantum algorithm determining whether an unknown function is constant, or has value 0 for half its arguments and 1 for the rest [263,264], and (6) implementation of a quantum algorithm for estimating the number of matching items in a search operation [265,266]. The NMR quantum computers have poor scaling with the number of qubits. The measured signal scales as 2^{-N}. This feature will likely limit NMR quantum computers to applications requiring only 10 to 20 qubits [267]. Other concepts involving the manipulation of spin states have also been proposed that use (1) electron spins [268], (2) atomic-force microscopy to manipulate nuclear spins [269,120], and (3) electron-nuclear spin interactions in the Hall regime [270].

NMR can be combined with semiconductor technology in a hybrid quantum computer implementation, the *silicon-based nuclear spin quantum computer*. Since it is unlikely that NMR quantum computers can be scaled up to produce large-scale quantum computations involving very large numbers of qubits, a hybrid concept has been proposed that uses semiconductor physics to deterministically manipulate nuclear spins [271]. The silicon-based nuclear spin quantum computer would consist of an equally spaced linear or planar array of dopant phosphorus nuclear spins implanted in a silicon semiconductor crystal, separated by an insulator layer from overlaying voltage-controlled metal gate electrodes; the gate electrodes would implement quantum logic operations by affecting the shape of the electron wavefunction surrounding each phosphorus nucleus. The qubits are the nuclear spins of the phosphorus nuclei embedded periodically in the silicon crystal, each located directly beneath a gate referred to as an "A gate." A phosphorus atom in a silicon host is an electron donor, and at room temperature one of its outer electrons can move freely in the crystal; however, at the very low temperature of operation of the device, the electron is weakly bound by the phosphorus ion, and the electron spin can interact with the nuclear spin. Thus, the weakly bound electron spin can affect the state of a qubit, since electron and nuclear spins are coupled by the hyperfine interaction [249]. Also, the electrons can mediate nuclear spin interactions and facilitate the measurement of nuclear spins. A voltage applied to an A gate can cause the wave function of the electron bound to the phosphorus nucleus beneath it to become altered, thereby changing its overlap with the nucleus. This electron-nucleus interaction affects the relative energies of the nuclear qubit, and therefore also the resonant rf frequency needed to cause a nuclear spin flip. This makes it possible for a resonant rf pulse to selectively change the state of only that nucleus. Between any two neighboring A gates is a "J gate," for affecting the overlap between two electron orbitals bound to neighboring phosphorus nuclei in the lattice. This J gate results in an indirect coupling between the two neighboring phosphorus qubits, and makes it possible to implement the quantum gates necessary for quantum computation. Since it is prohibitively difficult to directly measure the spin state of an individual nucleus, an indirect approach is implemented, involving a chain of interactions among the nuclear spin, its bound electron and a neighboring electron, the external magnetic field, and the J gate overlaying the two electron orbitals; these interactions affect the capacity between neighboring A gate electrodes, which can be measured. Normally, all electron spins are pointed in the direction of the external magnetic field, but if the overlap between two neighboring electron orbitals is sufficiently increased by an applied J-gate voltage, it may become energetically favorable for the pair of electrons to change their state so that their spins are opposite. Whether this happens depends on the direction of whichever phosphorus spin is coupled most strongly to its bound electron, and that depends on the A-gate voltage. The Pauli exclusion principle does not allow both electrons to hop into the same atom, unless their spins are opposite, and a hop changes the capacitance between the neighboring A electrodes. This same mechanism also enables qubit states to be initialized, since each can be measured individually, and the measured state can be reversed with an NMR pulse if necessary. Before such a device can be successfully implemented, many formidable technological problems must be overcome, including (1) emplacement of individual phosphorus atoms in a prescribed regular array in a perfect Silicon crystal, (2) development of defect-free semiconductor and overlaying layers, (3) limitation of the decoherence rate of the phosphorus qubits in the presence of electrode fluctuations, gate biasing, rf-induced eddy currents, charge fluctuations, spin impurities, and crystal defects, (4) sufficient limitation of the probability of error in each operation, and (5) nanoscale fabrication [271–273].

Another popular solid-state approach to quantum computer development is the *quantum-dot quantum computer*. Various approaches have been considered to the potential development of quantum dot quantum computers [274–286]. In one of the most interesting approaches [277–281], a qubit would be the two spin states of an electron in a single-electron quantum dot, and a quantum register would consist of an array of coupled single-electron quantum dots. Each semiconductor quantum dot would consist of one excess electron with spin $1/2$ in a potential well that confines the electron in all three dimensions. Quantum gate operations would be performed by gating of the tunneling barrier between neighboring dots, to produce controlled entanglements of the qubits. The tunnel barrier between dots could be raised or lowered by the application of a higher or lower gate voltage [277,279]. If the barrier is sufficiently reduced, virtual tunneling can occur, resulting in transient spin-spin coupling. Hopping to a neighboring auxiliary ferromagnetic dot can be used to implement single-qubit operations. Also, readout can be implemented through tunneling to a neighboring auxiliary paramagnetic dot, which can nucleate a ferromagnetic domain that could then be measured [277]. Alternatively, spin-dependent tunneling into another neighboring auxiliary dot can enable spin measurement by means of an electrometer [280]. Reversing this procedure might accomplish general-state preparation. Ground-state preparation can be accomplished by cryogenic cooling in a uniform applied magnetic field. In other approaches, the qubit does not consist of electron spin states, but instead of pseudo-spin states, corresponding to charged orbital degrees of freedom [282–285]. Real spin has the advantage of (1) permanent well-defined qubits with no extra dimensions for qubit state leakage, and (2) much longer dephasing times [277]. Although spin degrees of freedom are not significantly affected by fluctuations in electric potential, they are subject to decoherence arising from magnetic coupling to the environment. In other approaches, gate operation may be performed by spectroscopic manipulation. Coherent optical control of quantum-dot states is an important area of research. Picosecond optical excitation can be used to coherently control quantum-dot states on a time scale that is small compared to the decoherence time [286]. Generally, it is expected that solid-state systems, because of their complex internal field and many-particle environment, will subject qubit states to numerous possible mechanisms of quantum decoherence, presenting formidable obstacles to the development of a practical large-scale quantum computer based on the quantum-dot approach.

Another possible condensed-matter approach to quantum computer development involves macroscopic superconducting quantum states. One of these is the *Josephson junction quantum computer*, and efforts are under way to develop it [287–293]. In one relatively simple exploratory approach, a nanoelectronic device would consist of an array of low-capacitance Josephson junctions [287–291]. The device would exploit coherent tunneling in the superconducting state, with the possibility of controlling individual charges by means of Coulomb blockade effects. The Josephson junction qubit is implemented in a small superconducting island connected by a tunnel junction to a superconducting electrode. The qubit consists of two charge states of the superconducting island adjacent to the junction. The logical states differ by one Cooper-pair charge. The island is connected to an ideal voltage source with a gate capacitor between them. An array of such Josephson junction qubits, each with its own voltage source, are connected in parallel with each other and also with a mutual inductor. The array would serve as a quantum register. One- and two-qubit gates can be implemented by application of appropriate sequences of voltages across the junctions, and by the tuning of selected qubits to resonance. Readout can be accomplished by capacitively coupling a dissipative normal-metal single-electron transistor to a qubit [288]. It is encouraging that coherent tunneling of Cooper pairs, and related characteristics of quantum superposition of charge states, have already been theoretically investigated and experimentally demonstrated [294–298]. This simple design presents various challenges: it requires high-precision timing control, and it involves residual two-qubit interactions that will produce errors. An improved design is being considered, in which the Josephson junctions are replaced by SQUIDs (superconducting quantum interference devices) that can be controlled by magnetic fluxes [289]. This design would enable exact on-off switching of the two-qubit coupling, relaxation of the timing control and system parameter requirements, and complete control of two-qubit couplings. Parallel operations on different qubits may be achieved, in principle, by more advanced designs, including additional tunable SQUIDs to decouple different parts of the circuit. Scaling to large numbers of qubits with massively parallel operation will necessitate much more elaborate designs, significant progress in nanotechnology, reduction in working temperature, near-perfect control of time-dependent gate voltages, and much longer decoherence times. It is well also to mention another possible approach to quantum computation using Josephson junctions [293], which would exploit the quantized states of position of superconducting vortices [299–301]. The two basis states of a qubit might be a vortex positioned in one of two neighboring superconducting loops with a Josephson junction between them. A vortex would have to be capable of being in a quantum superposition of positions. In other work, high-temperature-superconductor Josephson junctions are being considered for possible use in the construction of a quantum computer [292].

Another approach to quantum computer development involves a *SQUID quantum computer*, which would exploit the physics of superconducting quantum interference. The quantized magnetic flux in a SQUID might also be used as the basic qubit, instead of the charge of a Josephson junction island. This would provide the basis for a potential SQUID quantum computer [302–304]. It is encouraging that experimental demonstrations have been made of quantum jumps in SQUID rings [305] and resonant tunneling of the flux between quantized energy levels in different flux states of a SQUID [306,307]. Although SQUIDs are macroscopic objects, and macroscopic objects generally suffer decoherence in the extreme, many of the dissipative mechanisms that normally operate in macroscopic systems can be eliminated in SQUID systems [37,308–312]. The Hamiltonian of an rf SQUID can be represented as a two-state system [310]. The SQUID consists of a single tunnel junction with critical current I_c shunted by an inductor L. If a magnetic flux of half the fundamental flux quantum,

$$(253) \qquad \phi_0 = \pi \hbar / e,$$

is applied to the loop (e is the magnitude of the charge of the electron), and if

$$(254) \qquad 1 < 2\pi L I_c / \phi_0 < 5\pi/2,$$

then a two-state system can be created, in which the two states correspond to the loop containing either one flux quantum or none at all [302]. A supercurrent then circulates the SQUID ring in either direction. In principle, a superposition state can also be created, but this has only recently been accomplished [313,314], despite concerted previous attempts. One must also be able to entangle multiple qubits with each other. No quantum logic gate has yet been experimentally demonstrated in the SQUID approach. Important issues in the SQUID approach to quantum computer development include the required operating temperature, required junction quality, suppression of competing modes, magnetic coupling of flux qubits to magnetic impurities, unidentified decoherence mechanisms, sufficiently small junction capacitances, and required fabrication technology.

Another innovative approach to quantum computer development involves trapped atoms in optical lattices. Lasers can be used to confine ultracold atoms in periodic lattices [315–325]. The atoms are held together with light (instead of chemical bonds, as in a solid). Laser cooling and trapping techniques, used in producing

Bose-Einstein condensation, are also used to form optical lattices. In an optical lattice, ultracold atoms can be arranged in a crystal-like array in an optical potential in which the intensity or polarization of light varies periodically. Near-zero-temperature webs of interfering laser beams can be used to cool a collection of atoms, and the atoms become suspended in well-defined positions in the interfering beams. The separations of the atoms in an optical lattice are hundreds of times that in an ordinary solid. Currently, only about one in ten lattice sites is filled. The potential well depth is $\sim 10^{-9}$ that in a solid, and the dynamical oscillations of the atoms (1 to 100 kHz) are $\sim 10^6$ to 10^9 lower in frequency than in a solid. Defects and impurities are absent in the optical lattice. Through changes in the polarization and the direction of propagation of the laser beams, many different crystalline structures can be created in one, two, or three dimensions. The optical potential seen by an atom in an optical lattice depends on the magnetic quantum number of the atom. Dissipation and decoherence in optical lattices occurs due to spontaneous emission [326]. Evaporative cooling may be able to produce lattices in which every site is filled. Atoms can be trapped in a two-dimensional lattice and cooled to the zero point of motion by resolved-sideband cooling [327]. These characteristics are important for initial state preparation and the manipulation of quantum states. An optical lattice may serve as the arena for *neutral atom quantum computers*, as discussed below. Also, it will likely be possible to create Bose-Einstein condensates trapped in optical lattices [321]. This would provide the arena for *Bose condensate quantum computers*.

It may become possible to implement quantum logic with neutral atoms trapped in an optical lattice, very far off resonance [328]. A qubit would consist of two states of an atom. If the lasers are detuned very far off resonance, photon scattering is made negligible, and high laser intensities will maintain substantial potential wells. By means of laser cooling, the atoms can be prepared in the ground state of the potential well. The lattice geometry can be varied dynamically: changing the angle between different laser polarizations can control the distance between wells. Two atoms trapped in neighboring wells can be forced into the same well by varying the polarization of the trapping lasers. An auxiliary laser can then induce a near-resonant electric dipole, and the electric dipole-dipole potential will provide the predominant interaction between the atoms. Following this, the atoms can be separated by adiabatic rotation of the laser polarization. Quantum gates would be implemented through the induced coherent dipole interactions. Single-qubit operations would be performed with polarized resonant Raman pulses. Two-qubit operations would require conditioning the state of one atom on that of the other. A controlled-NOT gate could be achieved by conditioning the target atomic resonance on a resolvable level shift induced by the control atom. The resonant dipoles would be conditionally turned on only during conditional logic operations, and environmental decoherence would thereby be suppressed. Large numbers of atoms could be entangled by a sequence of two-qubit interactions. If the atoms are lightly confined to separations small relative to the wavelength, then a coherent dipole-dipole interaction can be induced with negligible photon scattering. The coherent level shift can thereby be substantially enhanced, while the cooperative emission rate is substantially suppressed. The atoms couple very weakly to the environment and would interact only during two-qubit logical operations, and all manipulations would be performed rapidly relative to the photon scattering rate, thus impeding spontaneous emission, which is the main source of decoherence. Before an operational neutral atom quantum computer can be successfully developed to perform even elementary quantum computations, many issues must be explored, including (1) increasing the filling fraction of atoms in the lattice, (2) developing methods for addressing and reading out individual qubits, (3) investigating the effect of atomic collisions, and (4) implementing quantum error-correction methods.

Bose condensates in optical lattices provide another innovative approach to quantum computer development. Bose condensates can be confined in an optical dipole trap [329]. They can also be created in an optical lattice [330]; the theory of condensates in optical potentials has been investigated [331]. A very innovative scheme has been proposed [332] to fill an optical lattice with a Bose condensate, and exploit ideas related to Mott transitions in optical lattices [333]. A far-detuned optical lattice acts as a conservative potential and could be loaded with a Bose condensed atomic vapor, resulting in tens of atoms per lattice site. It has been argued [333] that the dynamics of bosonic atoms corresponds to that of a Bose-Hubbard model [334], which describes the hopping of bosonic atoms between the lowest vibrational states of lattice sites. The important system parameters can be controlled by appropriate laser parameters and configurations. The model predicts a phase transition from a superfluid phase to a Mott insulator phase at low temperature. This results in the formation of an optical crystal with long-range order and period controlled by the laser light. A finite gap is produced in the excitation spectrum. An optical crystal could be created with uniform lattice occupation, or tailored atomic patterns could be produced. This would occur at sufficiently low temperature that cold laser-controlled coherent interactions could implement conditional dynamics in moving trap potentials. Methods were investigated for producing two-qubit quantum gates and highly entangled states, and may provide the basis for a possible *Bose condensate quantum computer*.

23. Quantum robots

Since computer technology has often been closely linked with automata or robot technology, it is a natural progression to consider the possibility of a *quantum robot*. A quantum robot would be a mobile quantum system containing a quantum computer and other ancillary systems [335–338]. It would perform tasks such as measuring the environment and changing the state of the environment. The robot's quantum computer can be modeled as a cyclic network of quantum gates. Computations performed by the quantum robot would include determining its next action. Actions of the robot would include moving itself.

For analytical and numerical convenience, models of quantum robots have been developed based on discretized space and time. Environments can be open or closed, and are modeled to include arbitrary numbers and types of systems moving in one-, two-, and three-dimensional spatial lattices. The systems are characterized by some internal quantum numbers, and can interact with each other or be free.

Quantum robot dynamics is described in terms of performing tasks. Tasks are described by their goals, such as producing changes in the state of the environment, and making measurements by transfer of information from the environment to the robot.

An example of a task might be to move each system, located in a spatial region R of a one-dimensional lattice, two sites to the right if the destination site is unoccupied. The path taken by the robot and the criteria for determining when it is inside or outside the region R must be specified, and the robot must be able to make the required movements. If there are, for example, three systems in region R at locations x_1, x_2, and x_3, then the initial state of the environment in that region is

$$|x\rangle \equiv |x_1\rangle |x_2\rangle |x_3\rangle, \tag{255}$$

and the final state is

$$|x+2\rangle \equiv |x_1+2\rangle |x_2+2\rangle |x_3+2\rangle. \tag{256}$$

If the initial state of R is a linear superposition,

$$|\psi\rangle = \sum_x c_x |x\rangle, \tag{257}$$

of three position states $|x\rangle$ in R, then the final state is

$$|\psi'\rangle = \sum_x c_x |x+2\rangle. \tag{258}$$

This equation illustrates the fact that quantum robots can complete the same task simultaneously on many environments.

Another task might be to perform measurements or experiments on the environment: for example, to determine the distance between a particle and the quantum robot.

The dynamics of each task can be described as a sequence of alternating computation and action phases. The purpose of each computation phase is to determine the action of the robot in the subsequent action phase, and to record local environment information. The computational input to the robot's computer includes the local state of the environment and other useful information, such as the output of the previous computation phase. During each action phase, the state of the environment can be changed. What happens during an action phase depends on the state of the output system and the state of the local environment. (A simple example of an environment is a one-dimensional lattice of qubits, or a one-dimensional lattice containing a particle located at some site.)

A unitary step operator T is associated with each task and describes the task dynamics during one time step. The system dynamics for n future directed time steps is represented by T^n, and for n past directed time steps by $(T^\dagger)^n$. The step operator T has two components,

$$T = T_c + T_a, \tag{259}$$

where T_c and T_a represent the computation and action phases, respectively, of the quantum robot.

The robot's quantum computer includes a finite-state output system o. In the computation phase represented by the operator T_c, the action to be carried out is determined. The states of o and the nearby environment are computational inputs. The multistep computation determines a new state of o as output. The action phase represented by T_a is carried out based on the state of o, and includes motion of the quantum robot and local changes in the state of the environment. The computer also has a control qubit c to regulate whether T_c or T_a is active. If c is in state $|0\rangle$, then T_c is active. If c is in state $|1\rangle$, then T_a is active. The last step of T_c or T_a

includes the changes $|0\rangle \Rightarrow |1\rangle$ or $|1\rangle \Rightarrow |0\rangle$, respectively. The operators T_c and T_a must satisfy certain requirements.

The requirement that T_c not change the robot location or the state of the environment is expressed as follows [336]:

$$T_c = \sum_{xE} P^e_{xE} T_c P^e_{xE} P^c_o, \tag{260}$$

where

$$P^e_{xE} = |xE\rangle \langle xE| \tag{261}$$

is a projection operator for the quantum robot when it is located at site x in state $|x\rangle$ and the environment is in state $|E\rangle$, and

$$|xE\rangle = |x\rangle |E\rangle, \tag{262}$$

where the tensor product is implicit. Equation (260) ensures that iteration of T_c does not change the location of the robot or the state of the environment. Also in Eq. (260),

$$P^c_0 = |0\rangle \langle 0| \tag{263}$$

is the projection operator for the control qubit in state $|0\rangle$, and it ensures that T_c is inactive if the control qubit is in state $|1\rangle$. Another condition on T_c follows from the requirement that it depend on the environment only in the neighborhood of the quantum robot [336].

The action phase operator T_a depends on the states of o, but does not change them. This results in an algebraic condition on T_a similar to Eq. (260), namely [336],

$$T_a = \sum_{xx'l_1} P^{qr}_{x'} P^o_{l_1} T_a P^o_{l_1} P^{qr}_{x} P^c_1, \tag{264}$$

where $P^o_{l_1}$ is the projection operator for o in state $|l_1\rangle$, P^{qr}_x is the projection operator for the quantum robot at site x, P^c_1 is the projection operator for c in state $|1\rangle$, and the sum restricts any movement during a time step to a neighboring site only. Also, T_a is independent of the states of the computer and the states of distant environmental systems, and this results in another algebraic condition on T_a [336].

The time development of a task can be expressed in terms of a sum over paths [339]. The overall state $|\psi(n)\rangle$ of the system after n time steps is

$$|\psi(n)\rangle = T^n |\psi(0)\rangle, \tag{265}$$

where $|\psi(0)\rangle$ is the initial overall state of the system. If $|b\rangle$, $|l\rangle$, and $|i\rangle$ denote basis states for the computer, output, and control qubit, respectively, then a basis state for the overall system is

$$|blixE\rangle = |b\rangle |l\rangle |i\rangle |x\rangle |E\rangle, \tag{266}$$

and by completeness of the set of states, one has

$$\sum_{blixE} |blixE\rangle \langle blixE| = 1. \tag{267}$$

Substituting Eq. (267) in Eq. (265), one has

$$|\psi(n)\rangle = T^n \left(\sum_{blixE} |blixE\rangle \langle blixE| \right) |\psi(0)\rangle$$
$$(268) \qquad = \sum_{b_1 l_1 i_1 x_1 E_1} T^n |b_1 l_1 i_1 x_1 E_1\rangle \langle b_1 l_1 i_1 x_1 E_1 |\psi(0)\rangle,$$

in which dummy summation variables b, l, i, x, and E are replaced by b_1, l_1, i_1, x_1, and E_1, respectively. Therefore, the amplitude for ending in the state $|blixE\rangle$ is

$$(269) \qquad \langle blixE|\psi(n)\rangle = \sum_{b_1 l_1 i_1 x_1 E_1} \langle blixE| T^n |b_1 l_1 i_1 x_1 E_1\rangle \langle b_1 l_1 i_1 x_1 E_1 |\psi(0)\rangle.$$

The notation is simplified if we define

$$(270) \qquad |wi\rangle \equiv |blixE\rangle.$$

Then Eq. (269) can be written as

$$(271) \qquad \langle wi|\psi(n)\rangle = \sum_{w_1 i_1} \langle wi| T^n |w_1 i_1\rangle \langle w_1 i_1 |\psi(0)\rangle.$$

The matrix element $\langle wi| T^n |w_1 i_1\rangle$ appearing in Eq. (271) gives the amplitude for evolving from the state $|w_1 i_1\rangle$ to the state $|wi\rangle$ in n steps.

It is true that

$$(272) \qquad T^n = T\,1 T\,1 ... 1 T.$$

Therefore, substituting Eq. (272) and the completeness relation, Eq. (267), in the matrix element $\langle wi| T^n |w_1 i_1\rangle$, one obtains the following expression [336]:

$$(273) \qquad \langle wi| T^n |w_1 i_1\rangle = \sum_{w_2 i_2 ... w_n i_n} \langle wi| T |w_n i_n\rangle \langle w_n i_n| T |w_{n-1} i_{n-1}\rangle ... \langle w_2 i_2| T |w_1 i_1\rangle.$$

As in the path integral approach [339], Eq. (273) can also be written in terms of a sum over paths of states $\{|wi\rangle\}$ of length $n+1$, where the initial element is $|w_1 i_1\rangle$, and the final element is $|wi\rangle$, namely,

$$\langle wi| T^n |w_1 i_1\rangle$$
$$= \sum_{\substack{\text{paths } p \text{ of} \\ \text{length } n+1}} \langle wi|p_{n+1}\rangle \langle p_{n+1}| T |p_n\rangle \langle p_n| T |p_{n-1}\rangle ... \langle p_2| T |p_1\rangle \langle p_1|w_1 i_1\rangle$$
$$= \sum_{\substack{\text{paths } p \text{ of} \\ \text{length } n+1}} \langle p_{n+1}| T |p_n\rangle \langle p_n| T |p_{n-1}\rangle ... \langle p_2| T |p_1\rangle \langle p_{n+1}|wi\rangle^* \langle p_1|w_1 i_1\rangle.$$
(274)

The control qubit projection operators are given by Eq. (263), together with

$$(275) \qquad P_1^c = |1\rangle \langle 1|.$$

Also, by completeness, one has

$$(276) \qquad P_0^c + P_1^c = 1,$$

and one can therefore write

$$(277) \qquad T^n = (P_0^c + P_1^c)\, T\, (P_0^c + P_1^c)\, T\, (P_0^c + P_1^c) ... (P_0^c + P_1^c)\, T\, (P_0^c + P_1^c).$$

From Eqs. (259), (260), and (264), one has

(278) $$T P_1^c = (T_c + T_a) P_1^c = T_a P_1^c = T_a ,$$

since

(279) $$P_1^c P_0^c = 0, \qquad P_1^c P_1^c = P_1^c, \qquad P_0^c P_0^c = P_0^c .$$

Similarly,

(280) $$T P_0^c = T_c .$$

Using Eqs. (278) and (280) in Eq. (277), one obtains [336]

(281) $$T^n = \sum_{v_1=a,c} \sum_{t=1}^{n} \sum_{h_1,h_2,\ldots h_t=1}^{\delta(\Sigma,n)} (P_0^c + P_1^c) (T_{v_t})^{h_t} (T_{v_{t-1}})^{h_{t-1}} \ldots (T_{v_2})^{h_2} (T_{v_1})^{h_1} ,$$

where $v_{j+1} = a$ if $v_j = c$, and $v_{j+1} = c$ if $v_j = a$. The summation upper limit $\delta(\Sigma, n)$ means that the sum must satisfy $h_1 + h_2 + \ldots + h_t = n$. Equation (281) expresses T^n as a sum of alternating computation and action phase operators. The operators T_a and T_c do not commute, and the operators for each phase are time ordered, with $(T_{v_{j+1}})^{h_{j+1}}$ occurring after $(T_{v_j})^{h_j}$. With Eq. (281), it can be shown that [336]

$$\langle wi | T^n | w_1 0 \rangle = \sum_{t=1}^{n} \sum_{w_2 \ldots w_t} \sum_{h_1,h_2,\ldots h_t=1}^{\delta(\Sigma,n)} \langle wi | (T_{v_t})^{h_t} | w_t \rangle \ldots$$

(282) $$\times \langle w_3 | (T_a)^{h_2} | w_2 \rangle \langle w_2 | (T_c)^{h_1} | w_1 \rangle ,$$

where

(283) $$|w\rangle \equiv |b\rangle |l\rangle |x\rangle |E\rangle .$$

In terms of paths, the same matrix element becomes [336]

$$\langle wi | T^n | w_1 0 \rangle = \sum_{t=1}^{n} \sum_{\text{paths } p \text{ of length } t+1} \sum_{h_1,h_2,\ldots h_t=1}^{\delta(\Sigma,n)} \langle p_{t+1} i | (T_{v_t})^{h_t} | p_t \rangle \ldots$$

(284) $$\times \langle p_3 | (T_a)^{h_2} | p_2 \rangle \langle p_2 | (T_c)^{h_1} | p_1 \rangle \langle w | p_{t+1} \rangle \langle p_1 | w_1 \rangle ,$$

where

(285) $$|p_j\rangle \equiv |w_j\rangle \equiv |b_j\rangle |l_j\rangle |x_j\rangle |E_j\rangle$$

denotes the j-th state in path p.

The formalism given above was used in a simple example in which the environment consists of one particle on a one-dimensional spatial lattice, and the robot is to determine its distance from the particle [336]. To complete the task, the robot moves to the right on the lattice, and counts the number of lattice sites as it moves. When the particle is located, the number of steps is registered as the distance. The robot then returns to its initial position, and the task is complete. More complex tasks that result in entanglement can also be addressed: tasks involving decision trees of sequences of measurements of noncommuting observables, or even tasks of factoring numbers [139] or searching a database [340], could also be treated as tasks for the quantum robot.

The possibility of a qubit device quantum dynamically affecting and reacting to both itself and its quantum environment is interesting. However, the same decoherence issues facing quantum computer development will also apply to quantum robots. Notwithstanding these difficulties, heroic efforts have been successfully made in recent years to develop quantum error-correction methods to overcome quantum decoherence problems (see Sect. 21).

24. Conclusions

Revolutionary new advances have been made recently in the areas of quantum communication, quantum information processing, quantum computing, and quantum cryptography. These advances have led to a wide variety of potentially useful qubit devices. Several of these devices (such as interaction-free detectors, quantum key receivers, quantum games, all-optical quantum information processors, and various quantum gates) offer near-term opportunities for practical development. Other qubit devices must first overcome the obstacle of quantum decoherence. It is therefore important that the physics of quantum decoherence be much more extensively and thoroughly investigated, both experimentally and theoretically, if these qubit devices are to become a useful reality. Although heroic efforts have been made to develop both theory and methods of quantum error correction and decoherence control, many practical issues remain, and it is important that these theoretical methods be physically implemented and tested.

Qubits have been successfully implemented with photons, atoms, and nuclear spins, including the production of superposition states and qubit entanglements. However, the qubits that would form the basis for quantum computers based on quantum dots, Josephson junctions, SQUIDs, and Bose condensates remain to be experimentally demonstrated. It is important to stress that a viable qubit must not only be a two-state system, it must also be capable of existing in a superposition of Boolean states and being entangled with other qubits.

To test the validity of the various approaches to useful quantum information processing, it is important to implement a variety of qubit entanglers that can construct multiple-qubit entanglements using photons, atoms, nuclear spins, quantum dots, Josephson junctions, SQUIDs, and Bose condensates. For the full power of quantum technology to be exploited, large-scale controlled entanglements must be producible.

The quantum key receiver is ready for practical development, and should be useful both in quantum cryptography and in fundamental studies of the generalized measurement of photonic qubits, based on a POVM. Deterministic single-photon sources need development.

Although quantum game theory has only begun to develop, it is possible that interesting quantum games could be constructed with NMR or all-optical quantum information processors. Also, all the various quantum gates should be implemented with all the various types of qubits, and their engineering design should be optimized to meet the severe fidelity requirements for large-scale quantum computation.

A variety of EPR-pair sources should be developed, along with innovative methods for efficiently storing EPR correlated states. Also, Bell-state synthesizers, using various types of qubits, need much further development for use in quantum dense coding and quantum teleportation. Bell-state analyzers must be developed that can measure all four Bell states; for that purpose, it is important that gates be developed that enable robust nonlinear interaction between two qubits.

Robust Bell-state analyzers are essential for the useful implementation of entanglement swapping. Theoretical generalizations of entanglement swapping to manipulate entangled multiparticle systems remain to be experimentally demonstrated. Possible applications of entanglement swapping include cryptographic conferencing and quantum telephone exchanges. Decoherence is a critical issue also for the development of practical entanglement swappers and quantum teleporters. It is presently not possible to store separated EPR particles for periods of time that are sufficiently long for many applications. Still awaiting exploration is the possible implementation of quantum-error correction methods, which are needed for reliable quantum teleportation and for long-distance quantum key distribution.

Although quantum states cannot be cloned, it is important to thoroughly investigate methods for making approximate copies of arbitrary quantum states for possible information processing and quantum cryptographic applications. All-optical quantum information processors are ready for practical development and should be exploited for proof-of-principle demonstrations of quantum computation, and as small special-purpose quantum information processors.

The construction of practical universal quantum computers is presently not feasible , and must await further new revolutionary discoveries and advances. The successful development of quantum simulators could enable simulations of physical systems that are intractable on classical computers. Although a quantum factorizer does not presently exist that can factor even a 5-digit number, if one were successfully developed to factor numbers of more than 250 digits, many cryptosystems would be rendered insecure and obsolete.

Although ion-trap quantum computers are unlikely candidates for performing large-scale quantum computations involving more than 100 qubits, the ion-trap technology affords continuing opportunities for experimental investigations of many issues of quantum decoherence and quantum information processing, and may also lead to the development of special-purpose, small-scale quantum computers.

The cavity QED approach to quantum computing needs much more extensive experimental verification. Cavity QED continues to serve as a near-term arena for experimental investigations of the fundamental physics of quantum information processing and quantum decoherence, and may also be useful for small-scale quantum computation; however, it is unlikely that it will lead to a practical large-scale quantum computer. Cavity QED does offer interesting possibilities for the development of quantum communication networks.

The NMR quantum computer is the first small-scale proof-of-principle quantum computer, and it continues to furnish useful insights into practical quantum information processing; because of its poor scaling, however, it will likely be limited to computations requiring only 10 to 20 qubits. Nevertheless, the possibility of combining NMR with semiconductor technology is currently receiving considerable attention. It is hoped that the silicon-based, nuclear spin quantum computer can be scaled up to perform large-scale quantum computations involving large numbers of qubits. The quantum-dot quantum computer is another attempt to implement semiconductor technology in quantum computation. However, severe problems posed by quantum decoherence face any quantum computer based on the solid state.

Josephson junction and SQUID quantum computers would be based on the superconducting state, and for both types, robust viable qubits (which can exist in superpositions and be entangled with each other) remain to be demonstrated. Neutral atoms or Bose condensates trapped in optical lattices provide still another possible innovative approach to quantum computer development.

Quantum robots offer the exciting possibility of a qubit device that can quantum dynamically affect and react to both its environment and itself. The successful implementation of quantum robots, quantum computers, and most qubit devices will depend on incorporating effective quantum error-correction methods, together with effective decoherence avoidance and control.

The future of quantum information science and technology appears bright, as do the prospects for successful implementations of a broad spectrum of innovative qubit devices.

25. Acknowledgement

This work was sponsored by the U.S. Army Research Laboratory. The hospitality is gratefully acknowledged of the University of Cambridge Isaac Newton Institute for Mathematical Sciences, where some of this work was completed. The author thanks Prof. Samuel Lomonaco for the invitation to present this lecture as part of the short course, *Quantum Computation: The Grand Mathematical Challenge for the Twenty-first Century and the Millennium*, at the American Mathematical Society Annual Meeting, 17–18 January 2000, in Washington DC.

References

[1] H. E. Brandt, "Quantum Decoherence in Qubit Devices," Opt. Eng. **37**, 600–609 (1998).
[2] B. Schumacher, "Quantum Coding," Phys. Rev. A **51**, 2738–2747 (1995).
[3] B. Schumacher and M. D. Westmoreland, "Sending Classical Information via Noisy Quantum Channels," Phys. Rev. A **56**, 131–138 (1997).
[4] A. Barenco, C. H. Bennett, R. Cleve, D. P. DiVincenzo, N. Margolus, P. Shor, T. Sleator, J. A. Smolin , and H. Weinfurter, "Elementary Gates for Quantum Computation," Phys. Rev. A **52**, 3457–3467 (1995).
[5] H. E. Brandt, Qubit Devices and the Issue of Quantum Decoherence, Progr. Quantum Electron. **22**, 257–370 (1998).
[6] A. C. Elitzur and L. Vaidman, "Quantum Mechanical Interaction-Free Measurements," Found. Phys. **23**, 987–997 (1993).
[7] L. Vaidman, "On the Realization of Interaction-Free Measurements," Quant. Opt. **6**, 119–124 (1994).
[8] P. G. Kwait, "Experimental and Theoretical Progress in Interaction-Free Measurements," Physica Scripta **T76**, 115–121 (1998).
[9] P.A.M. Dirac, *The Principles of Quantum Mechanics,* Revised Fourth Edition, Oxford University Press (1967).
[10] R. P. Feynman, *The Theory of Fundamental Processes,* Addison-Wesley Publishing Company, Redwood City, California (1961).
[11] R. P. Feynman, R. B. Leighton, and M. L. Sands, *The Feynman Lectures on Physics, Volume III, Quantum Mechanics,* Addison-Wesley Publishing Company, Redwood City, California (1989).
[12] G. Baym, *Lectures on Quantum Mechanics,* Addison-Wesley Publishing Company, Reading, Massachusetts (1969).
[13] V. Degiorgio, "Phase Shift Between the Transmitted and the Reflected Optical Fields of a Semireflecting Lossless Mirror is $\pi/2$," Am. J. Phys. **48**, 81–82 (1980).
[14] A. Zeilinger, "General Properties of Lossless Beamsplitters in Interferometry," Am. J. Phys. **49**, 882–883 (1981).
[15] Z. Y. Ou and L. Mandel, "Derivation of Reciprocity Relations for a Beamsplitter from Energy Balance," Am J. Phys. **57**, 66–67 (1989).

[16] A. Peres, *Quantum Theory: Concepts and Methods,* Kluwer, Dordrecht (1993).
[17] E. B. Davies, *Quantum Theory of Open Systems,* Academic, New York (1976).
[18] C. W. Helstrom, *Quantum Detection and Estimation Theory,* Academic Press, New York (1976).
[19] A. K. Ekert, B. Huttner, G. M. Palma, and A. Peres, "Eavesdropping on Quantum-Cryptographical Systems," Phys. Rev. A **50**, 1047–1056 (1994).
[20] H. E. Brandt, "Inconclusive Rate with a Positive Operator Valued Measure," preprint NI99015-CCP, University of Cambridge Isaac Newton Institute for Mathematical Sciences (1999); to appear in Contemporary Math., American Mathematical Society.
[21] H. E. Brandt, J. M. Myers, and S. J. Lomonaco, Jr., "Aspects of Entangled Translucent Eavesdropping in Quantum Cryptography," Phys. Rev. A **56**, 4456–4465 (1997); **58**, 2617 (1998).
[22] H. E. Brandt and J. M. Myers, Invention Disclosure: POVM Receiver for Quantum Cryptography (U.S. Army Research Laboratory, Adelphi, MD, 1996); U.S. Patent Number 5,999,285, "Positive-Operator-Valued-Measure Receiver," Dec. 7, 1999.
[23] J. M. Myers and H. E. Brandt, "Converting a Positive Operator-Valued Measure to a Design for a Measuring Instrument on the Laboratory Bench," Meas. Sci. Technol. **8**, 1222–1227 (1997).
[24] H. E. Brandt, "Eavesdropping Optimization for Quantum Cryptography using a Positive Operator Valued Measure," Phys. Rev. A **59**, 2665–2669 (1999); "Inconclusive Rate as a Disturbance Measure in Quantum Cryptography," Phys. Rev. A **62**, 042310-1-14 (2000).
[25] C. H. Bennett, "Quantum Cryptography Using Any Two Nonorthogonal States," Phys. Rev. Lett. **68**, 3121–3124 (1992).
[26] C. H. Bennett, G. Brassard, and A. K. Ekert, "Quantum Cryptography," Sci. Am., 50–57 (October 1992).
[27] C. H. Bennett, F. Bessette, G. Brassard, L. Salvail, and J. Smolin, "Experimental Quantum Cryptography," J. Cryptology **5**, 3–28 (1992).
[28] J. D. Franson and H. Ilves, "Quantum Cryptography Using Optical Fibers," Appl. Opt. **33**, 2949–2954 (1994).
[29] R. J. Hughes, D. M. Alde, P. Dyer, G. G. Luther, G. L. Morgan, and M. Schauer, "Quantum Cryptography," Contemp. Phys. **36**, 149–163 (1995).
[30] S.J.D. Phoenix and P. D. Townsend, "Quantum Cryptography: How to Beat the Code Breakers Using Quantum Mechanics," Contemp. Phys. **36**, 165–195 (1995).
[31] A. E. Ekert, "From Quantum Code-Making to Quantum Code-Breaking," in *The Geometric Universe: Science, Geometry and the Work of Roger Penrose,* A. A. Huggett et al, editors, 195–214, Oxford University Press (1998).
[32] W. K. Wootters and W. H. Zurek, "A Single Quantum Cannot be Cloned," Nature (London) **299**, 802–803 (1982).
[33] D. Dieks, "Communication by EPR Devices," Phys. Lett. **92A**, 271–272 (1982).
[34] B.E.A. Saleh and M. C. Teich, *Fundamentals of Photonics,* John Wiley and Sons, Inc., New York (1991).
[35] P. Busch, M. Grabowski, and P. J. Lahti, *Operational Quantum Physics,* Springer, Berlin (1995).
[36] C. Cohen-Tannoudji, B. Diu, and F. Laloe, *Quantum Mechanics,* Volumes 1 and 2, Wiley, New York (1977).
[37] R. Omnes, *The Interpretation of Quantum Mechanics,* Princeton University Press, Princeton, NJ (1994).
[38] R. Omnes, *Understanding Quantum Mechanics,* Princeton University Press, Princeton, NJ (1999).
[39] D. Giulini, E. Joos, C. Kiefer, J. Kupsch, I. O. Stamatescu, and H. D. Zeh, *Decoherence and the Appearance of a Classical World in Quantum Theory,* Springer, Berlin (1996).
[40] W. H. Zurek, "Pointer Basis of Quantum Apparatus: Into What Mixture Does the Wave Packet Collapse?," Phys. Rev. D **24**, 1516–1525 (1981).
[41] W. H. Zurek, "Environment-Induced Superselection Rules," Phys. Rev. D **26**, 1862–1880 (1982).
[42] W. H. Zurek, "Decoherence and the Transition from Quantum to Classical," Phys. Today, 36–44 (October 1991).

[43] J. R. Anglin, J. P. Paz, and W. H. Zurek, "Deconstructing Decoherence," Phys. Rev. A **55**, 4041–4053 (1997).
[44] W. H. Zurek, "Decoherence, Einselection and the Existential Interpretation (The Rough Guide)," Phil. Trans. R. Soc. Lond. A **356**, 1793–1821 (1998).
[45] W. H. Zurek, "Decoherence, Chaos, Quantum-Classical Correspondence, and the Algorithmic Arrow of Time," Physica Scripta **T76**, 186–198 (1998).
[46] J. P. Paz and W. H. Zurek, "Quantum Limit of Decoherence: Environment Induced Superselection of Energy Eigenstates," Phys. Rev. Lett. **82**, 5181–5185 (1999).
[47] K. Hepp and E. H. Lieb, "Phase Transitions in Reservoir-Driven Open Systems with Applications to Lasers and Superconductors," Helv. Phys. Acta **46**, 573–603 (1973).
[48] R. Omnès, "General Theory of the Decoherence Effect in Quantum Mechanics," Phys. Rev. A **56**, 3383–3394 (1997).
[49] R. Omnès, "Theory of the Decoherence Effect," Fortschr. Phys. **46**, 771–777 (1998).
[50] C. Kiefer and E. Joos, "Decoherence: Concepts and Examples," quant-ph/9803052 (1998) to appear in *Proc. of 10th Born Symposium*.
[51] J. von Neumann and O. Morgenstern, *Theory of Games and Economic Behavior*, John Wiley and Sons, Inc., New York (1967).
[52] D. Blackwell and M. A. Girshick, *Theory of Games and Statistical Decisions*, Dover Publications, Inc., New York (1979).
[53] R. D. Luce and H. Raiffa, *Games and Decisions: Introduction and Critical Survey*, Dover Publications, Inc., New York (1989).
[54] M. D. Davis, *Game Theory: A Nontechnical Introduction*, Dover Publications, Inc., Mineola, New York (1997).
[55] A. P. Maitra and W. D. Sudderth, *Discrete Gambling and Stochastic Games*, Springer-Verlag, New York (1996).
[56] D. A. Meyer, "Quantum Strategies," Phys. Rev. Lett. **82**, 1052–1055 (1999).
[57] J. Eisert and M. Wilkens, "Quantum Games and Quantum Strategies," Phys. Rev. Lett. **83**, 3077–3080 (1999).
[58] L. Goldenberg. L Vaidman, and S. Wiesner, "Quantum Gambling," Phys. Rev. Lett. **82**, 3356–3359 (1999).
[59] R. F. Werner, "Optimal Cloning of Pure States," Phys. Rev. A **58**, 1827–1832 (1998).
[60] A. K. Ekert, "Quantum Cryptography Based on Bell's Theorem," Phys. Rev. Lett. **67**, 661–663 (1991).
[61] N. Gisin and B. Huttner, "Quantum Cloning, Eavesdropping and Bell's Inequality," Phys. Lett. A **228**, 13–21 (1997).
[62] R. Derka, V. Buzek, and A. K. Ekert, "Universal Algorithm for Optimal Estimation of Quantum States for Finite Ensembles via Realizable Generalized Measurement," Phys. Rev. Lett. **80**, 1571–1575 (1998).
[63] P. G. Kwait, K. Mattle, H. Weinfurter, and A. Zeilinger, "New High-Intensity Source of Polarization-Entangled Photon Pairs," Phys. Rev. Lett. **75**, 4337–4341 (1995).
[64] P. G. Kwait, E. Waks, A. G. White, I. Appelbaum and P. H. Eberhard, "Ultra-Bright Source of Polarization Entangled Photons," Phys. Rev. A **60**, R773–R776 (1999).
[65] C. H. Bennett, S. J. Wiesner, "Communication via One- and Two-Particle Operators on Einstein-Podolsky-Rosen States," Phys. Rev. Lett. **69**, 2881–2884 (1992).
[66] K. Mattle, H. Weinfurter, P. G. Kwait, and A. Zeilinger, "Dense Coding in Experimental and Quantum Communication," Phys. Rev. Lett. **76**, 4656–4659 (1996).
[67] A. Zeilinger, "Quantum Entanglement: A Fundamental Concept Finding its Applications," Physica Scripta **T-76**, 203–209 (1998).
[68] N. Bloembergen, *Nonlinear Optics*, World Scientific, Singapore (1996).
[69] *Selected Papers on Nonlinear Optics*, H. E. Brandt, editor, SPIE Optical Engineering Press, Bellingham, WA (1991).
[70] J. D. Franson, "Cooperative Enhancement of Optical Quantum Gates," Phys. Rev. Lett. **78**, 3852–3855 (1997).
[71] J. D. Franson and T. B. Pittman, "Quantum Logic Operations Based on Photon Exchange Interactions," Phys. Rev. A **60**, 917–936 (1999).
[72] J. D. Franson and T. B. Pittman, "Nonlocality in Quantum Computing," Fortschr. Phys. **46**, 697-705 (1998).

[73] Q. A. Turchette, C. J. Hood, W. Lange, H. Mabuchi, and H. J. Kimble, "Measurement of Conditional Phase Shifts for Quantum Logic," Phys. Rev. Lett. **75**, 4710–4713 (1995).

[74] M. Brune, P. Nussenzveig, F. Schmidt-Kaler, F. Bernardot, A. Maali, J. M. Raimond, and S. Haroche, "From Lamb Shift to Light Shifts: Vacuum and Subphoton Cavity Fields Measured by Atomic Phase Sensitive Detection," Phys. Rev. Lett. **72**, 3339–3342 (1994).

[75] *Atomic, Molecular, and Optical Physics Handbook,* G.W.F. Drake, editor, American Institute of Physics, Woodbury, New York (1996).

[76] A. Zeilinger, H. J. Bernstein, and M. A. Horne, "Information Transfer with Two-State Two-Particle Quantum Systems," J. Mod. Opt. **41**, 2375–2384 (1994).

[77] W. Pauli, "The Connection Between Spin and Statistics," Phys. Rev. **58**, 716–722 (1940).

[78] R. F. Streater and A. S. Wightman, *PCT, Spin and Statistics, and All That*, Addison-Wesley Publishing Company, Inc., Redwood City, CA (1989).

[79] C. K. Hong, Z. Y. Ou, and L. Mandel, "Measurement of Subpicosecond Time Intervals Between Two Photons by Interference," Phys. Rev. Lett. **59**, 2044–2046 (1987).

[80] N. Lutkenhaus, J. Culsamiglia, and K. A. Suominen, "Bell Measurements for Teleportation," Phys. Rev. A **59** 3295-3300 (1999).

[81] M. Zukowski, A. Zeilinger, M. A. Horne, and A. K. Ekert, "'Event-Ready-Detectors,' Bell Experiment via Entanglement Swapping," Phys. Rev. Lett. **71**, 4287–4290 (1993).

[82] C. H. Bennett, G. Brassard, C. Crépeau, R. Jozsa, A. Peres, and W. K. Wootters, "Teleporting an Unknown Quantum State via Dual Classical and Einstein-Podolsky-Rosen Channels," Phys. Rev. Lett. **70**, 1895–1899 (1993).

[83] C. H. Bennett, D. P. DiVincenzo, J. A. Smolin, and W. K. Wootters, "Mixed-State Entanglement and Quantum Error Correction," Phys. Rev. A **54**, 3824–3851 (1996).

[84] S. W. Bose, V. Vedral, and P. L. Knight, "Multiparticle Generalization of Entanglement Swapping," Phys. Rev. A **57**, 822–829 (1998).

[85] S. Bose, P. L. Knight, M. Murao, M. B. Plenio, and V. Vedral, "Implementations of Quantum Logic: Fundamental and Experimental Limits," Phil. Trans. R. Soc. Lond. A **356**, 1823–1839 (1998).

[86] J. W. Pan, D. Bouwmeester, H. Weinfurter, and A. Zeilinger, "Experimental Entanglement Swapping: Entangling Photons that Never Interacted," Phys. Rev. Lett. **80**, 3891–3894 (1998).

[87] D. Bouwmeester, J. W. Pan, K. Mattle, M. Eibl, H. Weinfurter and A. Zeilinger, "Experimental Quantum Teleportation," Nature **390**, 575–579 (1997).

[88] D. Boschi, S. Branca, F. De Martini, L. Hardy, and S. Popescu, "Experimental Realization of Teleportating an Unknown Pure Quantum State via Dual Classical and Einstein-Podolsky-Rosen Channels," Phys. Rev. Lett. **80**, 1121–1125 (1998).

[89] A. Furusawa, J. L. Sorensen, S. L. Braunstein, C. A. Fuchs, H. J. Kimble, and E. S. Polzik, "Unconditional Quantum Teleportation," Science **282**, 706–709 (1998).

[90] M. A. Nielsen, E. Knill, and R. Laflamme, "Complete Quantum Teleportation Using Nuclear Magnetic Resonance," Nature **396**, 52–55 (1998).

[91] 180. V. Buzek and M. Hillery, "Quantum Copying: Beyond the No-Cloning Theorem," Phys. Rev. A **54**, 1844–1852 (1996).

[92] V. Buzek, S. L. Braunstein, M. Hillery, and D. Bruss, "Quantum Copying: A Network," Phys. Rev. A **56**, 3446–3452 (1997).

[93] M. Hillery and V. Buzek, "Quantum Copying: Fundamental Inequalities," Phys. Rev. A **56**, 1212–1216 (1997).

[94] V. Buzek, V. Vedral, M. Plenio, P. L. Knight, and M. Hillery, "Broadcasting of Entanglement via local Copying," Phys. Rev. A **55**, 3327–3332 (1997).

[95] V. Buzek and M. Hillery, "Universal Optimal Cloning of Arbitrary Quantum States: From Qubits to Quantum Registers," Phys. Rev. Lett. **81**, 5003–5006 (1998).

[96] V. Buzek, M. Hillery, and P. L. Knight, "Flocks of Quantum Clones: Multiple Copying of Qubits," Fortschr. Phys. **46**, 521–533 (1998).

[97] N. Gisin and S. Massar, "Optimal Quantum Cloning Machines," Phys. Rev. Lett **79**, 2153–2156 (1997).

[98] D. Bruss, A. Ekert, and C. Macchiavello, "Optimal Universal Quantum Cloning and State Estimation," Phys. Rev. Lett. **81**, 2598–2601 (1998).

[99] M. Keyl and R. F. Werner, "Optimal Cloning of Pure States, Judging Single Clones," quant-ph/9807010 (1998); J. Math. Phys. **40**, 3283–3299 (1990).

[100] P. Masiek and P. L. Knight, "Copying of Entangled States and the Degradation of Correlations," quant-ph/9808043 (1998).
[101] D. Bruss, D. P. DiVincenzo, A. Ekert, C. A. Fuchs, C. Macchiavello, and J. A. Smolin, "Optimal Universal and State-Dependent Quantum Cloning," Phys. Rev. A **57**, 2368–2378 (1998).
[102] M. Murao, D, Jonathen, M. B. Plenio, and V. Vedral, "Quantum Telecloning and Multiparticle Entanglement," Phys. Rev. A **59**, 156–161 (1999).
[103] C. S. Niu and R. B. Griffiths, "Optimal Copying of One Quantum Bit," Phys. Rev. A **58**, 4377–4393 (1998).
[104] C. S. Niu and R. B. Griffiths, "Two Qubit Copy Machine for Economical Quantum Eavesdropping," Phys. Rev. A **60**, 2764–2776 (1999).
[105] C. Adami and N. J. Cerf, "Quantum Computation with Linear Optics," quant-ph/9806048 (1998).
[106] N. J. Cerf, C. Adami, and P. G. Kwait, "Optical Simulation of Quantum Logic," Phys. Rev. A **57**, R1477–1480 (1998).
[107] A. Ekert, "Quantum Interferometers as Quantum Computers," Physica Scripta **T76**, 218–222 (1998).
[108] I. L. Chuang and Y. Yamamoto, "Simple Quantum Computer," Phys. Rev. A **52**, 3489–3496 (1995).
[109] I. L. Chuang and Y. Yamamoto, "Quantum Bit Regeneration," Phys. Rev. Lett. **76**, 4281–4284 (1996).
[110] G. J. Milburn, "Quantum Optical Fredkin Gate," Phys. Rev. Lett. **62**, 2124–2127 (1989).
[111] P. Törmä and S. Stenholm, "Quantum Logic Using Polarized Photons," Phys. Rev. A **54**, 4701–4706 (1996).
[112] S. Lloyd, "Necessary and Sufficient Conditions for Quantum Computation," J. Mod. Optics **41**, 2503–2520 (1994).
[113] E. Bernstein and U. Vazirani, "Quantum Complexity Theory," SIAM J. Comput. **26**, 1411–1473 (1997).
[114] Deutsch, "Quantum Theory, the Church-Turing Principle and the Universal Quantum Computer," Proc. R. Soc. Lond. A **400**, 97–117 (1985).
[115] Deutsch, "Quantum Computational Networks," Proc. R. Soc. Lond. A **425**, 73–90 (1989).
[116] A. Steane, "Quantum Computing," Rep. Prog. Phys. **61**, 117–173 (1998).
[117] D. Deutsch, A. Barenco, and A. Ekert, "Universality in Quantum Computation," Proc. R. Soc. Lond. A **449**, 669–677 (1995).
[118] S. Lloyd, "Almost Any Quantum Logic Gate is Universal," Phys. Rev. Lett. **75**, 346–349 (1995).
[119] A. Barenco, "A Universal Two-Bit Gate for Quantum Computation," Proc. R. Soc. A **449**, 679–683 (1995).
[120] D. P. DiVincenzo, "Two-Bit Gates are Universal for Quantum Computation," Phys. Rev. A **51**, 1015–1022 (1995).
[121] J. M. Myers, "Can a Universal Quantum Computer be Fully Quantum?," Phys. Rev. Lett. **78**, 1823–1824 (1997).
[122] N. Linden and S. Popescu, "The Halting Problem for Quantum Computers," quant-ph/9806054 v2 (1998).
[123] T. D. Kieu and M. Danos, "The Halting Problem for Universal Quantum Computers," quant-ph/9811001 (1998).
[124] M. A. Nielsen, "Computable Functions, Quantum Measurements, and Quantum Dynamics," Phys. Rev. Lett. **79**, 2915–2918 (1997).
[125] R. P. Feynman, "Simulating Physics with Computers," Int. J. Theoret. Phys. **21**, 467–488 (1982).
[126] R. P. Feynman, "Quantum Mechanical Computers," Found. Phys. **16**, 507–531 (1986).
[127] S. Lloyd, "Universal Quantum Simulators," Science **273**, 1073–1078 (1996).
[128] C. Zalka, "Efficient Simulation of Quantum Systems by Quantum Computers," Fortschr. Phys. **46**, 877–879 (1998).
[129] S. Wiesner, "Simulations of Many-Body Quantum Systems by a Quantum Computer," quant-ph/9603028 (1996).
[130] D. A. Meyer, " Quantum Mechanics of Lattice Gas Automata: One-Particle Plane Waves and Potentials," Phys. Rev. E **55**, 5261–5269 (1997).

[131] D. A. Lidar and O. Biham, "Simulating Ising Spin Glasses on a Quantum Computer," Phys. Rev. E **56**, 3661–3681 (1997).

[132] D. S. Abrams and S. Lloyd, "Simulation of Many-Body Fermi Systems on a Universal Quantum Computer," Phys. Rev. Lett. **79**, 2586–2589 (1997).

[133] B. M. Boghosian and W. Taylor, "Simulating Quantum Mechanics on a Quantum Computer," Physica D **120**, 30–42 (1998).

[134] C. Zalka, "Simulating Quantum Systems on a Quantum Computer," Proc. R. Soc. Lond. A **454**, 313–322 (1998).

[135] B. Boghosian and W. Taylor, "A Quantum Lattice-Gas Model for the Many-Body Schrödinger Equation in d Dimensions," Phys. Rev. E **57**, 54–66 (1998).

[136] D. A. Meyer, "From Quantum Cellular Automata to Quantum Lattice Gases," J. Stat. Phys. **85**, 551–574 (1996).

[137] B. M. Terhal and D. P. DiVincenzo, "On the Problem of Equilibration and the Computation of Correlation Functions on a Quantum Computer," Phys. Rev. A **61**, 022301-1-22 (2000).

[138] R. Schack, "Using a Quantum Computer to Investigate Quantum Chaos," Phys. Rev. A **57**, 1634–1635 (1998).

[139] P. W. Shor, "Polynomial-Time Algorithms for Prime Factorization and Discrete Logarithms on a Quantum Computer," Proc. 35th Annual Symp. on Foundations of Computer Science (Sante Fe, NM: IEEE Computer Society Press, Los Alamitos, CA, 1994), 124–134; SIAM J. Comput. **26**, 1484–1509 (1997).

[140] I. L. Chuang, R. Laflamme, P. W. Shor, and W. H. Zurek, "Quantum Computers, Factoring and Decoherence," Science **270**, 1633–1635 (1995).

[141] A. Ekert and R. Jozsa, "Quantum Computation and Shor's Factoring Algorithm," Rev. Mod. Phys. **68**, 733–753 (1996).

[142] D. Beckman, A. N. Chari, S. Devabhaktuni, and J. Preskill, "Efficient Networks for Quantum Factoring," Phys. Rev. **54**, 1034–1063 (1996).

[143] C. Miquel, J. P. Paz, and R. Perazzo, "Factoring in a Dissipative Quantum Computer," Phys. Rev. A **54**, 2605–2613 (1996).

[144] P. W. Shor, "Scheme for Reducing Decoherence in Quantum Computer Memory," Phys. Rev. A **52**, R2493–2496 (1995).

[145] M. A. Nielsen, C. M. Caves, B. Schumacher, and H. Barnum, "Information-Theoretic Approach to Quantum Error Correction and Reversible Measurement," Proc. R. Soc. Lond. A **454**, 277–304 (1998).

[146] A. M. Steane, "Error Correcting Codes in Quantum Theory," Phys. Rev. Lett. **77**, 793–797 (1996).

[147] M. Steane, "Multiple Particle Interference and Quantum Error Correction," Proc. R. Soc. A **452**, 2551–2577 (1996).

[148] A. R. Calderbank and P. W. Shor, "Good Quantum Error-Correcting Codes Exist," Phys. Rev. A **54**, 1098–1105 (1996).

[149] E. Knill and R. Laflamme, "Theory of Quantum Error-Correcting Codes," Phys. Rev. A **55**, 900–911 (1997).

[150] A. Ekert and C. Macchiavello, "Quantum Error Correction for Communication," Phys. Rev. Lett. **77**, 2585–2588 (1996).

[151] R. Laflamme, C. Miquel, J. P. Paz, and W. H. Zurek, "Perfect Quantum Error Correcting Code," Phys. Rev. Lett. **77**, 198–201 (1996).

[152] E. M. Rains, R. H. Hardin, P. W. Shor, and N.J.A. Sloane, "Nonadditive Quantum Code," Phys. Rev. Lett **79**, 953–954 (1997).

[153] D. Gottesman, "Class of Quantum Error-Correcting Codes Saturating the Quantum Hamming Bound," Phys. Rev. A **54,** 1862–1868 (1996).

[154] A. R. Calderbank, E. M. Rains. P. W. Shor, and N.J.A. Sloane, "Quantum Error Correction and Orthogonal Geometry," Phys. Rev. Lett. **78**, 405–408 (1997).

[155] P. W. Shor and R. Laflamme, "Quantum Analog of the Mac Williams Identities for Classical Coding Theory," Phys. Rev. Lett. **78**, 1600–1602 (1997).

[156] P. W. Shor, "Fault Tolerant Quantum Computation," *Proc. 37th Annual Symp. on Foundations of Computer Science* (Los Alamitos, CA: IEEE Computer Society Press) 56–65 (1996).

[157] T. Pellizzari, "Quantum Computers, Error-Correction and Networking: Quantum Optical Approaches," in *Introduction to Quantum Computation and Information*, H. K. Lo et al, editors, 270–310, World Scientific, Singapore (1998).

[158] D. P. DiVincenzo and P. W. Shor, "Fault-Tolerant Error Correction with Efficient Quantum Codes," Phys. Rev. Lett. **77**, 3260–3263 (1996).
[159] A. M. Steane, "Active Stabilization, Quantum Computation, and Quantum State Synthesis," Phys. Rev. Lett. **78**, 2252–2255 (1997).
[160] A. M. Steane, "Space, Time, Parallelism and Noise Requirements for Reliable Quantum Computing," Fortschr. Phys. **46**, 443–457 (1998).
[161] E. Knill and R. Laflamme, "Concatenated Quantum Codes," quant-ph/9608012 (1996).
[162] E. Knill, R. Laflamme, and W. H. Zurek, "Threshold Accuracy for Quantum Computation," quant-ph/9610011 (1996).
[163] D. Aharonov and M. Ben-Or, "Fault-Tolerant Quantum Computation with Constant Error," quant-ph/9611025 (1996).
[164] A. M. Steane, "Simple Quantum Error-Correcting Codes," Phys. Rev. A **54**, 4741–4751 (1996).
[165] E. Knill, R. Laflamme, and W. H. Zurek, "Resilient Quantum Computation: Error Models and Thresholds," Proc. R. Soc. Lond. A **454**, 365–384 (1998).
[166] E. Knill, R. Laflamme, and W. H. Zurek, "Resilient Quantum Computation," Science **279**, 342–345 (1998).
[167] D. Gottesman, "Theory of Fault-Tolerant Quantum Computation," Phys. Rev. A **57**, 127–137 (1998).
[168] A. Kitaev, "Fault-Tolerant Quantum Computation by Anyons," quant-ph/9707021 (1997).
[169] A. Barenco, T. A. Brun, R. Schack, and T. P. Spiller, "Effects of Noise on Quantum Error Correction Algorithms," Phys. Rev. A **56**, 1177–1188 (1997).
[170] A. M. Steane, "Introduction to Quantum Error Correction," Phil. Trans. R. Soc. Lond. A **356**, 1739–1758 (1998).
[171] J. Preskill, "Reliable Quantum Computers," Proc. R. Soc. Lond. A **454**, 385–410 (1998).
[172] J. I. Cirac, T. Pellizari, and P. Zoller, "Enforcing Coherent Evolution in Dissipative Quantum Dynamics," Science **273**, 1207–1210 (1996).
[173] I. L. Chuang and Y. Yamamoto, "Creation of a Persistent Qubit Using Error Correction," Phys. Rev. A **55**, 114–127 (1997).
[174] D. Gottesman, "Stabilizer Codes and Quantum Error Correction," Ph.D. thesis, California Institute of Technology, quant-ph/9705052 (1997).
[175] J. Preskill, "Fault Tolerant Quantum Computation," in *Introduction to Quantum Computation and Information,* H. K. Lo et al, editors, 213–269, World Scientific, Singapore (1998).
[176] A. M. Steane, "Quantum Error Correction," in *Introduction to Quantum Computation and Information,* H. K. Lo et al, editors, 184–212, World Scientific, Singapore (1998).
[177] G. M. Palma, K. Suominen, and A. K. Ekert, "Quantum Computers and Dissipation," Proc. R. Soc. Lond. A **452**, 567–584 (1996).
[178] P. Zanardi and M. Rasetti, "Error Avoiding Quantum Codes," Mod. Phys. Lett. B **11**, 1085–1093 (1997).
[179] P. Zanardi and M. Rasetti, "Noiseless Quantum Codes," Phys. Rev. Lett. **79**, 3306–3309 (1997).
[180] P. Zanardi, "Dissipation and Decoherence in a Quantum Register," Phys. Rev. A **57**, 3276–3284 (1998).
[181] D. A. Lidar, I. L. Chuang, and K. B. Whaley, "Decoherence-Free Subspaces for Quantum Computation," Phys. Rev. Lett. **81**, 2594–2597 (1998).
[182] G. Lindblad, "On the Generators of Quantum Dynamical Semigroups," Commun. Math. Phys. **48**, 119–130 (1976).
[183] R. Alicki and K. Lendi, *Quantum Dynamical Semigroups and Applications*, in Lecture Notes in Physics, No. 286, Springer-Verlag, Berlin (1987).
[184] D. Vitali, P. Tombesi, and G. J. Milburn, "Controlling the Decoherence of a 'Meter' via Stroboscopic Feedback," Phys. Rev. Lett. **79**, 2442–2445 (1997).
[185] D. Vitali and P. Tombesi, "Decoherence Control for Optical Qubits," in *Quantum Computing and Quantum Communications*, C. P. Williams, editor, 402–412, Springer, NY (1999).
[186] S. Lloyd, "Quantum Controllers for Quantum Systems," quant-ph/9703042 (1997).
[187] L. Viola and S. Lloyd, "Dynamical Suppression of Decoherence in Two-State Quantum Systems," Phys. Rev. A **58**, 2733–2744 (1998).

[188] L. Viola and S. Lloyd, "Decoherence Control in Quantum Information Processing: Simple Models," quant-ph/9809058 (1998), to appear in *Proc. of 4th International Conference on Quantum Measurement, Communication, and Computing*, P. Kumar, editor, Northwestern.
[189] L. Viola, E. Knill, and S. Lloyd, "Dynamical Decoupling of Open Quantum Systems," Phys. Rev. Lett **82**, 2417–2421 (1999), quant-ph/9809071 (1998).
[190] Preskill, *Lecture Notes for Physics 229: Quantum Information and Computation*, http://www.theory.caltech.edu/people/preskill/ph229/#lecture (1998).
[191] D. Bouwmeester, A. Ekert, and A. Zeilinger, *The Physics of Quantum Information*, Springer Verlag, New York (2000).
[192] M. A. Nielson and I. L. Chuang, *Quantum Computation and Quantum Information*, Cambridge University Press (2000).
[193] C. P. Williams and S. H. Clearwater, *Explorations in Quantum Computing,"* Springer-Verlag, New York (1998).
[194] *Introduction to Quantum Computation and Information,* H. K. Lo, S. Popescu, and T. Spiller, editors, World Scientific, Singapore (1998).
[195] J. Gruska, *Quantum Computing*, McGraw-Hill, London (1999).
[196] J. I. Cirac and P. Zoller, "Quantum Computation with Cold Trapped Ions," Phys. Rev. Lett. **74**, 4091–4094 (1995).
[197] W. Paul and H. Steinwedel, "Ein neues Massenspektrometer ohne Magnetfeld," Z. Naturforsch. A **8**, 448–450 (1953).
[198] W. Paul, "Electromagnetic Trap for Charged and Neutral Particles," Rev. Mod. Phys. **62**, 531–540 (1990).
[199] M. G. Raizen, J. M. Gillligan, J. C. Bergquist, W. M. Itano, and D. J. Wineland, "Ionic Crystals in a Linear Paul Trap," Phys. Rev. A **45**, 6493–6501 (1992).
[200] P. K. Ghosh, *Ion Traps*, Clarendon Press, Oxford (1995).
[201] C. Monroe, D. M. Meekhof, B. E. King, W. M. Itano, and D. J. Wineland, "Demonstration of a Fundamental Quantum Logic Gate," Phys. Rev. Lett. **75**, 4714–4717 (1995).
[202] C. Monroe, D. M. Meekhof, B. E. King, and D. J. Wineland, "A 'Schrödinger Cat' Superposition State of an Atom," Science **272**, 1131–1136 (1996).
[203] Q. A. Turchette, C. S. Wood, B. E. King, C. J. Myatt, D. Leibfried, W. M. Itano, C. Monroe, and D. J. Wineland, "Deterministic Entanglement of Two Trapped Ions," Phys. Rev. Lett. **81**, 3631–3634 (1998).
[204] D. J. Wineland, C. Monroe, W. M. Itano, D. Leibfried, B. E. King, and D. M. Meekhof, "Experimental Issues in Coherent Quantum-State Manipulation of Trapped Atomic Ions," J. Res. Natl. Inst. Stand. Technol. **103**, 259–328 (1998).
[205] A. M. Steane, "The Ion Trap Quantum Information Processor," Appl. Phys. B **64**, 623–642 (1997).
[206] D. J. Wineland, C. Monroe, W. M. Itano, B. E. King, D. Leibfried, D. M. Meekhof, C. Myatt , and C. Wood, "Experimental Primer on the Trapped Ion Quantum Computer," Fortschr. Phys. **46**, 363–390 (1998).
[207] D. J. Wineland, C. Monroe, W. M. Itano, B. E. King, D. Leibfried, C. Myatt, and C. Wood, "Trapped-Ion Quantum Simulator," Phys. Scripta **T76**, 147–151 (1998).
[208] D. M. Meekhof, C. Monroe, B. E. King, W. M. Itano, and D. J. Wineland, "Generation of Nonclassical Motional States of a Trapped Atom," Phys. Rev. Lett. **76**, 1796–1799 (1996).
[209] D. J. Wineland, C. Monroe, D. M. Meekhof, B. E. King, D. Leibfried, W. M. Itano, J. C. Bergquist , D. Berkeland, J. J. Bollinger, and J. Miller, "Quantum State Manipulation of Trapped Atomic Ions," Proc. R. Soc. Lond. A **454**, 411–429 (1998).
[210] C. Monroe, D. Leibfried, B. E. King, D. M. Meekhof, W. M. Itano, and D. J. Wineland, "Simplified Quantum Logic with Trapped Ions," Phys. Rev. A **55**, R2489–2491 (1997).
[211] B. E. King, C. S. Wood, C. J. Myatt , Q. A. Turchette, D. Leibfried, W. M. Itano, C. Monroe , and D. J. Wineland, "Cooling the Collective Motion of Trapped Ions to Initialize a Quantum Register," Phys. Rev. Lett. **81**, 1525–1528 (1998).
[212] R. J. Hughes, "Cryptography, Quantum Computation and Trapped Ions," Phil. Trans. R. Soc. Lond. A **356**, 1853–1868 (1998).
[213] R. J. Hughes and D.F.V. James, "Prospects for Quantum Computation with Trapped Ions," Fortschr. Phys. **46**, 759–769 (1998).
[214] R. J. Hughes, D.F.V. James, J. J. Gomez, M. S. Gulley, M. H. Holzscheiter, P. G. Kwait, S. K. Lamoreaux, C. G. Peterson, V. D. Sandberg, M. M. Schauer, C. M. Simmons,

C. E. Thorburn, D. Tupa, P. Z. Wang, and A. G. White, "The Los Alamos Trapped Ion Quantum Computer Experiment," Fortschr. Phys. **46**, 329–361 (1998).

[215] D.F.V. James, M. S. Gulley, M. H. Holzscheiter, R. J. Hughes, P. G. Kwait, S. K. Lamoreaux, C. G. Peterson, V. D. Sanberg, M. M. Schauer, C. M. Simmons, D. Tupa, P. Z. Wang, and A. G. White, "Trapped Ion Quantum Computer Research at Los Alamos," quant-ph/9807071 (1998).

[216] R. J. Hughes, D.F.V. James, E. H. Knill, R. Laflamme, and A. G. Petschek, "Decoherence Bounds on Quantum Computation with Trapped Ions," Phys. Rev. Lett. **77**, 3240–3243 (1996).

[217] D.F.V. James, "Theory of Heating of the Ground State of Trapped Ions," Phys. Rev. Lett. **81**, 317–320 (1998).

[218] D. Stevens, J. Brochard, and A. M. Steane, "Simple Experimental Methods for Trapped-Ion Quantum Processors," Phys. Rev. A **58**, 2750–2759 (1998).

[219] J. Steinbach, J. Twamley, and P. L. Knight, "Engineering Two-Mode Interaction in Ion Trap," Phys. Rev. A **56**, 4815–4825 (1997).

[220] S. Schneider, H. M. Wiseman, W. J. Munro, and G. J. Milburn, "Measurement and State Preparation via Ion Trap Quantum Computing," Fortschr. Phys. **46**, 391–399 (1998).

[221] S. Schneider and G. J. Milburn, "Decoherence in Ion Traps Due to Laser Intensity and Phase Fluctuations," Phys. Rev. A **57**, 3748–3752 (1998).

[222] S. Schneider and G. J. Milburn, "Decoherence and Fidelity in Ion Traps with Fluctuating Trap Parameters," quant-ph/9812044 (1998).

[223] R. Onofrio and L. Viola, "Lindblad Approach to Nonlinear Jaynes-Cummings Dynamics of a Trapped Ion," Phys. Rev. A **56**, 39–43 (1997).

[224] J. F. Poyatos, J. I. Cirac, and P. Zoller, "Quantum Gates with 'Hot Trapped Ions'," Phys. Rev. Lett. **81** 1322–1325 (1998).

[225] S. Schneider, D.F.V. James, and G. J. Milburn, "Method of Quantum Computation with 'Hot' Trapped Ions," quant-ph/9808012 (1998).

[226] K. M. Obenland and A. M. Despain, "Simulating the Effect of Decoherence and Inaccuracies on a Quantum Computer," in *Quantum Computing and Quantum Communications*, C. P. Williams, editor, 447–459, Springer, NY (1999).

[227] M. B. Plenio and P. L. Knight, "Decoherence Limits to Quantum Computation Using Trapped Ions," Proc. R. Soc. Lond. A **453**, 2017–2041 (1997).

[228] M. B. Plenio and P. L. Knight, "Realistic Lower Bounds for the Factorization Time of Large Numbers on a Quantum Computer," Phys. Rev. A **53**, 2986–2990 (1996).

[229] M. Murao and P. L. Knight, "Decoherence in Nonclassical Motional States of a Trapped Ion," Phys. Rev. A **58**, 663–669 (1998).

[230] P. R. Berman, editor, *Cavity Quantum Electrodynamics*, Academic Press, Boston (1994).

[231] H. J. Kimble, "Strong Interaction of Single Atoms and Photons in Cavity QED," Physica Scripta **T76**, 127–137 (1998).

[232] M. Brune and S. Haroche, "Cavity Quantum Electrodynamics," in *Quantum Dynamics of Simple Systems*, G.-L. Oppo, S. M. Barnett, E. Riis, and M. Wilkinson, editors, Inst. of Physics Publishers, Bristol, 49–70 (1996).

[233] S. Haroche, "Mesoscopic Coherences in Cavity QED," Il Nuovo Cimento **110B**, 545–556 (1995).

[234] T. Pellizzari, S. A. Gardiner, J. I. Cirac, and P. Zoller, "Decoherence, Continuous Observation, and Quantum Computing: A Cavity QED Model," Phys. Rev. Lett. **75**, 3788–3791 (1995).

[235] H. J. Kimble, Q. A. Turchette, N. Ph. Georgiades, C. J. Hood, W. Lange, H. Mabuchi, E. S. Polzik, and D. W. Vernooy, "Cavity Quantum Electrodynamics with a Capital Q," in *Coherence and Quantum Optics VII*, J. H. Eberly, L. Mandel, and E. Wolf, editors, 203–310, Plenum Press, New York (1996).

[236] X. Maître, E. Hagley, G. Nogues, C. Wunderlich, P. Goy, M. Brune, J. M. Raimond, and S. Haroche, "Quantum Memory with a Single Photon in a Cavity," Phys. Rev. Lett. **79**, 769–772 (1997).

[237] T. Sleator and H. Weinfurter, "Realizable Universal Quantum Logic Gates," Phys. Rev. Lett. **74**, 4087–4090 (1995).

[238] P. Domokos, J. M. Raimond, M. Brune, and S. Haroche, "Simple Cavity-QED Two-Bit Quantum Logic Gate: The Principle and Expected Performances," Phys. Rev. A. **52**, 3554–3559 (1995).

[239] C. K. Law and H. J. Kimble, "Deterministic Generation of a Bit-Stream of Single-Photon Pulses," J. Mod. Opt. **44**, 2067–2074 (1997).
[240] K. M. Gheri, C. Saavedra, P. Törmä, J. I. Cirac, and P. Zoller, "Entanglement Engineering of One-Photon Wave Packets Using a Single-Atom Source," Phys. Rev. A **58**, R2627–2630 (1998).
[241] L. Davidovich, N. Zagury, M. Brune, J. M. Raimond, and S. Haroche, "Teleportation of an Atomic State Between Two Cavities Using Nonlocal Microwave Fields," Phys. Rev. A **50**, R895–898 (1994).
[242] J. I. Cirac, P. Zoller, H. J. Kimble, and H. Mabuchi, "Quantum State Transfer and Entanglement Distribution Among Distant Nodes in a Quantum Network," Phys. Rev. Lett. **78**, 3221–3224 (1997).
[243] H. J. Kimble, "Strong Interactions of Single Atoms and Photons in Cavity QED," Physica Scripta **T76**, 127–137 (1998).
[244] H.-J. Briegel, W. Dür, S. J. Van Enk, J. I. Cirac, and P. Zoller, "Quantum Communication and the Creation of Maximally Entangled Pairs of Atoms over a Noisy Channel," Phil Trans. R. Soc. Lond. A **356**, 1841–1851 (1998).
[245] J. I. Cirac, S. J. Van Enk, P. Zoller, H. J. Kimble, and H. Mabuchi, "Quantum Communication in a Quantum Network," Physica Scripta **T76**, 223–232 (1998).
[246] S. J. Van Enk, J. I. Cirac, and P. Zoller, "Photonic Channels for Quantum Communication," Science **279**, 205–208 (1998).
[247] S. J. Van Enk, J. I. Cirac, P. Zoller, H. J. Kimble, and H. Mabuchi, "Quantum State Transfer in a Quantum Network: a Quantum Optical Implementation," J. Mod. Opt. **44**, 1727–1736 (1997).
[248] S. J. van Enk, J. I. Cirac, and P. Zoller, "Ideal Quantum Communication over Noisy Channels: A Quantum Optical Implementation," Phys. Rev. Lett. **78**, 4293–4296 (1997).
[249] C. P. Slichter, *Principles of Magnetic Resonance,* 3rd Edition, Springer, New York (1996).
[250] M. Goldman, *Quantum Description of High-Resolution NMR in Liquids,* Oxford Scientific Publications, London (1988).
[251] N. A. Gershenfeld and I. L. Chuang, "Bulk Spin-Resonance Quantum Computation," Science **275**, 350–356 (1997).
[252] I. L. Chuang, N. Gershenfeld, M. G. Kubinec, and D. W. Leung, "Bulk Quantum Computation with Nuclear Magnetic Resonance: Theory and Experiment," Proc. R. Soc. Lond. A **454**, 447–467 (1998).
[253] I. L. Chuang, "Quantum Computation with Nuclear Magnetic Resonance," in *Introduction to Quantum Computation and Information,* H. K. Lo et al, editors, 311–339, World Scientific, Singapore (1998).
[254] D. G. Cory, A. F. Fahmy, and T. F. Havel, "Ensemble Quantum Computing by NMR Spectroscopy," Proc. Natl. Acad. Sci. USA **94**, 1634–1639 (1997); "Nuclear Magnetic Resonance Spectroscopy: An Experimentally Accessible Paradigm for Quantum Computing," in *Proc. 4th Workshop on Physics and Computation,* New England Complex Systems Institute, Boston, MA, 87–91 (1996).
[255] S. S. Somaroo, D. G. Cory, and T. F. Havel, "Expressing the Operations of Quantum Computing in Multiparticle Geometric Algebra," Phys. Lett. A **240**, 1–7 (1998).
[256] T. F. Havel, S. S. Somaroo, C.-H. Tseng, and D. G. Cory, "Principles and Demonstrations of Quantum Information Processing by NMR Spectroscopy," quant-ph/9812086 (1998) to appear in *Applicable Algebra and Engineering, Communication and Computing.*
[257] I. L. Chuang, N. Gershenfeld, and M. Kubinic, "Experimental Implementation of Fast Quantum Searching," Phys. Rev. Lett. **80**, 3408–3411 (1998).
[258] J. A. Jones, M. Mosca, and R. H. Hansen, "Implementation of a Quantum Search Algorithm on a Quantum Computer," Nature **393**, 344–346 (1998).
[259] J. A. Jones, "Fast Searches with Nuclear Magnetic Resonance Computers," Science **280**, 229 (1998).
[260] D. G. Cory, M. D. Price, T. F. Havel, "Nuclear Magnetic Resonance Spectroscopy: An Experimentally Accessible Paradigm for Quantum Computing," Physica D **120**, 82–101 (1998).
[261] R. Laflamme, E. Knill, W. H. Zurek, P. Catasti, and S.V.S. Mariappan, "NMR Greenberger-Horne-Zeiliger States," Phil. Trans. R. Soc. Lond. A **356**, 1941–1948 (1998).

[262] D. G. Cory, M. D. Price, W. Maas, E. Knill, R. Laflamme, W. H. Zurek, T. F. Havel, and S. S. Somaroo, "Experimental Quantum Error Correction," Phys. Rev. Lett. **81**, 2152–2155 (1998).

[263] I. L. Chuang. L.M.K. Vandersypen, Y. Zhou, D. W. Leung, and S. Lloyd, "Experimental Realization of a Quantum Algorithm," Nature **393**, 143–146 (1998).

[264] J. A. Jones, "Implementation of a Quantum Algorithm on a Nuclear Magnetic Resonance Quantum Computer," J. Chem. Phys. **109**, 1648–1653 (1998).

[265] N. Linden, H. Barjat, and R. Freeman, "An Implementation of the Deutsch-Jozsa Algorithm on a Three-Qubit NMR Quantum Computer," quant-ph/9808039 (1998).

[266] J. A. Jones and M. Mosca, "Approximate Quantum Counting on an NMR Ensemble Quantum Computer," Phys. Rev. Lett. **83**, 1050–1053 (1999).

[267] W. S. Warren, "The Usefulness of NMR Quantum Computing," Science **277**, 1688–1689 (1997).

[268] J. R. Friedman, M. P. Sarachik, J. Tejada, and R. Ziolo, "Macroscopic Measurement of Resonant Magnetization Tunneling in High-Spin Molecules," Phys. Rev. Lett. **76**, 3830–3833 (1996).

[269] D. P. DiVincenzo, "Quantum Computation," Science **270**, 255–261 (1995).

[270] V. Privman, I. D. Vagner, and G. Kventsel, "Quantum Computation in Quantum-Hall Systems," Phys. Lett. A **239**, 141–146 (1998).

[271] B. E. Kane, "A Silicon-Based Nuclear Spin Quantum Computer," Nature **393**, 133–137 (1998).

[272] J. W. Lyding, "UHV/STM Nanofabrication: Progress, Technology, Spin-offs, and Challenges," Proc. IEEE **85**, 589–600 (1997).

[273] D. P. DiVincenzo, "Real and Realistic Quantum Computers," Nature **393**, 113–114 (1998).

[274] S. Bandyopadhyay and V. Roychowdhury, "Computational Paradigms in Nanoelectronics: Quantum Coupled Electron Logic and Neuromorphic Networks," Jpn. J. Appl. Phys. **35**, 3350–3362 (1996).

[275] S. N. Molotkov, "Quantum Controlled-NOT Gate Based on a Single Quantum Dot," JETP Lett. **64**, 237–243 (1996).

[276] S. Bandyopadhyay, A. Balandin, V. P. Roychowdhury, and F. Vatan, "Nanoelectronic Implementation of Reversible and Quantum Logic," Superlat. and Microstr. **23**, 445–464 (1998).

[277] D. Loss and D. P. DiVincenzo, "Quantum Computation with Quantum Dots," Phys. Rev. A **57**, 120–126 (1998).

[278] D. P. DiVincenzo and D. Loss, "Quantum Information is Physical," Superlat. and Microstr. **23**, 419–432 (1998).

[279] G. Burkard, D. Loss, and D. P. DiVincenzo, "Coupled Quantum Dots as Quantum Gates," Phys. Rev. B **59**, 2070–2078 (1999).

[280] D. P. DiVincenzo, "Quantum Computing and Single-Qubit Measurement Using the Spin Filter Effect," J. Appl. Phys. **85**, 4785–4787 (1999).

[281] D. P. DiVincenzo and D. Loss, "Quantum Computers and Quantum Coherence," cond-mat/9901137 (1998), to appear in J. Magn. Matl. (1999).

[282] A. Barenco, D. Deutsch, A. Ekert, and R. Josza, "Conditioned Quantum Dynamics and Logic Gates," Phys. Rev. Lett. **74**, 4083–4086 (1995).

[283] J. A. Brum and P. Hawrylak, "Coupled Quantum Dots as Quantum Exclusive-OR Gate," Superlat. and Microstr. **22**, 431–436 (1997).

[284] P. Zanardi and F. Rossi, "Quantum Information in Semiconductors: Noiseless Encoding in a Quantum Dot Array," Phys. Rev. Lett. **81**, 4752–4755 (1998).

[285] P. Zanardi and F. Rossi, "Subdecoherent Information Encoding in a Quantum-Dot Array," Phys. Rev. B **59**, 8170–8181 (1999).

[286] N. H. Bonadeo, J. England, D. Gammon, D. Park, D. S. Katzer, and D. G. Steel, "Coherent Optical Control of the Quantum State of a Single Quantum Dot," Science **282**, 1473–1476 (1998); G. Chen, N. H. Bonadeo, D. G. Steel, D. Gammon, D. S. Katzer, D. Park, L. J. Sham, "Optically Induced Entanglement of Excitons in a Single Quantum Dot," Science **289**, 1906–1909 (2000).

[287] A. Shnirman, G. Schön, and Z. Hermon, "Quantum Manipulations of Small Josephson Junctions," Phys. Rev. Lett. **79**, 2371–2374 (1997).

[288] A. Shnirman and G. Schön, "Quantum Measurements Performed with a Single-Electron Transistor," Phys. Rev. B **57**, 15 400–15 407 (1998).

[289] Y. Makhlin, G. Schön, and A. Shnirman, "Josephson-Junction Qubits with Controlled Coupling ," Nature **398**, 305–307 (1999), cond-mat/9808067 (1998).
[290] G. Schön, A. Shnirman, and Y. Makhlin, "Josephson-Junction Qubits and the Readout Process by Single-Electron Transistors," cond-mat/9811029 (1998).
[291] D. V. Averin, "Adiabatic Quantum Computation with Cooper Pairs," Solid State Commun. **105**, 65 (1998).
[292] L. B. Ioffe, V. B. Geshkenbein, M. V. Feigel'man, A. L. Fauchère, and G. Blatter, "Quiet SDS Josephson Junctions for Quantum Computing," cond-mat/9809116 (1998); Nature **398**, 679 (1999).
[293] D. P. DiVincenzo, "Topics in Quantum Computers," in *Mesoscopic Electron Transport*, L. Sohn , L. Kouwenhoven, and G. Shoen, editors, 657–677, Kluwer (1997), cond-mat/9612126 v.2 (1996).
[294] V. Bouchiat, D. Vion, P.Joyez, D. Esteve, and M. H. Devoret, "Quantum Coherence with a Single Cooper Pair," Physica Scripta **T76**, 165–170 (1998).
[295] A. M. van den Brink, G. Schön, and L. J. Geerligs, "Combined Single-Electron and Coherent-Cooper-Pair Tunneling in Voltage-Biased Josephson Junctions," Phys. Rev. Lett. **67**, 3030–3033 (1991).
[296] M. T. Tuominen, J. M. Hergenrother, T. S. Tighe, and M. Tinkham, "Experimental Evidence for Parity-Based 2e Periodicity in a Superconducting Single-Electron Tunneling Transistor," Phys. Rev. Lett. **69,** 1997–2000 (1992).
[297] J. Siewert and G. Schön, "Charge Transport in Voltage-Biased Superconducting Single-Electron Transistors," Phys. Rev. B **54**, 7421–7424 (1996).
[298] Y. Nakamura, C. D. Chen, and J. S. Tsai, "Spectroscopy of Energy-Level Splitting Between Two Macroscopic Quantum States of Charge Coherently Superposed by Josephson Coupling," Phys. Rev. Lett. **79**, 2328–2331 (1997).
[299] A. Van Oudenaarden and J. E. Mooij, "One Dimensional Mott Insulator Formed by Quantum Vortices in Josephson Junction Arrays," Phys. Rev. Lett. **76**, 4947–4950 (1996).
[300] W. J. Elion, J. J. Wachters, L. L. Sohn, and J. E. Mooij, "Observation of the Aharonov-Casher Effect for Vortices in Josephson-Junction Arrays," Phys. Rev. Lett. **71**, 2311–2314 (1993).
[301] W. J. Elion, J. J. Wachters, L. L. Sohn, and J. E. Mooij, "The Aharonov-Casher Effect for Vortices in Josephson-Junction Arrays," Physica B **203**, 497–503 (1994).
[302] M. F. Bocko, A. M. Herr, and M. J. Feldman, "Prospect for Quantum Coherent Computation Using Superconducting Electrons," IEEE Trans. Appl. Superconductivity **7**, 3638–3641 (1997).
[303] B. Rosen, "Superconducting Circuit Implementation of Qubits and Quantum Computer Logic," preprint (1997).
[304] X. Xue and H. Wei, "Superconducting State Quantum Logic," quant-ph/9702041 v2 (1997).
[305] R. J. Prance, R. Whiteman, T. D. Clark, J. Diggins, H. Prance, J. F. Ralph, G. Buckling, G. Colyer, C. Vittoria, A. Widom, and Y. Srivastava, "Observation of Quantum Jumps in SQUID Rings," in *Quantum Communications and Measurement,* V. P. Belavkin et al, editors, Plenum Press, New York, 299–306 (1995).
[306] R. Rouse, S. Y. Han, and J. E. Lukens, "Observation of Resonant Tunneling Between Macroscopically Distinct Quantum Levels," Phys. Rev. Lett. **75**, 1614–1617 (1995).
[307] R. Rouse, S. Han, and J. E. Lukens, "Photon Assisted Resonant Tunneling Between Macroscopically Distinct States of a SQUID," in *Quantum Coherence and Decoherence,* K. Fujikawa and Y. A. Ono, editors, Elsevier, Amsterdam, 179–182 (1996).
[308] L. Viola, R. Onofrio, and T. Calarco, "Macroscopic Quantum Damping in SQUID Rings," Phys. Lett. A **229**, 23–31 (1997).
[309] T. D. Clark, J. Diggins, J. F. Ralph, M. Everitt, R. J. Prance, H. Prance, R. Whiteman, A. Widom, and Y. N. Srivastava, "Coherent Evolution and Quantum Transitions in a Two Level Model of a SQUID Ring," Annals of Phys. **268**, 1–30 (1998).
[310] A. J. Leggett, S. Chakravarty, A. T. Dorsey, M.P.A. Fisher, A. Garg, and W. Zwerger, "Dynamics of the Dissipative Two-State System," Rev. Mod. Phys. **59**, 1–85 (1987).
[311] A. O. Caldiera and A. J. Leggett, "Quantum Tunneling in a Dissippative System," Ann. Phys. **149**, 374–456 (1983).
[312] U. Weiss, H. Grabert, and S. Linkwitz, "Influence of Temperature and Friction on Macroscopic Quantum Coherence in SQUID-Rings," Jap. J. Appl. Phys. **26**, Suppl 26-3, 1391 (1987).

[313] J. R. Friedman, V. Patel, W. Chen, S. K. Tolpygo, and J. E. Lukens, "Quantum Superposition of Distinct Macroscopic States," Nature **406**, 43–46 (2000).

[314] C. H. van der Wal, A.C.J. ter Haar, F. K. Wilhelm, R. N. Schouten, C.J.P.M. Harmans, T. P. Orlando, S. Lloyd, and J. E. Mooij, "Quantum Superposition of Macroscopic Persistent Current States," Science **290**, 773–777 (2000).

[315] P. Verkerk, B. Luonis, C. Salomon, C. Cohen-Tannoudji, J.-Y. Courtois, and G. Grynberg, "Dynamics and Spatial Order of Cold Cesium Atoms in a Periodic Optical Potential," Phys. Rev. Lett. **68**, 3861–3864 (1992).

[316] P. S. Jessen, C. Gerz, P. Lett, W. D. Phillips, S. L. Rolston, R.J.C. Spreeuw, and C. I. Westbrook , "Observation of Quantized Motion of Rb Atoms in an Optical Field," Phys. Rev. Lett. **69**, 49–52 (1992).

[317] I. H. Deutsch and P. S. Jessen, "Quantum-State Control in Optical Lattices," Phys. Rev. A **57** 1972–1986 (1998).

[318] A. Hemmerich and T. W. Hänsch, "Two-Dimensional Atomic Crystal Bound by Light," Phys. Rev. Lett. **70**, 410–413 (1993).

[319] G. Grynberg, B. Luonis, P. Verkerk, J.-Y. Courtois, and C. Salomon, "Quantized Motion of Cold Cesium Atoms in Two- and Three-Dimensional Optical Potentials," Phys. Rev. Lett. **70**, 2249–2252 (1993).

[320] A. Görlitz, M. Weidenmüller, T. W. Hänsch, and A. Hemmerich, "Observing the Position Spread of Atomic Wave Packets," Phys. Rev. Lett. **78**, 2096–2099 (1997).

[321] S. Rolston, "Optical Lattices," Physics World, 27–32 (October 1998).

[322] D. R. Meacher, "Optical Lattices—Crystalline Structures Bound by Light," Contemp. Phys. **39**, 329–350 (1998).

[323] C. S. Adams and E. Riis, "Laser Cooling and Trapping of Neutral Atoms," Prog. Quant. Electr. **21**, 1–79 (1997).

[324] D. L. Haycock, S. E. Hamann, G. Klose, and P. S. Jessen, "Atomic Trapping in Deeply Bound States of a Far-Off Resonance Optical Lattice," Phys. Rev. A **55**, R3991–3994 (1997).

[325] P. S. Jessen and I. H. Deutsch, "'Optical Lattices," in *Advances in Atomic, Molecular, and Optical Physics* **37**, B. Bederson and H. Walther, editors, 95, Cambridge (1996).

[326] G. Raithel, W. D. Phillips, and S. L. Rolston, "Collapse and Revivals of Wave Packets in Optical Lattices," Phys. Rev. Lett. **81**, 3615–3618 (1998).

[327] S. E. Hamann, D. L. Haycock, G. Klose, P. H. Pax, I. H. Deutsch, and P. S. Jessen, "Resolved-Sideband Raman Cooling to the Ground State of an Optical Lattice," Phys. Rev. Lett. **80**, 4149–4152 (1998).

[328] G. K. Brennen, C. M. Caves, P. S. Jessen, and I. H. Deutsch, "Quantum Logic Gates in Optical Lattices," Phys. Rev. Lett. **82**, 1060–1063 (1999).

[329] D. M. Stamper-Kurn, M. R. Andrews, A. P. Chikkatur, S. Inouye, H.-J. Miesner, J. Stenger, and W. Ketterle, "Optical Confinement of a Bose-Einstein Condensate," Phys. Rev. Lett. **80**, 2027–2030 (1998).

[330] B. P. Anderson and M. A. Kasevich, "Macroscopic Quantum Interference from Atomic Tunnel Arrays," Science **282**, 1686–1689 (1998).

[331] K. Marzlin and W. Zhang, "Photonic Band Gap and Defect States Induced by Excitations of Bose-Einstein Condensates in Optical Lattices," Phys. Rev. A **59**, 2982–2989 (1999).

[332] D. Jaksch, H.-J. Briegel, J. I. Cirac, C. W. Gardiner, and P. Zoller, "Entanglement of Atoms via Cold Controlled Collisions," Phys. Rev. Lett. **82**, 1975–1978 (1999), quant-ph/9810087 (1998).

[333] D. Jaksch, C. Bruder, J. I. Cirac, C. W. Gardiner, and P. Zoller, "Cold Bosonic Atoms in Optical Lattices," Phys. Rev. Lett. **81**, 3108–3111 (1998).

[334] M.P.A. Fisher, P. B. Weichman, G. Grinstein, and D. S. Fisher, "Boson Localization and the Superfluid-Insulator Transition," Phys. Rev. B **40**, 546–570 (1989).

[335] P. Benioff, "Quantum Robots," in *Feynman and Computation: Exploring the Limits*, A.J.G. Hey, editors, Perseus Books, Reading, Massachusetts, 155–175 (1999).

[336] P. Benioff, "Quantum Robots and Environments," Phys. Rev. A **58**, 893–904 (1998).

[337] P. Benioff, "Some Foundational Aspects of Quantum Computers and Quantum Robots," Superlatt. and Microstr. **23**, 407–417 (1998).

[338] P. Benioff, "Quantum Robots Plus Environments," quant-ph/9807032 (1998), to appear in *Proc. of 4th International Conference on Quantum Measurement, Communication and Computing*, P. Kumar, editor, Northwestern.

[339] R. P. Feynman and A. R. Hibbs, *Quantum Mechanics and Path Integrals*, McGraw-Hill, New York (1965).

[340] L. K. Grover, "Quantum Mechanics Helps in Searching for a Needle in a Haystack," Phys. Rev. Lett. **79**, 325–328 (1997).

U.S. ARMY RESEARCH LABORATORY, ADELPHI, MD 20783, AND UNIVERSITY OF CAMBRIDGE ISAAC NEWTON INSTITUTE FOR MATHEMATICAL SCIENCES, CAMBRIDGE, UK
E-mail address: hbrandt@arl.army.mil

Chapter II

Quantum Algorithms and Quantum Complexity Theory

Introduction to Quantum Algorithms

Peter W. Shor

ABSTRACT. These notes discuss the quantum algorithms we know of that can solve problems significantly faster than the corresponding classical algorithms. So far, we have only discovered a few techniques which can produce speed up versus classical algorithms. It is not clear yet whether the reason for this is that we do not have enough intuition to discover more techniques, or that there are only a few problems for which quantum computers can significantly speed up the solution.

In the first section of these notes, I try to explain why the recent results about quantum computing have been so surprising. This section comes from a talk I have been giving for several years now, and discusses the history of quantum computing and its relation to the mathematical foundations of computer science. In Sections 2 and 3, I talk about the quantum computing model and its relationship to physics. These sections rely heavily on two of my papers [SIAM J. Comp. **26** (1997), 1484–1509; Doc. Math. **Extra Vol. ICM I** (1998), 467–486]. Sections 4 and 5 illustrate the general technique of using quantum Fourier transforms to find periodicity. Section 4 contains an algorithm of Dan Simon showing that quantum computers are likely to be exponentially faster than classical computers for some problems. Section 5 discusses my factoring algorithm, which was inspired in part by Dan Simon's paper. In the final section, I discuss Lov Grover's search algorithm, which illustrates a different technique for speeding up classical algorithms. These techniques for constructing faster algorithms for classical problems on quantum computers are the only two significant ones which have been discovered so far.

Contents

1. History and Foundations
2. The Quantum Circuit Model
3. Relation of the Model to Quantum Physics
4. Simon's Algorithm
5. The Factoring Algorithm
6. Grover's Algorithm

References

2000 *Mathematics Subject Classification.* [2000]Primary 81P68.
Key words and phrases. Quantum computing, prime factorization, period finding, searching.

1. History and Foundations

The first results in the mathematical theory of theoretical computer science appeared before the discipline of computer science existed; in fact, even before electronic computers existed. Shortly after Gödel proved his famous incompleteness result, several papers [**13, 27, 32, 41**] were published that drew a distinction between computable and non-computable functions. These papers showed that there are some mathematically defined functions which are impossible to compute algorithmically. Of course, proving such a theorem requires a mathematical definition of what it means to compute a function. These papers contained several distinct definitions of computation. What was observed was that, despite the fact that these definitions appear quite different, they all result in the same class of computable functions. This led to the proposal of what is now called the Church-Turing thesis, after two of its proponents. This thesis says that any function that is computable by any means, can also be computed by a Turing machine. This is not a mathematical theorem, because it does not give a mathematical precise definition of computable; it is rather a statement about the real world. In fact, many such mathematical theorems have been proven for various definitions of computation. What was not widely appreciated until recently is that, since the Church-Turing thesis implicitly refers to the physical world, it is in fact a statement about physics. In the sixty years since Church proposed his thesis, nobody has discovered any counterexamples to it and it is now widely accepted. The current theories of physics appear to support this thesis, although as we do not yet have a comprehensive theory of physical laws, we must wait until we make a final judgment on this thesis.

The model that the majority of these early papers used for intuition about computation does not appear to have been a digital computer, as these did not yet exist. Rather, they appear to have been inspired by considering a mathematician scribbling on sheets of paper. Less than a decade after 1936, the first digital computers were built. As the Church-Turing thesis asserts, the class of functions computable by digital machines with arbitrarily large amounts of time and memory is indeed those functions computable by a Turing machine.

With the advent of practical digital computers, it became clear that the distinction between computable and non-computable was much too course for practical use, as actual computers do not have an arbitrary amount of time and memory. After all, it doesn't do much good in practice to know that a function is computable if the sun will burn out long before any conceivable computer could reach the end of the computation. What was needed was some classification of functions as efficiently or inefficiently computable, based on their computational difficulty. In the late 1960's and early 1970's theoretical computer scientists came up with an asymptotic classification that reflects this distinction moderately well in practice, and is also tractable to work with theoretically, that is, useful for proving theorems about the difficulty of computation. Computer scientists call an algorithm polynomial-time if the running time grows polynomially in the input size, and they say that a problem is in the complexity class P if there is a polynomial-time algorithm solving it. This does not capture the intuitive notion of efficient perfectly — hardly anybody would claim that an algorithm with an n^{30} running time is feasible — but it works reasonably well in practice. Experience seems to show that most natural problems in P tend to have reasonably efficient algorithms, and most natural problems not in P tend not to be solvable much faster than exponential time. Further, the complexity class P has been very useful for proving theorems, an advantage which is unlikely to hold for any definition which differentiates between $O(n^3)$ and $O(n^{30})$ algorithms.

For the definition of P to make sense, you need to know that it does not depend on the exact type of computer used for the computation. This led to a "folk" thesis, which we call the polynomial Church's thesis, whose origins appear to be impossible to pin down, but which has nevertheless been widely referred to in the literature. This thesis says that any physically computable function can be computed on a Turing machine with at most a polynomial increase in the running time. That is, if a function can be computed on a physical computer in time T, it can be computed on a Turing machine in time $O(T^c)$ for some constant c depending only on the class of computing machine used.

Why might this folk thesis be true? One explanation might be that the physical laws of our universe are efficiently simulable by computers. This would explain it via the following argument: if we have some physical machine that solves a problem, then we can simulate the physical laws driving this machine, and by our hypothesis this simulation runs in polynomial time. Conversely, if we are interested in counterexamples to the polynomial Church's thesis, we should look at physical systems which appear to be very difficult to simulate on a digital computer. Two classes of physical systems immediately spring to mind for which simulation currently consumes vast amounts of computer time, even while trying to solve relatively simple problems. One of these is turbulence, about which I unfortunately have nothing further to say. The other is quantum mechanics.

In 1982, Feynman [19] argued that simulating quantum mechanics inherently required an exponential amount of overhead, so that it must take enormous amounts of computer time no matter how clever you are. This realization was come to independently, and somewhat earlier, in 1980, in the Soviet Union by Yuri Manin [30]. It is not true that all quantum mechanical systems are difficult to simulate; some of them have exact solutions and others have very clever computational shortcuts, but it does appear to be true when simulating a generic quantum mechanics system. Another thing Feynman suggested in this paper was the use of quantum computers to get around this. That is, a computer based on fundamentally quantum mechanical phenomena might be used to simulate quantum mechanics much more efficiently. In much the same spirit, you could think of a wind tunnel as a "turbulence computer". Benioff [5] had already showed how quantum mechanical processes could be used as the basis of a classical Turing machine. Feynman [20] refined these ideas in a later paper.

In 1985, David Deutsch [15] gave an abstract model of quantum computation, and also raised the question of whether quantum computers might actually be useful for classical problems. Subsequently, he and a number of other people [16, 8, 39] came up with rather contrived-appearing problems for which quantum computers seemed to work better than classical computers. It was by studying these algorithms, especially Dan Simon's [39], that I figured out how to design the factoring algorithm.

2. The Quantum Circuit Model

In this section we discuss the *quantum circuit model* [44] for quantum computation. This is a rigorous mathematical model for a quantum computer. It is not the only mathematical model that has been proposed for quantum computation; there are also the *quantum Turing machine model* [8, 44] and the *quantum cellular automata model* [31, 42]. All these models result in the same class of polynomial-time quantum computable functions. These are, of course, not the only potential models for quantum computation, and some of the assumptions made in these models, such as unitarity of all gates, and the lack of fermion/boson particle statistics, clearly are not physically realistic in that it is easy to conceive of machines that do not conform to the above assumptions. However, there do not seem to be any physically realistic models which have more computational power than the ones listed above. Neither non-unitarity [3] nor fermions [9] add significant power to the mathematical model. Of these models, the quantum circuit model is possibly the simplest to describe. It is also easier to connect with possible physical implementations of quantum computers than the quantum Turing machine model. A disadvantage of this model is that it is not naturally a *uniform* model. Uniformity is a technical condition arising in complexity theory, and to make the quantum circuit model uniform, additional constraints must be imposed on it. This issue is discussed later.

In analogy with a classical bit, a two-state quantum system is called a *qubit*, or quantum bit. Mathematically, a qubit takes a value in the vector space $\mathbb{C}^2$. We single out two orthogonal basis vectors in this space, and label these V_0 and V_1. In Dirac's "bra-ket" notation, which comes from physics and is commonly used in the quantum computing field, these are represented as $|0\rangle$ and $|1\rangle$. More precisely, quantum states are invariant under multiplication by scalars, so a qubit lives in two-dimensional complex projective space. To conform with physics usage, we treat qubits as column vectors and operate on them by left multiplication.

One of the fundamental principles of quantum mechanics is that the joint quantum state space of two systems is the tensor product of their individual quantum state spaces. Thus, the quantum state space of n qubits is the space $\mathbb{C}^{2^n}$. The basis vectors of this space are parameterized by binary strings of length n. We make extensive use of the tensor decomposition of this space into n copies of $\mathbb{C}^2$, where we represent a basis state V_b corresponding to the binary string $b_1 b_2 \cdots b_n$ by

$$V_{b_1 b_2 \cdots b_n} = V_{b_1} \otimes V_{b_2} \otimes \ldots \otimes V_{b_n}.$$

In "bra-ket" notation, this state is written as $|b_1 b_2 b_3 \cdots b_n\rangle$ or equivalently, as the tensor product $|b_1\rangle|b_2\rangle|b_3\rangle \cdots |b_n\rangle$. Generally, we use position to distinguish the n different qubits. Occasionally we need some other notation for distinguishing them, in which case we denote the i'th qubit by $V^{[i]}$. Since quantum states are invariant under multiplication by scalars, they can without loss of generality be normalized to be unit length vectors; except where otherwise noted, quantum states in this paper will be assumed to be normalized. Quantum computation takes place in the quantum state space of n qubits $\mathbb{C}^{2^n}$, and obtains extra computational power from its exponential dimensionality.

In a usable computer, we need some means of giving it the problem we want solved (input), some means of extracting the answer from it (output), and some means of manipulating the state of the computer to transform the input into the desired output (computation). We next briefly describe input and output for the quantum circuit model. We then take a brief detour to describe the classical circuit model; this will motivate the rules for performing the computation on a quantum computer.

Since we are comparing quantum computers to classical computers, and solving classical problems on a quantum computer, in this paper the input to a quantum computer will always be classical information. It can thus can be expressed as a binary string S of some length k. We need to encode this in the initial quantum state of the computer, which must be a vector in $\mathbb{C}^{2^n}$. The way we do this is to concatenate the bit string S with $n - k$ 0's to obtain the length n string $S0\ldots0$. We then initialize the quantum computer in the state $V_{S0\ldots0}$. Note that the number of qubits is in general larger than the input. These extra qubits, which we can take to be initialized to 0, are often required for workspace in implementing quantum algorithms.

At the end of a computation, the quantum computer is in a state which is a unit vector in $\mathbb{C}^{2^n}$. This state can be written explicitly as $W = \sum_s \alpha_s V_s$ where s ranges over binary strings of length n, $\alpha_s \in \mathbb{C}$, and $\sum_s |\alpha_s|^2 = 1$. These α_s are called *probability amplitudes,* and we say that W is a *superposition* of basis vectors V_s. In quantum mechanics, the Heisenberg uncertainty principle tells us that we cannot measure the complete quantum state of this system. There are a large number of permissible measurements; for example, any orthogonal basis of $\mathbb{C}^{2^n}$ defines a measurement whose possible outcomes are the elements of this basis. However, we assume that the output is obtained by projecting each qubit onto the basis $\{V_0, V_1\}$. This measurement has the great advantage of being simple, and it appears that any physically reasonable measurements can be accomplished by first doing some precomputation and then making the above canonical measurement.

When applied to a state $\sum_s \alpha_s V_s$, this projection produces the string s with probability $|\alpha_s|^2$. The quantum measurement process is inherently probabilistic. Thus we do not require that the computation gives the right answer all the time; but that we obtain the right answer at least 2/3 of the time. Here, the probability 2/3 can be replaced by any number strictly between 1/2 and 1 without altering the class of functions that can be computed in polynomial time by quantum computers—if the probability of obtaining the right answer is strictly larger than 1/2, it can be amplified by running the computation several times and taking the majority vote of the results of these separate computations.

In order to motivate the rules for state manipulation in a quantum circuit, we now take a brief detour and describe the classical circuit model. Recall that a classical circuit can always be written solely with the three gates AND ($\wedge$), OR ($\vee$) and NOT ($\neg$). These three gates are thus said to form a *universal* set of gates. Besides these three gates, note that we also need elements which duplicate the values on wires. It is arguable that these elements should also be classified as gates. These duplicating "gates" are not possible in the domain of quantum computing, because of the theorem that an arbitrary quantum state cannot be cloned (duplicated) [**17, 43**].

A quantum circuit is similarly built out of logical quantum wires carrying qubits, and quantum gates acting on these qubits. Each wire corresponds to one of the n qubits. We assume each gate acts on either one or two wires. The possible physical transformations of a quantum system are unitary transformations, so each quantum gate can be described by a unitary matrix. A quantum gate on one qubit is then described by a 2×2 matrix, and a quantum gate on two qubits by a 4×4 matrix. Note that since unitary matrices are invertible, the computation is reversible; thus starting with the output and working backwards one obtains the input. Further note that for quantum gates, the dimension of the output space is equal to that of the input space, so at all times during the computation we have n qubits carried on n quantum wires.

It should be noted that these requirements of unitary and of maintaining only the original n qubits at all times need to be revised for dealing with noisy gates, an area not covered in this paper. In fact, it can be shown that with these requirements, noisy unitary gates make it impossible to carry out long computations [2]; some means of eliminating noise by resetting qubits to values near 0 is required.

Quantum gates acting on one or two qubits ($\mathbb{C}^2$ or $\mathbb{C}^4$) naturally induce a transformation on the state space of the entire quantum computer ($\mathbb{C}^{2^n}$). For example, if A is a 4×4 matrix acting on qubits i and j, the induced action on a basis vector of $\mathbb{C}^{2^n}$ is

$$(2.1) \qquad A^{[i,j]} V_{b_1 b_2 \cdots b_n} = \sum_{s=0}^{1} \sum_{t=0}^{1} A_{b_i b_j\, st}\, V_{b_1 b_2 \cdots b_{i-1} s b_{i+1} \cdots b_{j-1} t b_{j+1} \cdots b_n}.$$

This is the tensor product of A (acting on qubits i and j) with $n-2$ identity matrices (acting on each of the remaining qubits). When we multiply a general vector by a quantum gate, it can have negative and positive coefficients which cancel out, leading to quantum interference.

As there are for classical circuits, there are universal sets of gates for quantum circuits; such a universal set of gates is sufficient to build circuits for any quantum computation. One particularly useful universal set of gates is the set of all one-bit gates and a specific two-bit gate called the Controlled NOT (CNOT). These gates can efficiently simulate any quantum circuits whose gates act on only a constant number of qubits [4]. On basis vectors, the CNOT gate negates the second (target) qubit if and only if the first (control) qubit is 1. In other words, it takes V_{XY} to V_{XZ} where $Z = X + Y$ (mod 2). This corresponds to the unitary matrix

$$\begin{pmatrix} 1 & 0 & 0 & 0 \\ 0 & 1 & 0 & 0 \\ 0 & 0 & 0 & 1 \\ 0 & 0 & 1 & 0 \end{pmatrix}$$

Note that the CNOT is a classical reversible gate. To obtain a universal set of classical reversible gates, you need at least one reversible three-bit gate, such as a Toffoli gate; otherwise you can only perform linear Boolean computations. A Toffoli gate is a doubly controlled NOT, which negates the 3rd bit if and only if the first two are both 1. By itself the Toffoli gate is universal for reversible classical computation, as it can simulate both AND and NOT gates [21]. Thus, if you can make a Toffoli gate, you can perform any reversible classical computation. Further, as long as the input is not erased, any classical computation can be efficiently performed reversibly [6], and thus implemented efficiently by Toffoli gates. The matrix corresponding to a Toffoli gate is

(2.3)
$$\begin{pmatrix} 1 & 0 & 0 & 0 & 0 & 0 & 0 & 0 \\ 0 & 1 & 0 & 0 & 0 & 0 & 0 & 0 \\ 0 & 0 & 1 & 0 & 0 & 0 & 0 & 0 \\ 0 & 0 & 0 & 1 & 0 & 0 & 0 & 0 \\ 0 & 0 & 0 & 0 & 1 & 0 & 0 & 0 \\ 0 & 0 & 0 & 0 & 0 & 1 & 0 & 0 \\ 0 & 0 & 0 & 0 & 0 & 0 & 0 & 1 \\ 0 & 0 & 0 & 0 & 0 & 0 & 1 & 0 \end{pmatrix}$$

We now define the complexity class BQP, which stands for bounded-error quantum polynomial time. This is the class of languages which can be computed on a quantum computer in polynomial time, with the computer giving the correct answer at least 2/3 of the time.

To give a rigorous definition of this complexity class using quantum circuits, we need to impose uniformity conditions. Any specific quantum circuit can only compute a function whose domain (input) is binary strings of a specific length. To use the quantum circuit model to implement functions taking arbitrary length binary strings for input, we need a family of quantum circuits that contains one circuit for inputs of each length. Without any further conditions on this family of circuits, the designer of this circuit family could hide an uncomputable function in the design of the circuits for each input length. This definition would thus result in the unfortunate inclusion of uncomputable functions in the complexity class BQP. One should note that there is a name for this nonuniform class of functions. It is called BQP/poly, meaning that there can be at most a polynomial amount of extra information included in the circuit design.

To exclude this possibility of including non-computable information in the circuit, we require *uniformity* conditions on the circuit family. The easiest way of doing this is to require a classical Turing machine that on input n outputs a description of the circuit for length n inputs, and which runs in time polynomial in n. For quantum computing, we need an additional uniformity condition on the circuits. It is also be possible for the circuit designer to hide uncomputable (or hard-to-compute) information in the unitary matrices corresponding to quantum gates. We thus require that the k'th digit of the entries of these matrices can be computed by a second Turing machine in time polynomial in k and n. Although we do not have space to discuss this fully, the power of the classical machines designing the circuit family can actually be varied over a wide range; they can be varied from classes much smaller than P to the classical randomized class BPP. This helps us convince ourselves that we have the right definition of BQP.

The definition of polynomial time computable functions on a quantum computer is thus those functions computable by a *uniform* family of circuits whose size (number of gates) is polynomial in the length of the input, and which for any input gives the right answer at least 2/3 of the time. The corresponding set of languages (languages are functions with values in $\{0,1\}$) is called BQP.

3. Relation of the Model to Quantum Physics

The quantum circuit model of the previous section is much simplified from the realities of quantum physics. There are operations possible in physical quantum systems which do not correspond to any simple operation allowable in the quantum circuit model, and complexities that occur when performing experiments that are not reflected in the quantum circuit model. This section contains a brief discussion of these issues, some of which are discussed more thoroughly in [8, 18].

In everyday life, objects behave very classically, and on large scales we do not see any quantum mechanical behavior. This is due to a phenomenon called decoherence, which makes superpositions of states decay, and makes large-scale superpositions of states decay very quickly. A thorough, elementary, discussion of decoherence can be found in [47]; one reason it occurs is that we are dealing with open systems rather than closed ones. Although closed systems quantum mechanically undergo unitary evolution, open systems need not. They are subsystems of systems undergoing unitary evolution, and the process of taking subsystems does not preserve unitarity.

However hard we may try to isolate quantum computers from the environment, it is virtually inevitable that they will still undergo some decoherence and errors. We need to know that these processes do not fundamentally change their behavior. Using no error correction, if each gate results in an amount of decoherence and error of order $1/t$, then $O(t)$ operations can be performed before the quantum state becomes so noisy as to usually give the wrong answer [8]. Active error correction can improve this situation substantially; this is discussed in Gottesman's notes for this course [24].

In some proposed physical architectures for quantum computers, there are restrictions that are more severe than the quantum circuit model given in the preceding section. Many of these restrictions do not change the class BQP. For example, it might be the case that a gate could only be applied to a pair of *adjacent* qubits. We can still operate on a pair of arbitrary qubits: by repeatedly exchanging one of these qubits with a neighbor we can bring this pair together. If there are n qubits in the computer, this can only increase the computation time by a factor of n, preserving the complexity class BQP.

The quantum circuit model described in the previous section postpones all measurements to the end, and assumes that we are not allowed to use probabilistic steps. Both of these possibilities are allowed in general by quantum mechanics, but neither possibility makes the complexity class BQP larger [8]. For fault-tolerant quantum computing, however, it is very useful to permit measurements in the middle of the computation, in order to measure and correct errors.

The quantum circuit model also assumes that we only operate on a constant number of qubits at a time. In general quantum systems, all the qubits evolve simultaneously according to some Hamiltonian describing the system. This simultaneous evolution of many qubits cannot be described by a single gate in our model, which only operates on two qubits at once. In a realistic model of quantum computation, however, we cannot allow general Hamiltonians, since they are not experimentally realizable. Some Hamiltonians that act on all the qubits at once are experimentally realizable. It would be nice to know that even though these Hamiltonians cannot be directly described by our model, they cannot be used to compute functions not in BQP in polynomial time. This could be accomplished by showing that systems with such Hamiltonians can be efficiently simulated by a quantum computer. Some work has been done on simulating Hamiltonians on quantum computers [1, 29, 45], but I do not believe this question has been completely addressed yet.

4. Simon's Algorithm

In this section, we give Dan Simon's algorithm [39] for a problem that takes exponential time on a classical computer, but quadratic time on a quantum computer. This is an "oracle" problem, in that there is a function f given as a "black box" subroutine, and the computer is allowed to compute f, but is not allowed to look at the code for f. In fact, to prove the lower bound on a classical computer, we must permit the computer to use functions f which are not efficiently computable.

We now describe Simon's problem. The computer is given a function f mapping $\mathbf{F}_2^n$ to $\mathbf{F}_2^n$ which has the property that there is a c such that

(4.1) $$f(x) = f(y) \longleftrightarrow x \equiv y + c \pmod{\mathbf{F}_2^n}$$

Here, the addition is bitwise binary addition. Essentially, this is a function which is periodic over $\mathbf{F}_2^n$ with period c.

We now describe the lower bound for a classical computer. Suppose that the function f is chosen at random from all functions with property (4.1). We show that you need to compute $O(2^{n/2})$ function evaluations to find c. Suppose that you have evaluated s values of f. You have then eliminated at most one value of c for each pair of the s values of f computed, but c is equally likely to be any of the remaining possibilities. Thus, after computing s values of f, you will have eliminated at most $s(s-1)/2$ values of c. At least half the time, you must try more than half the possibilities for c, and this takes $O(2^{n/2})$ function evaluations.

We now describe Simon's algorithm for finding the period on a quantum computer. To do this, we need to introduce the Hadamard gate,

$$H = \frac{1}{\sqrt{2}} \begin{pmatrix} 1 & 1 \\ 1 & -1 \end{pmatrix}.$$

Now, suppose that we apply the Hadamard transformation to each of k qubits. We obtain, for a vector a in $\mathbf{F}_2^k$,

(4.2) $$H^{\otimes k}(V_a) = \frac{1}{2^{k/2}} \sum_{b=0}^{2^k-1} (-1)^{a \cdot b} V_b.$$

It is easy to see that each entry of the matrix $H^{\otimes k}$ is $\pm 2^{-k/2}$. Further, the (a,b) entry picks up a factor of -1 for each position which is 1 in both a and b, giving a sign of $(-1)^{a \cdot b}$. Here,

$$a \cdot b = \sum_i a_i b_i \pmod{2}$$

is the binary inner product of a and b. This is in fact the Fourier transform over $\mathbf{F}_2^k$.

We are now ready to describe Simon's algorithm. We will use two registers, both with n qubits. We start with the state $V_0 \otimes V_0$. The first step is to take each qubit in the first register to $\frac{1}{\sqrt{2}}(V_0 + V_1)$, putting the first register in an equal superposition o all binary strings of length n. The computer is now in the state

$$2^{-n/2} \sum_{x=0}^{2^n-1} V_x \otimes V_0.$$

The second step is to compute $f(x)$ in the second register. We now obtain the state

$$2^{-n/2} \sum_{x=0}^{2^n-1} V_x \otimes V_{f(x)}.$$

Note that since the input x of the function $f(x)$ is kept in memory, this is a reversible classical transformation, and thus unitary. The third step is to take the Fourier transform of the first register. This leaves the first register in the state

$$2^{-n} \sum_{x=0}^{2^n-1} \sum_{y=0}^{2^n-1} (-1)^{x \cdot y} V_y \otimes V_{f(x)}.$$

Finally, we observe the state of the computer in the basis $V_i \otimes V_j$. We see the state $V_y \otimes V_{f(x)}$ with probability equal to the square of its amplitude in the above sum. There are exactly two x which give the value $f(x)$, namely x and $x + c$. The probability of observing $V_y \otimes V_{f(x)}$ is thus

$$2^{-2n} \left((-1)^{x \cdot y} + (-1)^{(x+c) \cdot y} \right)^2.$$

This probability is either 2^{2n-2} or 0, depending on whether $y \cdot c$ is 0 or 1. The above measurement thus produces a random y with $c \cdot y = 0$. It is straightforward to show that $O(n)$ such y's chosen at random will be of full rank in $c^\perp$, the $n-1$ dimensional space perpendicular to c, and thus determine c uniquely. Thus, if we repeat the above procedure $O(n)$ times, we will be able to deduce c. Since each of these repetitions takes $O(n)$ steps on the quantum computer, we obtain the answer in $O(n^2 + nF)$ time, where F is the cost of the evaluating the function f.

Simon's algorithm is at least a moderately convincing argument that BQP is strictly larger than BPP, although it is not a rigorous proof. However, Simon's problem is contrived in that it does not seem to have arisen in any other context. It did point the way to my discovery of the factoring algorithm, which will be discussed in the next section. The factoring algorithm is a much less convincing argument that BQP is larger than BPP, as nobody really knows the complexity of factoring. However, as factoring is a widely studied problem that is fundamental for public key cryptography [35], the quantum factoring algorithm brought widespread attention to the field of quantum computing.

5. The Factoring Algorithm

For factoring an L-bit number N, the best classical algorithm known is the number field sieve [28]; this algorithm asymptotically takes time $O(\exp(cL^{1/3} \log^{2/3} L))$. On a quantum computer, the quantum factoring algorithm takes asymptotically $O(L^2 \log L \log \log L)$ steps. The key idea of the quantum factoring algorithm is the use of a Fourier transform to find the period of the sequence $u_i = x^i \pmod{N}$, from which period a factorization of N can be obtained. The period of this sequence is exponential in L, so this approach is not practical on a digital computer. On a quantum computer, however, we can find the period in polynomial time by exploiting the 2^{2L}-dimensional state space of $2L$ qubits, and taking a Fourier transform over this space. The exponential dimensionality of this space permits us to take the Fourier transform of an exponential length sequence. How this works will be made clearer by the following sketch of the algorithm, the full details of which are in [36], along with a quantum algorithm for finding discrete logarithms.

The idea behind all the fast factoring algorithms (classical or quantum) is fairly simple. To factor N, find two residues mod N such that

(5.1) $$s^2 \equiv t^2 \pmod{N}$$

but $s \not\equiv \pm t \pmod{N}$. We now have

(5.2) $$(s+t)(s-t) \equiv 0 \pmod{N}$$

and neither of these two factors is $0 \pmod{N}$. Thus, $s+t$ must contain one factor of N (and $s-t$ another). We can extract this factor by finding the greatest common divisor of $s+t$ and N; this computation can be done in polynomial time using Euclid's algorithm.

In the quantum factoring algorithm, we find the multiplicative period r of a residue $x \pmod{N}$. This period r satisfies $x^r \equiv 1 \pmod{N}$. If we are lucky and r is even, then both sides of this congruence are squares and we can try the above factorization method. If we are just a little bit more lucky, then $x^{r/2} \not\equiv -1 \pmod{N}$, and we obtain a factor by computing $\gcd(x^{r/2}+1, N)$. The greatest common divisor can be computed in polynomial time on a classical computer using Euclid's algorithm.

It is a relatively simple exercise in number theory to show that for large N with two or more prime factors, at least half the residues $x \pmod{N}$ produce prime factors using this technique, and that for most large N the fraction of good residues x is much higher; thus, if we try several different values for x, we have to be particularly unlucky not to obtain a factorization using this method.

We now need to explain what the quantum Fourier transform is. The quantum Fourier transform on k qubits maps the state V_a, where a is considered as an integer between 0 and $2^k - 1$, to a superposition of the states V_b as follows:

(5.3) $$V_a \to \frac{1}{2^{k/2}} \sum_{b=0}^{2^k-1} \exp(2\pi i ab/2^k) V_b$$

It is easy to check that this transformation defines a unitary matrix. It is not as straightforward to implement this Fourier transform as a sequence of one- and two-bit quantum gates. However, an adaption of the Cooley-Tukey algorithm decomposes this transformation into a sequence of $k(k-1)/2$ one- and two-bit gates. More generally, the discrete Fourier transform over any product Q of small primes (each of size at most $\log Q$) can be performed in polynomial time on a quantum computer. We will show how to break the above Fourier transform of Eq. (5.3) into this product of two-bit gates at the end of this section.

We are now ready to give the quantum algorithm for factoring. What we do is design a polynomial-size circuit which starts in the quantum state $V_{00...0}$ and whose output, with reasonable probability, lets us factor an L-bit number N in polynomial time using a digital computer. This circuit has two main registers, the first of which is composed of $2L$ qubits and the second of L qubits. It also requires a few extra qubits of work space, which we do not mention in the summary below but which are required for implementing the step (5.5) below.

We start by putting the computer into the state representing the superposition of all possible values of the first register:

(5.4) $$\frac{1}{2^L} \sum_{a=0}^{2^{2L}-1} V_a \otimes V_0.$$

This can easily be done using $2L$ gates by putting each of the qubits in the first register into the state $\frac{1}{\sqrt{2}}(V_0 + V_1)$.

We next use the value of a in the first register to compute the value $x^a \pmod{N}$ in the second register. This can be done using a reversible classical circuit for computing $x^a \pmod{N}$ from a. Computing $x^a \pmod{N}$ using repeated squaring takes $O(L^3)$ quantum gates using the grade school multiplication algorithm, and asymptotically $O(L^2 \log L \log \log L)$ gates using fast integer multiplication (which is actually faster only for moderately large values of L). This leaves the computer in the state

$$(5.5) \qquad \frac{1}{2^L} \sum_{a=0}^{2^{2L}-1} V_a \otimes V_{x^a (\bmod\ N)}.$$

The next step is to take the discrete Fourier transform of the first register, as in Equation (5.3). This puts the computer into the state

$$(5.6) \qquad \frac{1}{2^{2L}} \sum_{a=0}^{2^{2L}-1} \sum_{c=0}^{2^{2L}-1} \exp(2\pi i a b / 2^{2L}) V_c \otimes V_{x^a (\bmod\ N)}.$$

Finally, we measure the state of our machine. This yields the output $V_c \otimes V_{x^j (\bmod\ N)}$ with probability equal to the square of the coefficient on this state in the sum (5.6). Since many values of $x^a \pmod{N}$ are equal, many terms in this sum contribute to each coefficient. All these a's giving the same value of $x^a \pmod{N}$ can be represented as

$$a = a_0 + br,$$

where a_0 is the smallest of these a's and b is some integer between 0 and $\lceil 2^{2L}/r \rceil$. Explicitly, this probability is:

$$(5.7) \qquad \frac{1}{2^{4L}} \left| \exp(2\pi i a_0 c / 2^{2L}) \sum_{b=0}^{\lfloor 2^{2L}/r \rfloor + \eta} \exp(2\pi i b r c / 2^{2L}) \right|^2.$$

where η is either 0 or 1, depending on the values of $2^{2L} \pmod{r}$ and a_0. This sum in Eq. (5.7) is a geometric sum of unit complex numbers equally spaced around the unit circle, and it is straightforward to check that this sum is small except when these complex numbers point predominantly in the same direction. For this to happen, we need that the angle between the two complex phases for b and $b+1$ is on the order of the reciprocal of the number of possible b's, i.e., that

$$(5.8) \qquad rc/2^{2L} = d + O(r/2^{2L})$$

for some integer d. We thus are likely to observe only values of b satisfying (5.8). Recalling that $2^{2L} \approx N^2$, we can rewrite this equation to obtain

$$(5.9) \qquad \frac{c}{2^{2L}} = \frac{d}{r} + O(1/N^2).$$

We know c and 2^{2L}, and we want to find r. Since both d and r are less than N, if the $O(1/N^2)$ in Eq. (5.9) were exactly $1/2N^2$, we would have

$$\left| \frac{c}{2^{2L}} - \frac{d}{r} \right| \leq \frac{1}{2N^2}$$

and $\frac{d}{r}$ would be the closest fraction to $c/2^{2L}$ with numerator and denominator less than N. In actuality, it is likely to be one of the closest ones. Thus, all we need do to find r is to round $c/2^{2L}$ to find all close fractions with denominators less than N. This can be done in polynomial time using a continued fraction expansion, and since we can check whether we have obtained the right value of r, we can search the close fractions until we have obtained the correct one. We chose $2L$ as the size of the first register in order to make d/r likely to be the closest fraction to $c/2^{2L}$ with numerator and denominator at most N.

More details of this algorithm can be found in [36]. Recently, Zalka [46] has analyzed the resources required by this algorithm much more thoroughly, improving upon their original values in many respects. For example, he shows that you can use only $3L + o(L)$ qubits, whereas the original algorithm required $2L$ extra qubits for workspace, giving a total of $5L$ qubits. He also shows how to efficiently parallelize the algorithm to run on a parallel quantum computer.

5.1. Implementing the Quantum Fourier Transform. We now show how to break the discrete Fourier transform (Eq. 5.3) into a product of two-bit gates, a step which we previously postponed to this subsection. Let us consider the Fourier transform on $k + 1$ bits.

$$(5.10) \qquad V_a \to \frac{1}{2^{(k+1)/2}} \sum_{b=0}^{2^{k+1}-1} \exp(2\pi i a b / 2^{k+1}) V_b$$

We will assume that we have an expression for the Fourier transform on k qubits, and show how to obtain an expression for the Fourier transform on $k + 1$ qubits using only $k + 1$ additional gates.

We break the input space V_a on $k+1$ qubits into the tensor product of a k-qubit space and a 1-qubit space, so that $V_a = V_{a_-} \otimes V_{a_0}$, where the $(k+1)$-bit string a is the concatenation of the k-bit string a_- and the one-bit string a_0. Thus, a_0 is the rightmost bit of the binary number a, i.e., the units bit. We similarly break the output space V_b into the tensor product of a 1-qubit space and a k-qubit space, but this time we choose the first bit as the 1-qubit space, so $V_b = V_{b_k} \otimes V_{b_-}$, where b_k is the leftmost bit of b, i.e. the bit with value 2^k, and b_- comprises the k rightmost bits. Now, the Fourier transform becomes

$$(5.11) \quad V_{a_-} V_{a_0} \to 2^{-\frac{k+1}{2}} \sum_{\substack{a_k=0 \\ b_0=0}}^{1} \sum_{\substack{a_-=0 \\ b_-=0}}^{2^k-1} \exp\left(2\pi i \left(\frac{a_0 b_k}{2} + \frac{a_0 b_-}{2^{k+1}} + a_- b_k + \frac{a_- b_-}{2^k}\right)\right) V_{b_k} V_{b_-}.$$

We now analyze this expression. First, the term $\exp(2\pi i a_- b_k)$ is always 1, and thus can be dropped. The term $\exp(2\pi i a_- b_- / 2^k)$ is the phase factor in the quantum Fourier transform on k qubits. Thus, if we first perform the Fourier transform on k qubits (which we can do by the induction hypothesis), we take V_{a_-} to V_{b_-} and obtain this phase factor. The term $\exp(2\pi i a_0 b_- / 2^{k+1})$ can be expressed as the product of k gates, by letting the gate

$$T_{j,k} = \begin{pmatrix} 1 & 0 & 0 & 0 \\ 0 & 1 & 0 & 0 \\ 0 & 0 & 1 & 0 \\ 0 & 0 & 0 & \exp\left(\frac{2\pi i}{2^{k+1-j}}\right) \end{pmatrix}$$

operate on the qubits corresponding to a_0 and b_j, by which we mean the bit of b_- with value 2^j, i.e., the $j+1$'st bit from the right. This gate applies the phase factor

of $\exp(2\pi i/2^{k+1-j})$ if and only if both the bits a_0 and b_j are 1. Finally, the term
$$\exp(2\pi i a_0 b_k/2) = (-1)^{a_0 \cdot b_k}$$
is the unitary gate
$$H = \frac{1}{\sqrt{2}} \begin{pmatrix} 1 & 1 \\ 1 & -1 \end{pmatrix}$$
which takes V_{a_0} to V_{b_k} with the phase factor $(-1)^{a_0 \cdot b_k}$. We now see that we can obtain the Fourier transform on $k+1$ qubits by first applying the Fourier transform on k qubits, taking V_{a_-} to $\sum \exp(2\pi i a_- b_-/2^k) V_{b_-}$, next applying the gate $T_{j,k}$ on the qubits V_{a_0} and V_{b_j} for $j = 0$ to $k-1$, and finally by applying the gate H on the qubit V_{a_0} (yielding the qubit V_{b_k}). For those readers who are familiar with the Cooley-Tukey fast Fourier transform, this is almost a direct translation of it to a quantum algorithm. Multiplying the gates $T_{j,k}$ for a fixed k gives the "twiddle factor" of the Cooley-Tukey FFT.

One objection that might be raised to this expansion of the Fourier transform is that it requires gates with exponentially small phases, which could not possibly be implemented with any physical accuracy. In fact, one can omit these gates and obtain an approximate Fourier transform which is close enough to the actual Fourier transform that it barely changes the probability that the factoring algorithm succeeds [14]. This reduces the number of gates required for the quantum Fourier transform from $O(k^2)$ to $O(k \log k)$.

6. Grover's Algorithm

Another very important algorithm in quantum computing is L. K. Grover's search algorithm, which searches an unordered list of N items (or the range of an efficiently computable function) for a specific item in time $O(\sqrt{N})$, an improvement on the optimal classical algorithm, which must look at $N/2$ items on average before finding a specific item [25]. The technique used in this algorithm can be applied to a number of other problems to also obtain a square root speed-up [26]. If you are searching an unordered database, this square root speed-up is as good as a quantum computer can do; this is proved using techniques developed in [7]. Finally, a generalization of both Grover's search algorithm and the lower bound above gives tight bounds on how much a quantum computer can amplify a quantum procedure that has a given probability of success [10]. The quantum search algorithm can be thought of in these terms; the procedure is just that of choosing a random element of the N-element list, so the probability of success is $1/N$. A quantum computer can amplify this probability to near-unity by using $O(\sqrt{N})$ iterations while a classical computer requires order N iterations. I sketch Grover's search algorithm below.

Grover's algorithm uses only three transformations. The first is the transformation $W = H^{\otimes k}$, which is the transformation obtained by applying the matrix
$$H = \frac{1}{\sqrt{2}} \begin{pmatrix} 1 & 1 \\ 1 & -1 \end{pmatrix}$$
to each qubit. It is easy to check that $W^2 = \text{Id}$, because $H^2 = \text{Id}$. The second transformation is Z_0, which takes the basis vector V_0 to $-V_0$ and leaves V_i unchanged for $i \neq 0$. The third is Z_t, which takes V_t to $-V_t$ and leaves V_i unchanged for $i \neq t$, where the t'th element of the list is the one we are trying to find. At first glance, it might seem that we need to know t to apply Z_t; however, if we can

design a quantum circuit that tests whether an integer i is equal to t, than we can use it to perform the transformation Z_t. For example, if we are searching for a specific element in an unordered list, it is fairly straightforward to write a program that tests whether the i'th element of the list is indeed the desired element, and negates the phase if it is, without knowing where the desired element is in the list. Similarly, if we are searching for a solution to some mathematical problem, we need only to be able to efficiently test whether a given integer i encodes a solution to the problem.

Suppose that we are searching among $N = 2^k$ items, which are encoded by the integers 0 to $N - 1$. Here we use k qubits to keep track of the items. We will now calculate that if we start in the superposition

$$\Sigma_{i=0}^{N-1} \alpha_i V_i$$

then the transformation $-WZ_0W$ leaves us in the state

$$\Sigma_{i=0}^{N-1} (2m - \alpha_i) V_i$$

where $m = \frac{1}{N} \Sigma_0^{N-1} \alpha_i$ is the mean of all the amplitudes. The proof of this follows from the observation that after the transform W, the amplitude of V_0 is $\sqrt{N}m$. Recall that $W^2 = \text{Id}$. These two observations can be used to show that the transformation WZ_0W extracts the mean m in the amplitude of V_0, negates it, and redistributes it negated over all the basis states V_i. The transformation WZ_0W thus takes $\Sigma_i \alpha_i V_i$ to $\Sigma_i (\alpha_i - 2m) V_i$.

We are now in a position to describe Grover's algorithm in detail. We start in the equal superposition of all V_i, i.e. the state $\frac{1}{\sqrt{N}} \Sigma_{i=0}^{N-1} V_i$. We then repeat the transformation $Z_t W Z_0 W$ for $c\sqrt{N}$ iterations, for the appropriately chosen constant c. What this accomplishes is to gradually increase the amplitude on V_t at the expense of all the other amplitudes, until after $c\sqrt{N}$ iterations the amplitude on V_t is nearly unity. Suppose that we have reached a point where the amplitude on V_i is α for all $i \neq t$ and β for V_t. It is easy to see that in the next step, these amplitudes are $2m - \alpha$ and $\beta + 2m$, respectively, where $m = (\beta + (N-1)\alpha)/N$ is the mean amplitude. When β is small, $m \approx \alpha \approx 1/\sqrt{N}$, and thus the amplitudes on V_i, $i \neq t$ decrease slightly and the amplitude on V_t increases by approximately $2/\sqrt{N}$. I will not go into the details in this write-up, but this at least gives the intuition that, after $c\sqrt{N}$ steps, we obtain a state very close to V_t. There are many variations of this algorithm, including ones that work when there is more than one desired solution. For more details, I recommend reading Grover's paper [**25**].

Finally, as Feynman suggested, it appears that quantum computing is good at computing simulations of quantum mechanical dynamics. I will not be discussing this. Some work in this regard has appeared in [**1, 29, 45**], but much remains to be done.

References

[1] D. S. Abrams and S. Lloyd, Simulation of many-body Fermi systems on a universal quantum computer, *Phys. Rev. Lett.* **79** (1997), 2586–2589.

[2] D. Aharonov, M. Ben-Or, R. Impagliazzo and N. Nisan, Limitations of noisy reversible computation, (1996), http://xxx.lanl.gov/abs/quant-ph/9611028.

[3] D. Aharonov, A. Kitaev and N. Nisan, Quantum circuits with mixed states, in *Proceedings of the Thirtieth Annual ACM Symposium on Theory of Computation*, ACM Press, New York (1998), 20–30. Also http://xxx.lanl.gov/abs/quant-ph/9806029.

[4] A. Barenco, C. H. Bennett, R. Cleve, D. P. DiVincenzo, N. Margolus, P. Shor, T. Sleator, J. A. Smolin, and H. Weinfurter, Elementary gates for quantum computation, *Phys. Rev. A* **52** (1995), 3457–3467.
[5] P. Benioff, The computer as a physical system: A microscopic quantum mechanical Hamiltonian model of computers as represented by Turing machines, *J. Statist. Phys.* **22** (1980), 563–591.
[6] C. H. Bennett, Logical reversibility of computation, *IBM J. Res. Develop.* **17** (1973), 525–532.
[7] C. Bennett, E. Bernstein G. Brassard and U. Vazirani, Strengths and weaknesses of quantum computing, *SIAM J. Computing* **26** (1997), 1510–1523.
[8] E. Bernstein and U. Vazirani, Quantum complexity theory, *SIAM J. Computing* **26** (1997), 1411–1473.
[9] S. Bravyi and A. Kitaev, Fermionic quantum computation, (2000), http://xxx.lanl.gov/abs/quant-ph/0003137.
[10] H. Burhman, R. Cleve, R. de Wolfe, and C. Zalka, Bounds for small-error and zero-error quantum algorithms, *Proceedings of the Fortieth Annual Symposium on Foundations of Computer Science,* IEEE Computer Society, Los Alamitos, CA (1999), 358–368.
[11] A. R. Calderbank, E. M. Rains, P. W. Shor and N. J. A. Sloane, Quantum error correction via codes over GF(4), *IEEE Transactions on Information Theory* **44** (1998), 1369–1387.
[12] A. R. Calderbank and P. W. Shor, Good quantum error-correcting codes exist, *Phys. Rev. A* **54** (1995), 1098–1106.
[13] A. Church (1936), An unsolvable problem of elementary number theory, *Amer. J. Math.* **58,** (1936) 345–363.
[14] R. Coppersmith, An approximate Fourier transform useful in quantum factoring, IBM Research Report RC 19642 (1994).
[15] D. Deutsch, Quantum theory, the Church–Turing principle and the universal quantum computer, *Proc. Roy. Soc. London Ser. A* **400** (1985), 96–117.
[16] D. Deutsch and R. Jozsa, Rapid solution of problems by quantum computation, *Proc. Roy. Soc. London Ser. A* **439,** (1992), 553–558.
[17] D. Dieks, Communication by EPR devices, *Phys. Lett. A* **92**, 271–272 (1982).
[18] D. P. DiVincenzo, The physical implementation of quantum computation, *Fortsch. Phys.* **48** (2000), 771–783. Also http://xxx.lanl.gov/abs/quant-ph/0002077.
[19] R. Feynman, Simulating physics with computers, *Internat. J. Theoret. Phys.* **21** (1982), 467–488.
[20] R. Feynman, Quantum mechanical computers, *Found. Phys.* **16** (1986), 507–531; originally in *Optics News* (February 1985), 11–20.
[21] E. Fredkin and T. Toffoli, Conservative logic, *Internat. J. Theoret. Phys.* **21** (1982), 219–253.
[22] D. Gottesman, A class of quantum error-correcting codes saturating the quantum Hamming bound, *Phys. Rev. A* **54** (1996), 1862–1868.
[23] D. Gottesman, *Stabilizer Codes and Quantum Error Correction*, Ph.D. Thesis, California Institute of Technology (1997). Also http://xxx.lanl.gov/abs/quant-ph/9705052.
[24] D. Gottesman, An introduction to quantum error correction, this AMS Proceedings of Symposia in Applied Mathematics (PSAPM).
[25] L. K. Grover, Quantum mechanics helps in searching for a needle in a haystack, *Phys. Rev. Lett.* **78** (1997), 325–328. Also http://xxx.lanl.gov/abs/quant-ph/9706033.
[26] L. K. Grover, A framework for fast quantum mechanical algorithms, in *Proceedings of the 30th Annual ACM Symposium on Theory of Computing*, ACM Press, New York (1998), 53–62.
[27] S. C. Kleene, General recursive functions of natural numbers, *Mathematische Annalen* **112** (1936), pp. 727–742.
[28] A. K. Lenstra and H. W. Lenstra, Jr., editors, *The Development of the Number Field Sieve,* Lecture Notes in Mathematics 1554, Springer Verlag, Berlin (1993).
[29] S. Lloyd, Universal quantum simulators, *Science* **273** (1996), 1073–1078.
[30] Yu. Manin, *Computable and Uncomputable* (in Russian), Sovetskoye Radio, Moscow (1980).
[31] N. Margolus, Parallel quantum computation, in *Complexity, Entropy, and the Physics of Information*, edited by W. Zurek, Addison-Wesley (1990) 273–287.
[32] E. Post, Finite combinatory processes. Formulation I, *J. Symbolic Logic* **1** (1936) 103–105.
[33] J. Preskill, Fault-tolerant quantum computation, in *Introduction to Quantum Computation*, edited by H.-K. Lo, S Popescu and T. P. Spiller, World Scientific, Singapore (1998), 213–269. Also http://xxx.lanl.gov/abs/quant-ph/9712048.

[34] J. Preskill, Lecture notes for Physics 219/Computer Science 219: Quantum Computation (1999), available online at http://www.theory.caltech.edu/people/preskill/ph229.
[35] R. L. Rivest, A. Shamir and L. Adleman, A method of obtaining digital signatures and public-key cryptosystems, *Comm. Assoc. Comput. Mach.* **21** (1978), 120–126.
[36] P. W. Shor, Polynomial-time algorithms for prime factorization and discrete logarithms on a quantum computer, *SIAM J. Computing* **26** (1997), 1484–1509.
[37] P. W. Shor, Fault-tolerant quantum computation, in *Proc. 37nd Annual Symposium on Foundations of Computer Science*, IEEE Computer Society Press, Los Alamitos, CA (1996), 56–65.
[38] P. W. Shor, Quantum computing, *Documenta Mathematica* **Extra Vol. ICM I** (1998), 467–486.
[39] D. R. Simon, On the power of quantum computation, *SIAM J. Computing* **26** (1997), 1474–1483.
[40] A. Steane, Multiple particle interference and quantum error correction, *Proc. Roy. Soc. London Ser. A* **452** (1996), 2551–2577.
[41] A. M. Turing, On computable numbers, with an application to the Entscheidungsproblem, *Proc. London Math. Soc. (2)* **42**(1936), 230–265; Corrections in *Proc. London Math. Soc. (2)* **43** (1937), 544–546.
[42] W. van Dam, A universal quantum cellular automaton, in *Proceedings of PhysComp96*, edited by T. Toffoli, M. Biafore and J. Leão, New England Complex Systems Institute (1996), 323-331.
[43] W. Wootters and W. H. Zurek, A single quantum cannot be cloned, *Nature* **299** (1982), 802–803.
[44] A. Yao, Quantum circuit complexity, in *Proceedings of the 34th Annual Symposium on Foundations of Computer Science*, IEEE Computer Society Press, Los Alamitos, CA (1993), 352–361.
[45] C. Zalka, Efficient simulation of quantum systems by quantum computers, *Proc. Roy. Soc. London Ser. A* **454** (1998), 313–322.
[46] C. Zalka, Fast versions of Shor's quantum factoring algorithm, http://xxx.lanl.gov/abs/quant-ph/9806084 (1998).
[47] W. H. Zurek, Decoherence and the transition from quantum to classical, *Physics Today* **44** (1991), 36–44.

AT&T LABS – RESEARCH, FLORHAM PARK, NJ 07932, USA
E-mail address: shor@research.att.com
URL: http://research.att.com/~shor

Shor's Quantum Factoring Algorithm

Samuel J. Lomonaco, Jr.

ABSTRACT. This paper is a written version of a one hour lecture given on Peter Shor's quantum factoring algorithm. It is based on [4], [6], [7], [9], and [15].

CONTENTS

1. Preamble to Shor's algorithm
2. Number theoretic preliminaries
3. Overview of Shor's algorithm
4. Preparations for the quantum part of Shor's algorithm
5. The quantum part of Shor's algorithm
6. Peter Shor's stochastic source $\mathcal{S}$
7. A momentary digression: Continued fractions
8. Preparation for the final part of Shor's algorithm
9. The final part of Shor's algorithm
10. An example of Shor's algorithm

References

1. Preamble to Shor's algorithm

2000 *Mathematics Subject Classification.* Primary 81P68, 81-01, 81-02,68Q25, 68W40; Secondary 68Q05.

Key words and phrases. Shor's algorithm, factoring, quantum computation, quantum algorithms, algorithms, quantum mechanics.

This work was partially supported by Army Research Office (ARO) Grant #P-38804-PH-QC, by the National Institute of Standards and Technology (NIST), by the Defense Advanced Research Projects Agency (DARPA) and Air Force Materiel Command USAF under agreement number F30602-01-0522, and by the L-O-O-P Fund. The author gratefully acknowledges the hospitality of the University of Cambridge Isaac Newton Institute for Mathematical Sciences, Cambridge, England, where some of this work was completed. I would also like to thank the other AMS Short Course lecturers, Howard Brandt, Dan Gottesman, Lou Kauffman, Alexei Kitaev, Peter Shor, Umesh Vazirani and the many Short Course participants for their support.

© 2002 American Mathematical Society

There are cryptographic systems (such as RSA[1]) that are extensively used today (e.g., in the banking industry) which are based on the following questionable assumption, i.e., conjecture:

Conjecture(Assumption). *Integer factoring is computationally much harder than integer multiplication. In other words, while there are obviously many polynomial time algorithms for integer multiplication, there are no polynomial time algorithms for integer factoring. I.e., integer factoring computationally requires super-polynomial time.*

This assumption is based on the fact that, in spite of the intensive efforts over many centuries of the best minds to find a polynomial time factoring algorithm, no one has succeeded so far. As of this writing, the most asymptotically efficient *classical* algorithm is the number field sieve [**10**], [**11**], which is heuristically estimated[2] to factor an integer N in time $O\left(\exp\left[c\left(\lg N\right)^{1/3}\left(\lg \lg N\right)^{2/3}\right]\right)$. Thus, this is a super-polynomial time algorithm in the number $O\left(\lg N\right)$ of digits in N.

However, ... Peter Shor suddenly changed the rules of the game.

Hidden in the above conjecture is the unstated, but implicitly understood, assumption that all algorithms run on computers based on the principles of classical mechanics, i.e., on **classical computers**. But what if a computer could be built that is based not only on classical mechanics, but on quantum mechanics as well? I.e., what if we could build a **quantum computer**?

Shor, starting from the works of Benioff, Bennett, Deutsch, Feynman, Simon, and others, created an algorithm to be run on a quantum computer, i.e., a **quantum algorithm**, that factors integers in polynomial time! Shor's algorithm takes asymptotically $O\left(\left(\lg N\right)^2 \left(\lg \lg N\right)\left(\lg \lg \lg N\right)\right)$ steps on a quantum computer, which is polynomial time in the number of digits $O\left(\lg N\right)$ of N.

2. Number theoretic preliminaries

Since the time of Euclid, it has been known that every positive integer N can be uniquely (up to order) factored into the product of primes. Moreover, it is a computationally easy (polynomial time) task to determine whether or not N is a prime or composite number. For the primality testing algorithm of Miller-Rabin[**14**] makes such a determination at the cost of $O\left(s \lg N\right)$ arithmetic operations $[O\left(s \lg^3 N\right)$ bit operations] with probability of error $Prob_{Error} \leq 2^{-s}$.

---

[1]RSA is a public key cryptographic system invented by Rivest, Shamir, Adleman. Hence the name. For more information, please refer to [**17**].

[2]This heuristic estimate is based on unproven theorems in number theory. This estimate appears to match reasonably well with experimentally observed running times.

However, once an odd positive integer N is known to be composite, it does not appear to be an easy (polynomial time) task on a classical computer to determine its prime factors. As mentioned earlier, so far the most asymptotically efficient *classical* algorithm known is the number field sieve [10], [11], which is heuristically estimated to factor an integer N in time $O\left(c \exp\left[(\lg N)^{1/3} (\lg \lg N)^{2/3}\right]\right)$.

Prime Factorization Problem. *Given a composite odd positive integer N, find its prime factors.*

It is well known[14] that factoring N can be reduced to the task of choosing at random an integer m relatively prime to N, and then determining its modulo N multiplicative order P, i.e., to finding the smallest positive integer P such that

$$m^P = 1 \bmod N.$$

It was precisely this approach to factoring that enabled Shor to construct his factoring algorithm.

3. Overview of Shor's algorithm

But what is Shor's quantum factoring algorithm?

Let $\mathbb{N} = \{0, 1, 2, 3, \ldots\}$ denote the set of natural numbers.

Shor's algorithm provides a solution to the above problem. His algorithm consists of the five steps (**steps 1** through **5**), with only STEP **2** requiring the use of a quantum computer. The remaining four other steps of the algorithm are to be performed on a classical computer.

We begin by briefly describing all five steps. After that, we will then focus in on the quantum part of the algorithm, i.e., STEP **2**.

Step 1. Choose a random positive integer m. Use the polynomial time Euclidean algorithm[3] to compute the greatest common divisor $\gcd(m, N)$ of m and N. If the greatest common divisor $\gcd(m, N) \neq 1$, then we have found a non-trivial factor of N, and we are done. If, on the other hand, $\gcd(m, N) = 1$, then proceed to STEP **2**.

---

[3]The Euclidean algorithm is $O\left(\lg^2 N\right)$. For a description of the Euclidean algorithm, see for example [3] or [2].

STEP 2. Use a QUANTUM COMPUTER to determine the unknown period P of the function
$$\begin{array}{rcl} \mathbb{N} & \xrightarrow{f_N} & \mathbb{N} \\ a & \longmapsto & m^a \bmod N \end{array}$$

Step 3. If P is an odd integer, then go to **Step 1**. [The probability of P being odd is $(\frac{1}{2})^k$, where k is the number of distinct prime factors of N.] If P is even, then proceed to **Step 4**.

Step 4. Since P is even,
$$\left(m^{P/2} - 1\right)\left(m^{P/2} + 1\right) = m^P - 1 = 0 \bmod N \ .$$
If $m^{P/2} + 1 = 0 \bmod N$, then go to **Step 1**. If $m^{P/2} + 1 \neq 0 \bmod N$, then proceed to **Step 5**. It can be shown that the probability that $m^{P/2} + 1 = 0 \bmod N$ is less than $(\frac{1}{2})^{k-1}$, where k denotes the number of distinct prime factors of N.

Step 5. Use the Euclidean algorithm to compute $d = \gcd\left(m^{P/2} - 1, N\right)$. Since $m^{P/2} + 1 \neq 0 \bmod N$, it can easily be shown that d is a non-trivial factor of N. Exit with the answer d.

Thus, the task of factoring an odd positive integer N reduces to the following problem:

Problem. *Given a periodic function*
$$f : \mathbb{N} \longrightarrow \mathbb{N} \ ,$$
find the period P of f.

4. Preparations for the quantum part of Shor's algorithm

Choose a power of 2
$$Q = 2^L$$
such that
$$N^2 \leq Q = 2^L < 2N^2 \ ,$$
and consider f restricted to the set
$$S_Q = \{0, 1, \ldots, Q-1\}$$
which we also denote by f, i.e.,
$$f : S_Q \longrightarrow S_Q \ .$$

In preparation for a discussion of STEP 2 of Shor's algorithm, we construct an L-qubit quantum registers REGISTER1 and a $\lceil \lg N \rceil$-qubit quantum REGISTER2 to hold respectively the arguments and the values of the function f, i.e.,

$$|\text{REG1}\rangle |\text{REG2}\rangle = |a\rangle |f(a)\rangle = |a\rangle |b\rangle = |a_0 a_1 \cdots a_{L-1}\rangle |b_0 b_1 \cdots b_{\lceil \lg N \rceil -1}\rangle$$

In doing so, we have adopted the following convention for representing integers in these registers:

Notation Convention. In a quantum computer, *we represent an integer a with radix 2 representation*

$$a = \sum_{j=0}^{L-1} a_j 2^j ,$$

as a quantum register consisting of the 2^n qubits

$$|a\rangle = |a_0 a_1 \cdots a_{L-1}\rangle = \bigotimes_{j=0}^{L-1} |a_j\rangle$$

For example, the integer 23 is represented in our quantum computer as n qubits in the state:

$$|23\rangle = |10111000\cdots 0\rangle$$

Before continuing, we remind the reader of the classical definition of the Q-point Fourier transform.

DEFINITION 1. *Let ω be a primitive Q-th root of unity, e.g., $\omega = e^{2\pi i/Q}$. Then the Q-point Fourier transform is the map*

$$Map(S_Q, \mathbb{C}) \xrightarrow{\mathcal{F}} Map(S_Q, \mathbb{C})$$
$$[f : S_Q \longrightarrow \mathbb{C}] \longmapsto \left[\widehat{f} : S_Q \longrightarrow \mathbb{C}\right]$$

where

$$\widehat{f}(y) = \frac{1}{\sqrt{Q}} \sum_{x \in S_Q} f(x) \omega^{xy}$$

We implement the Fourier transform $\mathcal{F}$ as a unitary transformation, which in the standard basis

$$|0\rangle, |1\rangle, \ldots, |Q-1\rangle$$

is given by the $Q \times Q$ unitary matrix

$$\mathcal{F} = \frac{1}{\sqrt{Q}} \left(\omega^{xy}\right) .$$

This unitary transformation can be factored into the product of $O\left(\lg^2 Q\right) = O\left(\lg^2 N\right)$ sufficiently local unitary transformations. (See [15], [6].)

5. The quantum part of Shor's algorithm

The quantum part of Shor's algorithm, i.e., STEP 2, is the following:

$\boxed{\text{STEP 2.0}}$ Initialize registers 1 and 2, i.e.,

$$|\psi_0\rangle = |\text{REG1}\rangle |\text{REG2}\rangle = |0\rangle |1\rangle = |00\cdots 00\rangle |00\cdots 01\rangle$$

$\boxed{\text{STEP 2.1}}$ [4]Apply the Q-point Fourier transform $\mathcal{F}$ to REGISTER1.

$$|\psi_0\rangle = |0\rangle |1\rangle \stackrel{\mathcal{F}\otimes I}{\longmapsto} |\psi_1\rangle = \frac{1}{\sqrt{Q}}\sum_{x=0}^{Q-1} \omega^{0\cdot x} |x\rangle |0\rangle = \frac{1}{\sqrt{Q}}\sum_{x=0}^{Q-1} |x\rangle |1\rangle$$

REMARK 1. *Hence, REGISTER1 now holds all the integers*

$$0, 1, 2, \ldots, Q-1$$

in superposition.

$\boxed{\text{STEP 2.2}}$ Let U_f be the unitary transformation that takes $|x\rangle|0\rangle$ to $|x\rangle|f(x)\rangle$. Apply the linear transformation U_f to the two registers. The result is:

$$|\psi_1\rangle = \frac{1}{\sqrt{Q}}\sum_{x=0}^{Q-1} |x\rangle|1\rangle \stackrel{U_f}{\longmapsto} |\psi_2\rangle = \frac{1}{\sqrt{Q}}\sum_{x=0}^{Q-1} |x\rangle|f(x)\rangle$$

REMARK 2. *The state of the two registers is now more than a superposition of states. In this step, we have quantum entangled the two registers.*

$\boxed{\text{STEP 2.3.}}$ Apply the Q-point Fourier transform $\mathcal{F}$ to REG1. The resulting state is:

$$|\psi_2\rangle = \frac{1}{\sqrt{Q}}\sum_{x=0}^{Q-1}|x\rangle|f(x)\rangle \stackrel{\mathcal{F}\otimes I}{\longmapsto} |\psi_3\rangle = \frac{1}{Q}\sum_{x=0}^{Q-1}\sum_{y=0}^{Q-1}\omega^{xy}|y\rangle|f(x)\rangle$$

$$= \frac{1}{Q}\sum_{y=0}^{Q-1} \||\Upsilon(y)\rangle\| \cdot |y\rangle \frac{|\Upsilon(y)\rangle}{\||\Upsilon(y)\rangle\|},$$

where

$$|\Upsilon(y)\rangle = \sum_{x=0}^{Q-1} \omega^{xy} |f(x)\rangle.$$

---

[4]In this step we could have instead applied the Hadamard transform to REGISTER1 with the same result, but at the computational cost of $O(\lg N)$ sufficiently local unitary transformations. The phrase "sufficiently local unitary transformation" is defined in the last part of section 7.7 of [13].

STEP 2.4. Measure REG1, i.e., perform a measurement with respect to the orthogonal projections

$$|0\rangle\langle 0| \otimes I,\ |1\rangle\langle 1| \otimes I,\ |2\rangle\langle 2| \otimes I,\ \ldots,\ |Q-1\rangle\langle Q-1| \otimes I,$$

where I denotes the identity operator on the Hilbert space of the second register REG2.

As a result of this measurement, we have, with probability

$$Prob(y_0) = \frac{\||\Upsilon(y_0)\rangle\|^2}{Q^2},$$

moved to the state

$$|y_0\rangle \frac{|\Upsilon(y_0)\rangle}{\||\Upsilon(y_0)\rangle\|}$$

and measured the value

$$y_0 \in \{0, 1, 2, \ldots, Q-1\}.$$

If after this computation, we ignore the two registers REG1 and REG2, we see that what we have created is nothing more than a classical probability distribution $\mathcal{S}$ on the sample space

$$\{0, 1, 2, \ldots, Q-1\}.$$

In other words, the sole purpose of executing STEPS 2.1 to 2.4 is to create a classical finite memoryless stochastic source $\mathcal{S}$ which outputs a symbol $y_0 \in \{0, 1, 2, \ldots, Q-1\}$ with the probability

$$Prob(y_0) = \frac{\||\Upsilon(y_0)\rangle\|^2}{Q^2}.$$

(For more details, please refer to section 8.1 of [13].)

As we shall see, the objective of the remainder of Shor's algorithm is to glean information about the period P of f from the just created stochastic source $\mathcal{S}$. The stochastic source was created exactly for that reason.

6. Peter Shor's stochastic source $\mathcal{S}$

Before continuing to the final part of Shor's algorithm, we need to analyze the probability distribution $Prob(y)$ a little more carefully.

PROPOSITION 1. *Let q and r be the unique non-negative integers such that $Q = Pq + r$, where $0 \leq r < P$; and let $Q_0 = Pq$. Then*

$$Prob(y) = \begin{cases} \frac{r \sin^2\left(\frac{\pi P y}{Q} \cdot \left(\frac{Q_0}{P}+1\right)\right) + (P-r)\sin^2\left(\frac{\pi P y}{Q} \cdot \frac{Q_0}{P}\right)}{Q^2 \sin^2\left(\frac{\pi P y}{Q}\right)} & \text{if } Py \neq 0 \bmod Q \\ \frac{r(Q_0+P)^2 + (P-r)Q_0^2}{Q^2 P^2} & \text{if } Py = 0 \bmod Q \end{cases}$$

PROOF. We begin by deriving a more usable expression for $|\Upsilon(y)\rangle$.

$$\begin{aligned}
|\Upsilon(y)\rangle &= \sum_{x=0}^{Q-1} \omega^{xy} |f(x)\rangle = \sum_{x=0}^{Q_0-1} \omega^{xy} |f(x)\rangle + \sum_{x=Q_0}^{Q-1} \omega^{xy} |f(x)\rangle \\
&= \sum_{x_0=0}^{P-1} \sum_{x_1=0}^{\frac{Q_0}{P}-1} \omega^{(Px_1+x_0)y} |f(Px_1+x_0)\rangle + \sum_{x_0=0}^{r-1} \omega^{[P(\frac{Q_0}{P})+x_0]y} |f(Px_1+x_0)\rangle \\
&= \sum_{x_0=0}^{P-1} \omega^{x_0 y} \cdot \left(\sum_{x_1=0}^{\frac{Q_0}{P}-1} \omega^{Pyx_1} \right) |f(x_0)\rangle + \sum_{x_0=0}^{r-1} \omega^{x_0 y} \cdot \omega^{Py(\frac{Q_0}{P})} |f(x_0)\rangle \\
&= \sum_{x_0=0}^{r-1} \omega^{x_0 y} \cdot \left(\sum_{x_1=0}^{\frac{Q_0}{P}} \omega^{Pyx_1} \right) |f(x_0)\rangle + \sum_{x_0=r}^{P-1} \omega^{x_0 y} \cdot \left(\sum_{x_1=0}^{\frac{Q_0}{P}-1} \omega^{Pyx_1} \right) |f(x_0)\rangle
\end{aligned}$$

where we have used the fact that f is periodic of period P.

Since f is one-to-one when restricted to its period $0, 1, 2, \ldots, P-1$, all the kets
$$|f(0)\rangle, \; |f(1)\rangle, \; |f(2)\rangle, \; \ldots, \; |f(P-1)\rangle,$$
are mutually orthogonal. Hence,

$$\langle \Upsilon(y) \mid \Upsilon(y) \rangle = r \left| \sum_{x_1=0}^{\frac{Q_0}{P}} \omega^{Pyx_1} \right|^2 + (P-r) \left| \sum_{x_1=0}^{\frac{Q_0}{P}-1} \omega^{Pyx_1} \right|^2 .$$

If $Py = 0 \bmod Q$, then since ω is a Q-th root of unity, we have

$$\langle \Upsilon(y) \mid \Upsilon(y) \rangle = r \left(\frac{Q_0}{P} + 1 \right)^2 + (P-r) \left(\frac{Q_0}{P} \right)^2 .$$

On the other hand, if $Py \neq 0 \bmod Q$, then we can sum the geometric series to obtain

$$\langle \Upsilon(y) \mid \Upsilon(y) \rangle = r \left| \frac{\omega^{Py \cdot (\frac{Q_0}{P}+1)} - 1}{\omega^{Py} - 1} \right|^2 + (P-r)) \left| \frac{\omega^{Py \cdot (\frac{Q_0}{P})} - 1}{\omega^{Py} - 1} \right|^2$$

$$= r \left| \frac{e^{\frac{2\pi i}{Q} \cdot Py \cdot (\frac{Q_0}{P}+1)} - 1}{e^{\frac{2\pi i}{Q} \cdot Py} - 1} \right|^2 + (P-r)) \left| \frac{e^{\frac{2\pi i}{Q} \cdot Py \cdot (\frac{Q_0}{P})} - 1}{e^{\frac{2\pi i}{Q} \cdot Py} - 1} \right|^2$$

where we have used the fact that ω is the primitive Q-th root of unity given by
$$\omega = e^{2\pi i/Q} .$$

The remaining part of the proposition is a consequence of the trigonometric identity
$$\left| e^{i\theta} - 1 \right|^2 = 4 \sin^2 \left(\frac{\theta}{2} \right) .$$

As a corollary, we have

COROLLARY 1. *If P is an exact divisor of Q, then*
$$Prob(y) = \begin{cases} 0 & if \quad Py \neq 0 \bmod Q \\ \frac{1}{P} & if \quad Py = 0 \bmod Q \end{cases}$$

7. A momentary digression: Continued fractions

We digress for a moment to review the theory of continued fractions. (For a more in-depth explanation of the theory of continued fractions, please refer to [5] and [12].)

Every positive rational number ξ can be written as an expression in the form
$$\xi = a_0 + \cfrac{1}{a_1 + \cfrac{1}{a_2 + \cfrac{1}{a_3 + \cfrac{1}{\cdots + \cfrac{1}{a_N}}}}},$$
where a_0 is a non-negative integer, and where $a_1, \ldots, a_N$ are positive integers. Such an expression is called a (finite, simple) **continued fraction**, and is uniquely determined by ξ provided we impose the condition $a_N > 1$. For typographical simplicity, we denote the above continued fraction by
$$[a_0, a_1, \ldots, a_N] \ .$$

The continued fraction expansion of ξ can be computed with the following recurrence relation, which always terminates if ξ is rational:

$$\begin{cases} a_0 = \lfloor \xi \rfloor \\ \xi_0 = \xi - a_0 \end{cases}, \text{ and if } \xi_n \neq 0, \text{ then } \begin{cases} a_{n+1} = \lfloor 1/\xi_n \rfloor \\ \xi_{n+1} = \frac{1}{\xi_n} - a_{n+1} \end{cases}$$

The n-th **convergent** ($0 \leq n \leq N$) of the above continued fraction is defined as the rational number ξ_n given by
$$\xi_n = [a_0, a_1, \ldots, a_n] \ .$$
Each convergent ξ_n can be written in the form, $\xi_n = \frac{p_n}{q_n}$, where p_n and q_n are relatively prime integers ($\gcd(p_n, q_n) = 1$). The integers p_n and q_n are determined by the recurrence relation

$$p_0 = a_0, \quad p_1 = a_1 a_0 + 1, \quad p_n = a_n p_{n-1} + p_{n-2},$$
$$q_0 = 1, \quad q_1 = a_1, \quad q_n = a_n q_{n-1} + q_{n-2} \ .$$

8. Preparation for the final part of Shor's algorithm

DEFINITION 2. [5]*For each integer a, let $\{a\}_Q$ denote the **residue** of a modulo Q **of smallest magnitude**. In other words, $\{a\}_Q$ is the unique integer such that*

$$\begin{cases} a = \{a\}_Q \bmod Q \\ -Q/2 < \{a\}_Q \leq Q/2 \end{cases}.$$

PROPOSITION 2. *Let y be an integer lying in S_Q. Then*

$$Prob(y) \geq \begin{cases} \frac{4}{\pi^2} \cdot \frac{1}{P} \cdot \left(1 - \frac{1}{N}\right)^2 & \text{if } 0 < \left|\{Py\}_Q\right| \leq \frac{P}{2} \cdot \left(1 - \frac{1}{N}\right) \\ \frac{1}{P} \cdot \left(1 - \frac{1}{N}\right)^2 & \text{if } \{Py\}_Q = 0 \end{cases}$$

PROOF. We begin by noting that

$$\left|\frac{\pi\{Py\}_Q}{Q} \cdot \left(\frac{Q_0}{P} + 1\right)\right| \leq \frac{\pi}{Q} \cdot \frac{P}{2} \cdot \left(1 - \frac{1}{N}\right) \cdot \left(\frac{Q_0 + P}{P}\right) \leq \frac{\pi}{2} \cdot \left(1 - \frac{1}{N}\right) \cdot \left(\frac{Q+P}{Q}\right)$$

$$\leq \frac{\pi}{2} \cdot \left(1 - \frac{1}{N}\right) \cdot \left(1 + \frac{P}{Q}\right) \leq \frac{\pi}{2} \cdot \left(1 - \frac{1}{N}\right) \cdot \left(1 + \frac{N}{N^2}\right) < \frac{\pi}{2},$$

where we have made use of the inequalities

$$N^2 \leq Q < 2N^2 \quad \text{and} \quad 0 < P \leq N .$$

It immediately follows that

$$\left|\frac{\pi\{Py\}_Q}{Q} \cdot \frac{Q_0}{P}\right| < \frac{\pi}{2} .$$

As a result, we can legitimately use the inequality

$$\frac{4}{\pi^2}\theta^2 \leq \sin^2\theta \leq \theta^2, \text{ for } |\theta| < \frac{\pi}{2}$$

to simplify the expression for $Prob(y)$.

Thus,

$$Prob(y) = \frac{r\sin^2\left(\frac{\pi\{Py\}_Q}{Q} \cdot \left(\frac{Q_0}{P}+1\right)\right) + (P-r)\sin^2\left(\frac{\pi\{Py\}_Q}{Q} \cdot \frac{Q_0}{P}\right)}{Q^2 \sin^2\left(\frac{\pi Py}{Q}\right)}$$

$$\geq \frac{r \cdot \frac{4}{\pi^2} \cdot \left(\frac{\pi\{Py\}_Q}{Q} \cdot \left(\frac{Q_0}{P}+1\right)\right)^2 + (P-r) \cdot \frac{4}{\pi^2} \cdot \left(\frac{\pi\{Py\}_Q}{Q} \cdot \frac{Q_0}{P}\right)^2}{Q^2 \left(\frac{\pi\{Py\}_Q}{Q}\right)^2}$$

$$\geq \frac{4}{\pi^2} \cdot \frac{P \cdot \left(\frac{Q_0}{P}\right)^2}{Q^2} = \frac{4}{\pi^2} \cdot \frac{1}{P} \cdot \left(\frac{Q-r}{Q}\right)^2$$

$$= \frac{4}{\pi^2} \cdot \frac{1}{P} \cdot \left(1 - \frac{r}{Q}\right)^2 \geq \frac{4}{\pi^2} \cdot \frac{1}{P} \cdot \left(1 - \frac{1}{N}\right)^2$$

---

[5]$\{a\}_Q = a - Q \cdot \text{round}\left(\frac{a}{Q}\right) = a - Q \cdot \left\lfloor \frac{a}{Q} + \frac{1}{2} \right\rfloor.$

The remaining case, $\{Py\}_Q = 0$ is left to the reader. □

LEMMA 1. Let
$$Y = \left\{ y \in S_Q \mid \left|\{Py\}_Q\right| \leq \frac{P}{2} \right\} \quad \text{and} \quad S_P = \{d \in S_Q \mid 0 \leq d < P\} \ .$$
Then the map
$$\begin{aligned} Y &\longrightarrow S_P \\ y &\longmapsto d = d(y) = round\left(\frac{P}{Q} \cdot y\right) \end{aligned}$$
is a bijection with inverse
$$y = y(d) = round\left(\frac{Q}{P} \cdot d\right) \ .$$
Hence, Y and S_P are in one-to-one correspondence. Moreover,
$$\{Py\}_Q = P \cdot y - Q \cdot d(y) \ .$$

REMARK 3. Moreover, the following two sets of rationals are in one-to-one correspondence
$$\left\{ \frac{y}{Q} \mid y \in Y \right\} \longleftrightarrow \left\{ \frac{d}{P} \mid 0 \leq d < P \right\}$$

As a result of the measurement performed in STEP 2.4, we have in our possession an integer $y \in Y$. We now show how y can be use to determine the unknown period P.

We now need the following theorem[6] from the theory of continued fractions:

THEOREM 1. Let ξ be a real number, and let a and b be integers with $b > 0$. If
$$\left|\xi - \frac{a}{b}\right| \leq \frac{1}{2b^2} \ ,$$
then the rational number a/b is a convergent of the continued fraction expansion of ξ.

As a corollary, we have:

COROLLARY 2. If $\left|\{Py\}_Q\right| \leq \frac{P}{2}$, then the rational number $\frac{d(y)}{P}$ is a convergent of the continued fraction expansion of $\frac{y}{Q}$.

PROOF. Since
$$Py - Qd(y) = \{Py\}_Q \ ,$$
we know that
$$|Py - Qd(y)| \leq \frac{P}{2},$$
which can be rewritten as
$$\left|\frac{y}{Q} - \frac{d(y)}{P}\right| \leq \frac{1}{2Q} \ .$$

---

[6]See [5, Theorem 184, Section 10.15].

But, since $Q \geq N^2$, it follows that
$$\left| \frac{y}{Q} - \frac{d(y)}{P} \right| \leq \frac{1}{2N^2} .$$
Finally, since $P \leq N$ (and hence $\frac{1}{2N^2} \leq \frac{1}{2P^2}$), the above theorem can be applied. Thus, $\frac{d(y)}{P}$ is a convergent of the continued fraction expansion of $\xi = \frac{y}{Q}$. □

Since $\frac{d(y)}{P}$ is a convergent of the continued fraction expansion of $\frac{y}{Q}$, it follows that, for some n,
$$\frac{d(y)}{P} = \frac{p_n}{q_n} ,$$
where p_n and q_n are relatively prime positive integers given by a recurrence relation found in the previous subsection. So it would seem that we have found a way of deducing the period P from the output y of STEP 2.4, and so we are done.

Not quite!

We can determine P from the measured y produced by STEP 2.4, only if
$$\begin{cases} p_n = d(y) \\ q_n = P \end{cases} ,$$
which is true only when $d(y)$ and P are relatively prime.

So what is the probability that the $y \in Y$ produced by STEP 2.4 satisfies the additional condition that
$$\gcd(P, d(y)) = 1 \; ?$$

PROPOSITION 3. *The probability that the random y produced by STEP 2.4 is such that $d(y)$ and P are relatively prime is bounded below by the following expression*
$$\text{Prob}\{y \in Y \mid \gcd(d(y), P) = 1\} \geq \frac{4}{\pi^2} \cdot \frac{\phi(P)}{P} \cdot \left(1 - \frac{1}{N}\right)^2 ,$$
where $\phi(P)$ denotes Euler's totient function, i.e., $\phi(P)$ is the number of positive integers less than P which are relatively prime to P.

The following theorem can be found in [5, Theorem 328, Section 18.4]:

THEOREM 2.
$$\liminf \frac{\phi(N)}{N/\ln \ln N} = e^{-\gamma},$$
where γ denotes Euler's constant $\gamma = 0.57721566490153286061\ldots$, and where $e^{-\gamma} = 0.5614594836\ldots$.

As a corollary, we have:

COROLLARY 3.

$$Prob\{y \in Y \mid \gcd(d(y), P) = 1\} \geq \frac{4}{\pi^2 \ln 2} \cdot \frac{e^{-\gamma} - \epsilon(P)}{\lg \lg N} \cdot \left(1 - \frac{1}{N}\right)^2,$$

where $\epsilon(P)$ is a monotone decreasing sequence converging to zero. In terms of asymptotic notation,

$$Prob\{y \in Y \mid \gcd(d(y), P) = 1\} = \Omega\left(\frac{1}{\lg \lg N}\right).$$

Thus, if STEP 2.4 is repeated $O(\lg \lg N)$ times, then the probability of success is $\Omega(1)$.

PROOF. From the above theorem, we know that

$$\frac{\phi(P)}{P/\ln \ln P} \geq e^{-\gamma} - \epsilon(P).$$

where $\epsilon(P)$ is a monotone decreasing sequence of positive reals converging to zero. Thus,

$$\frac{\phi(P)}{P} \geq \frac{e^{-\gamma} - \epsilon(P)}{\ln \ln P} \geq \frac{e^{-\gamma} - \epsilon(P)}{\ln \ln N} = \frac{e^{-\gamma} - \epsilon(P)}{\ln \ln 2 + \ln \lg N} \geq \frac{e^{-\gamma} - \epsilon(P)}{\ln 2} \cdot \frac{1}{\lg \lg N}$$

□

REMARK 4. $\Omega(\frac{1}{\lg \lg N})$ denotes an asymptotic lower bound. Readers not familiar with the big-oh $O(*)$ and big-omega $\Omega(*)$ notation should refer to [2, Chapter 2] or [1, Chapter 2].

REMARK 5. For the curious reader, lower bounds $LB(P)$ of $e^{-\gamma} - \epsilon(P)$ for $3 \leq P \leq 841$ are given in the following table:

P	LB(P)
3	0.062
4	0.163
5	0.194
7	0.303
13	0.326
31	0.375
61	0.383
211	0.411
421	0.425
631	0.435
841	0.468

Thus, if one wants a reasonable bound on the $Prob\{y \in Y \mid \gcd(d(y), P) = 1\}$ before continuing with Shor's algorithm, it would pay to first use a classical algorithm to verify that the period P of the randomly chosen integer m is not too small.

9. The final part of Shor's algorithm

We are now prepared to give the last step in Shor's algorithm. This step can be performed on a classical computer. (We remind the reader that ξ_n denotes the n-th convergent of $\frac{y}{Q}$.)

Step 2.5 Compute the period P from the integer y produced by STEP 2.4.

- LOOP FOR EACH n FROM $n = 1$ UNTIL $\xi_n = 0$.

  - Use the recurrence relations given in subsection 13.7, to compute the p_n and q_n of the n-th convergent $\xi_n = \frac{p_n}{q_n}$ of $\frac{y}{Q}$.

  - Test to see if $q_n = P$ by computing[7]
    $$m^{q_n} = \prod_i \left(m^{2^i}\right)^{q_{n,i}} \bmod N \;,$$
    where $q_n = \sum_i q_{n,i} 2^i$ is the binary expansion of q_n.
    If $m^{q_n} = 1 \bmod N$, then exit with the answer $P = q_n$, and proceed to **Step 3**. If not, then continue to the loop.

- END OF LOOP

- If you happen to reach this point, you are a very unlucky quantum computer scientist. You must start over by returning to STEP 2.0. But don't give up hope! The probability that the integer y produced by STEP 2.4 will lead to a successful completion of Step 2.5 is bounded below by
  $$\frac{4}{\pi^2 \ln 2} \cdot \frac{e^{-\gamma} - \epsilon(P)}{\lg \lg N} \cdot \left(1 - \frac{1}{N}\right)^2 > \frac{0.232}{\lg \lg N} \cdot \left(1 - \frac{1}{N}\right)^2 ,$$
  provided the period P is greater than 3. [γ denotes Euler's constant.]

---

[7]The indicated algorithm for computing $m^{q_n} \bmod N$ requires $O(\lg q_n)$ arithmetic operations.

10. An example of Shor's algorithm

Let us now show how $N = 91 \ (= 7 \cdot 13)$ can be factored using Shor's algorithm.

We choose $Q = 2^{14} = 16384$ so that $N^2 \leq Q < 2N^2$.

Step 1 Choose a random positive integer m, say $m = 3$. Since $\gcd(91, 3) = 1$, we proceed to STEP 2 to find the period of the function f given by

$$f(a) = 3^a \bmod 91$$

REMARK 6. *Unknown to us, f has period $P = 6$. For,*

a	0	1	2	3	4	5	6	7	⋯
$f(a)$	1	3	9	27	81	61	1	3	⋯
∴ *Unknown period $P = 6$*									

STEP 2.0 Initialize registers 1 and 2. Thus, the state of the two registers becomes:

$$|\psi_0\rangle = |0\rangle |1\rangle$$

STEP 2.1 Apply the Q-point Fourier transform $\mathcal{F}$ to register #1, where

$$\mathcal{F}|k\rangle = \frac{1}{\sqrt{16384}} \sum_{x=0}^{16383} \omega^{0 \cdot x} |x\rangle ,$$

and where ω is a primitive Q-th root of unity, e.g., $\omega = e^{\frac{2\pi i}{16384}}$. Thus the state of the two registers becomes:

$$|\psi_1\rangle = \frac{1}{\sqrt{16384}} \sum_{x=0}^{16383} |x\rangle |1\rangle$$

STEP 2.2 Apply the unitary transformation U_f to registers #1 and #2, where

$$U_f |x\rangle |\ell\rangle = |x\rangle | f(x) - \ell \bmod 91\rangle .$$

(Please note that $U_f^2 = I$.) Thus, the state of the two registers becomes:

$$|\psi_2\rangle = \tfrac{1}{\sqrt{16384}} \sum_{x=0}^{16383} |x\rangle |3^x \bmod 91\rangle$$

$$= \tfrac{1}{\sqrt{16384}}(\ |0\rangle|1\rangle + |1\rangle|3\rangle + |2\rangle|9\rangle + |3\rangle|27\rangle + |4\rangle|81\rangle + |5\rangle|61\rangle$$

$$+ |6\rangle|1\rangle + |7\rangle|3\rangle + |8\rangle|9\rangle + |9\rangle|27\rangle + |10\rangle|81\rangle + |11\rangle|61\rangle$$

$$+ |12\rangle|1\rangle + |13\rangle|3\rangle + |14\rangle|9\rangle + |15\rangle|27\rangle + |16\rangle|81\rangle + |17\rangle|61\rangle$$

$$+ \ldots$$

$$+ |16380\rangle|1\rangle + |16381\rangle|3\rangle + |16382\rangle|9\rangle + |16383\rangle|27\rangle \ )$$

REMARK 7. *The state of the two registers is now more than a superposition of states. We have in the above step quantum entangled the two registers.*

STEP 2.3 Apply the Q-point $\mathcal{F}$ again to register #1. Thus, the state of the system becomes:

$$|\psi_3\rangle = \tfrac{1}{\sqrt{16384}} \sum_{x=0}^{16383} \tfrac{1}{\sqrt{16384}} \sum_{y=0}^{16383} \omega^{xy} |y\rangle |3^x \bmod 91\rangle$$

$$= \tfrac{1}{16384} \sum_{x=0}^{16383} |y\rangle \sum_{x=0}^{16383} \omega^{xy} |3^x \bmod 91\rangle$$

$$= \tfrac{1}{16384} \sum_{x=0}^{16383} |y\rangle |\Upsilon(y)\rangle \ ,$$

where

$$|\Upsilon(y)\rangle = \sum_{x=0}^{16383} \omega^{xy} |3^x \bmod 91\rangle$$

Thus,

$$|\Upsilon(y)\rangle = \quad |1\rangle + \omega^y |3\rangle + \omega^{2y}|9\rangle + \omega^{3y}|27\rangle + \omega^{4y}|81\rangle + \omega^{5y}|61\rangle$$

$$+ \omega^{6y}|1\rangle + \omega^{7y}|3\rangle + \omega^{8y}|9\rangle + \omega^{9y}|27\rangle + \omega^{10y}|81\rangle + \omega^{11y}|61\rangle$$

$$+ \omega^{12y}|1\rangle + \omega^{13y}|3\rangle + \omega^{14y}|9\rangle + \omega^{15y}|27\rangle + \omega^{16y}|81\rangle + \omega^{17y}|61\rangle$$

$$+ \ldots$$

$$+ \omega^{16380y}|1\rangle + \omega^{16381y}|3\rangle + \omega^{16382y}|9\rangle + \omega^{16383y}|27\rangle$$

STEP 2.4 Measure REG1. The result of our measurement just happens to turn out to be
$$y = 13453$$

Unknown to us, the probability of obtaining this particular y is:
$$0.3189335551 \times 10^{-6} .$$

Moreover, unknown to us, we're lucky! The corresponding d is relatively prime to P, i.e.,
$$d = d(y) = round(\frac{P}{Q} \cdot y) = 5$$

However, we do know that the probability of $d(y)$ being relatively prime to P is greater than
$$\frac{0.232}{\lg \lg N} \cdot \left(1 - \frac{1}{N}\right)^2 \approx 8.4\% \text{ (provided } P > 3) .$$

We also know that
$$\frac{d(y)}{P}$$
is a convergent of the continued fraction expansion of
$$\xi = \frac{y}{Q} = \frac{13453}{16384}$$

So with a reasonable amount of confidence, we proceed to **Step 2.5**.

Step 2.5 Using the recurrence relations found in subsection 13.7 of this paper, we successively compute (beginning with $n = 0$) the a_n's and q_n's for the continued fraction expansion of
$$\xi = \frac{y}{Q} = \frac{13453}{16384} .$$

For each non-trivial n in succession, we check to see if
$$3^{q_n} = 1 \bmod 91.$$

If this is the case, then we know $q_n = P$, and we immediately exit from **Step 2.5** and proceed to **Step 3**.

- In this example, $n = 0$ and $n = 1$ are trivial cases.

- For $n = 2$, $a_2 = 4$ and $q_2 = 5$. We test q_2 by computing
$$3^{q_2} = 3^5 = \left(3^{2^0}\right)^1 \cdot \left(3^{2^1}\right)^0 \cdot \left(3^{2^0}\right)^1 = 61 \neq 1 \bmod 91 .$$
Hence, $q_2 \neq P$.

- We proceed to $n = 3$, and compute

$$a_3 = 1 \text{ and } q_3 = 6.$$

We then test q_3 by computing

$$3^{q_3} = 3^6 = \left(3^{2^0}\right)^0 \cdot \left(3^{2^1}\right)^1 \cdot \left(3^{2^2}\right)^1 = 1 \bmod 91 \ .$$

Hence, $q_3 = P$. Since we now know the period P, there is no need to continue to compute the remaining a_n's and q_n's. We proceed immediately to **Step 3**.

To satisfy the reader's curiosity we have listed in the table below all the values of a_n, p_n, and q_n for $n = 0, 1, \ldots, 14$. But it should be mentioned again that we need only to compute a_n and q_n for $n = 0, 1, 2, 3$, as indicated above.

n	0	1	2	3	4	5	6	7	8	9	10	11	12	13	14
a_n	0	1	4	1	1	2	3	1	1	3	1	1	1	1	3
p_n	0	1	4	5	9	23	78	101	179	638	817	1455	2272	3727	13453
q_n	1	1	5	6	11	28	95	123	218	777	995	1772	2767	4539	16384

Step 3. Since $P = 6$ is even, we proceed to **Step 4**.

Step 4. Since

$$3^{P/2} = 3^3 = 27 \neq -1 \bmod 91,$$

we go to **Step 5**.

Step 5. With the Euclidean algorithm, we compute

$$\gcd\left(3^{P/2} - 1, 91\right) = \gcd\left(3^3 - 1, 91\right) = \gcd(26, 91) = 13 \ .$$

We have succeeded in finding a non-trivial factor of $N = 91$, namely 13. We exit Shor's algorithm, and proceed to celebrate!

We leave the reader with the following instructive exercise:

EXERCISE 1 (Exercise for the reader). *Go through the above example assuming that **Step. 1** produces $m = 5$ instead of $m = 3$.*

References

[1] Brassard, Gilles, and Paul Bratley, "**Algorithmics: Theory and Practice,**" Printice-Hall, (1988).
[2] Cormen, Thomas H., Charles E. Leiserson, and Ronald L. Rivest, "**Introduction to Algorithms,**" McGraw-Hill, (1990).
[3] Cox, David, John Little, and Donal O'Shea, "Ideals, Varieties, and Algorithms," (second edition), Springer-Verlag, (1996).
[4] Ekert, Artur K.and Richard Jozsa, **Quantum computation and Shor's factoring algorithm**, Rev. Mod. Phys., 68,(1996), pp 733-753.
[5] Hardy, G.H., and E.M. Wright, "**An Introduction to the Theory of Numbers,**" Oxford Press, (1965).
[6] Hoyer, Peter, **Efficient quantum transforms**, (1997), http://xxx.lanl.gov/abs/quant-ph/9702028.

[7] Jozsa, Richard, **Quantum algorithms and the Fourier transform**, (1997), http://xxx.lanl.gov/abs/quant-ph/9707033.

[8] Jozsa, Richard, Proc. Roy. Soc. London Soc., Ser. A, 454, (1998), 323 - 337.

[9] Kitaev, A., **Quantum measurement and the abelian stabiliser problem**, (1995), http://xxx.lanl.gov/abs/quant-ph/9511026.

[10] Lenstra, A.K., and H.W. Lenstra, Jr., eds., "**The Development of the Number Field Sieve**," Lecture Notes in Mathematics, Vol. 1554, Springer-Velag, (1993).

[11] Lenstra, A.K., H.W. Lenstra, Jr., M.S. Manasse, and J.M. Pollard, **The number field sieve**. Proc. 22nd Annual ACM Symposium on Theory of ComputingACM, New York, (1990), pp 564 - 572. (See exanded version in Lenstra & Lenstra, (1993), pp 11 - 42.)

[12] LeVeque, William Judson, "**Topics in Number Theory: Volume I**," Addison-Wesley, (1958).

[13] Lomonaco, Samuel J., Jr., **A Rosetta Stone for quantum mechanics with an introduction to quantum computation,** this AMS Proceedings of Symposia in Applied Mathematics (PSAPM). (http://xxx.lanl.gov/abs/quant-ph/0007045).

[14] Miller, G. L., **Riemann's hypothesis and tests for primality**, J. Comput. System Sci., 13, (1976), pp 300 - 317.

[15] Shor, Peter W., **Polynomial time algorithms for prime factorization and discrete logarithms on a quantum computer**, SIAM J. on Computing, 26(5) (1997), pp 1484 - 1509. (1995), (http://xxx.lanl.gov/abs/quant-ph/9508027)

[16] Shor, Peter W., **Introduction to quantum algorithms**, this AMS Proceedings of Symposia in Applied Mathematics (PSAPM). (http://xxx.lanl.gov/abs/quant-ph/0005003)

[17] Stinson, Douglas R., "**Cryptography: Theory and Practice**," CRC Press, Boca Raton, (1995).

UNIVERSITY OF MARYLAND BALTIMORE COUNTY, BALTIMORE, MD 21250
E-mail address: Lomonaco@UMBC.EDU
URL: http://www.csee.umbc.edu/~lomonaco

Grover's Quantum Search Algorithm

Samuel J. Lomonaco, Jr.

ABSTRACT. This paper is a written version of a one hour lecture given on Lov Grover's quantum database search algorithm. It is based on [5], [6], and [11].

CONTENTS

1. Problem definition
2. Two examples
3. The quantum mechanical perspective
4. Properties of the inversion $I_{|\psi\rangle}$
5. The method in Lov's "madness"
6. Summary of Grover's algorithm
7. An example of Grover's algorithm

References

1. Problem definition

We consider the problem of searching an unstructured database of $N = 2^n$ records for exactly one record which has been specifically marked. This can be rephrased in mathematical terms as an oracle problem as follows:

2000 *Mathematics Subject Classification.* Primary 81P68; Secondary 81-01.

Key words and phrases. Grover's algorithm, database search, quantum computation, quantum algorithms.

This work was partially supported by Army Research Office (ARO) Grant #P-38804-PH-QC, by the National Institute of Standards and Technology (NIST), by the Defense Advanced Research Projects Agency (DARPA) and Air Force Materiel Command USAF under agreement number F30602-01-0522, and by the L-O-O-P Fund. The author gratefully acknowledges the hospitality of the University of Cambridge Isaac Newton Institute for Mathematical Sciences, Cambridge, England, where some of this work was completed. I would also like to thank the other AMS Short Course lecturers, Howard Brandt, Dan Gottesman, Lou Kauffman, Alexei Kitaev, Peter Shor, Umesh Vazirani and the many Short Course participants for their support.

© 2002 American Mathematical Society

Label the records of the database with the integers

$$0, 1, 2, \ldots, N-1,$$

and denote the label of the unknown marked record by x_0. We are given an **oracle** which computes the n bit binary function

$$f : \{0,1\}^n \longrightarrow \{0,1\}$$

defined by

$$f(x) = \begin{cases} 1 & \text{if } x = x_0 \quad (\text{"Yes"}) \\ 0 & \text{otherwise} \quad (\text{"No"}) \end{cases}$$

By calling the function f an **oracle** we mean that we have neither access to the internal workings of the function f, nor immediate access to all argument-function pairs $(x, f(x))$. The oracle f operates simply as a blackbox function, which we can query as many times as we like. But with each such query comes an associated computational cost.

Search Problem for an Unstructured Database. *Find the record labeled as x_0 with the minimum amount of computational work, i.e., with the minimum number of queries of the oracle f.*

From elementary probability theory, we know that if we examine k random[1] records, i.e., if we compute the oracle f for k randomly chosen records, then the probability of finding the record labeled as x_0 is k/N. Hence, *on a classical computer*, finding the unknown record label x_0 comes with the computational price tag of $O(N) = O(2^n)$ computational steps.

2. Two examples

The above search problem for an unstructured database is a fundamental and practical problem which appears in many different guises.

2.1. Example 1. Searching a city phone book.

For example, consider a city phone book containing N phone numbers [3]. Find the name associated with the phone number

$$\boxed{x_0 = (123) \quad 456 - 7890}$$

As mentioned earlier, the best classical algorithm for finding the associated name, say Jane Doe, would search through $N/2$ phone numbers (i.e., $O(N)$ phone numbers) on average before finding the name Jane Doe.

---

[1] We are assuming the uniform probability distribution.

2.2. Example 2. A plaintext/ciphertext attack on DES.

As another example, consider a **plaintext/ciphertext attack** by brute force key search on a message encrypted with the **Data Encryption Standard (DES)**, where the key K is a 56 bit number.

Given the plaintext/ciphertext pair

PlainText	At0500BlowUpTheEmbassyAt
CipherText	xjejpwvziderkqldievmsfkfdlqye

,

crack the entire cipher by encrypting the PlainText

At0500BlowUpTheEmbassyAt

with each of the $N = 2^{56}$ keys

$$K = 0, 1, 2, \ldots, 2^{56} - 1 ,$$

in turn, until the key K_0 is found that produces the CipherText

xjejpwvziderkqldievmsfkfdlqye

In other words, if

$$(P, C)$$

denotes the available plaintext/ciphertext pair, and if

$$K_0$$

denotes the key such that

$$DES(P, K_0) = C ,$$

then the **oracle** is

$$f(K) = \begin{cases} 1 & \text{if } K = K_0 \quad (\text{"Yes"}) \\ 0 & \text{otherwise} \quad (\text{"No"}) \end{cases}$$

3. The quantum mechanical perspective

As we have seen, any classical algorithm for searching an unstructured database of N records must take on average at least $O(N)$ computational steps. However, much to everyone's surprise, Lov Grover actually found a non-classical quantum algorithm for searching such a database even faster, with an average of $O\left(\sqrt{N}\right)$ steps, and with an average total computational cost of $O\left(\sqrt{N} \lg N\right)$. Although this is not exponentially faster, it is indeed a significant speedup.

Let $\mathcal{H}_2$ be a 2 dimensional Hilbert space with orthonormal basis

$$\{|0\rangle, |1\rangle\} ;$$

and let
$$\{|0\rangle, |1\rangle, \ldots, |N-1\rangle\}$$
denote the induced orthonormal basis of the Hilbert space
$$\mathcal{H} = \bigotimes_{0}^{n-1} \mathcal{H}_2 .$$

From the quantum mechanical perspective, the oracle function f is given as a blackbox unitary transformation U_f, i.e., by

$$\mathcal{H} \otimes \mathcal{H}_2 \xrightarrow{U_f} \mathcal{H} \otimes \mathcal{H}_2$$

$$|x\rangle \otimes |y\rangle \longmapsto |x\rangle \otimes |f(x) \oplus y\rangle$$

where '$\oplus$' denotes exclusive 'OR', i.e., addition modulo 2.[2]

Instead of U_f, we will use the computationally equivalent unitary transformation

$$I_{|x_0\rangle}(|x\rangle) = (-1)^{f(x)} |x\rangle = \begin{cases} -|x_0\rangle & \text{if } x = x_0 \\ |x\rangle & \text{otherwise} \end{cases}$$

That $I_{|x_0\rangle}$ is computationally equivalent to U_f follows from the easily verifiable fact that

$$U_f \left(|x\rangle \otimes \frac{|0\rangle - |1\rangle}{\sqrt{2}} \right) = \left(I_{|x_0\rangle}(|x\rangle) \right) \otimes \frac{|0\rangle - |1\rangle}{\sqrt{2}} ,$$

and also from the fact that U_f can be constructed from a controlled-$I_{|x_0\rangle}$ and two one qubit Hadamard transforms. (For details, please refer to [**12**], [**13**].)

The unitary transformation $I_{|x_0\rangle}$ is actually an **inversion** [**1**] in $\mathcal{H}$ about the hyperplane perpendicular to $|x_0\rangle$. This becomes evident when $I_{|x_0\rangle}$ is rewritten in the form
$$I_{|x_0\rangle} = I - 2|x_0\rangle\langle x_0| ,$$
where 'I' denotes the identity transformation. More generally, for any unit length ket $|\psi\rangle$, the unitary transformation
$$I_{|\psi\rangle} = I - 2|\psi\rangle\langle\psi|$$
is an inversion in $\mathcal{H}$ about the hyperplane orthogonal to $|\psi\rangle$.

---

[2] Please note that $U_f = (\nu \circ \iota)(f)$, as defined in sections 10.3 and 10.4 of [**14**].

4. Properties of the inversion $I_{|\psi\rangle}$

We digress for a moment to discuss the properties of the unitary transformation $I_{|\psi\rangle}$. To do so, we need the following definition.

DEFINITION 1. *Let $|\psi\rangle$ and $|\chi\rangle$ be two kets in $\mathcal{H}$ for which the bracket product $\langle \psi \mid \chi \rangle$ is a real number. We define*

$$\mathcal{S}_{\mathbb{C}} = Span_{\mathbb{C}}(|\psi\rangle, |\chi\rangle) = \{\alpha |\psi\rangle + \beta |\chi\rangle \in \mathcal{H} \mid \alpha, \beta \in \mathbb{C}\}$$

as the sub-Hilbert space of $\mathcal{H}$ spanned by $|\psi\rangle$ and $|\chi\rangle$. We associate with the Hilbert space $\mathcal{S}_{\mathbb{C}}$ a real inner product space lying in $\mathcal{S}_{\mathbb{C}}$ defined by

$$\mathcal{S}_{\mathbb{R}} = Span_{\mathbb{R}}(|\psi\rangle, |\chi\rangle) = \{a |\psi\rangle + b |\chi\rangle \in \mathcal{H} \mid a, b \in \mathbb{R}\} \ ,$$

where the inner product on $\mathcal{S}_{\mathbb{R}}$ is that induced by the bracket product on $\mathcal{H}$. If $|\psi\rangle$ and $|\chi\rangle$ are also linearly independent, then $\mathcal{S}_{\mathbb{R}}$ is a 2 dimensional real inner product space (i.e., the 2 dimensional Euclidean plane) lying inside of the complex 2 dimensional space $\mathcal{S}_{\mathbb{C}}$.

PROPOSITION 1. *Let $|\psi\rangle$ and $|\chi\rangle$ be two linearly independent unit length kets in $\mathcal{H}$ with real bracket product; and let $\mathcal{S}_{\mathbb{C}} = Span_{\mathbb{C}}(|\psi\rangle, |\chi\rangle)$ and $\mathcal{S}_{\mathbb{R}} = Span_{\mathbb{R}}(|\psi\rangle, |\chi\rangle)$. Then*

1) *Both $\mathcal{S}_{\mathbb{C}}$ and $\mathcal{S}_{\mathbb{R}}$ are invariant under the transformations $I_{|\psi\rangle}$, $I_{|\chi\rangle}$, and hence $I_{|\psi\rangle} \circ I_{|\chi\rangle}$, i.e.,*

$I_{|\psi\rangle}(\mathcal{S}_{\mathbb{C}}) = \mathcal{S}_{\mathbb{C}}$	and	$I_{|\psi\rangle}(\mathcal{S}_{\mathbb{R}}) = \mathcal{S}_{\mathbb{R}}$
$I_{|\chi\rangle}(\mathcal{S}_{\mathbb{C}}) = \mathcal{S}_{\mathbb{C}}$	and	$I_{|\chi\rangle}(\mathcal{S}_{\mathbb{R}}) = \mathcal{S}_{\mathbb{R}}$
$I_{|\psi\rangle}I_{|\chi\rangle}(\mathcal{S}_{\mathbb{C}}) = \mathcal{S}_{\mathbb{C}}$	and	$I_{|\psi\rangle}I_{|\chi\rangle}(\mathcal{S}_{\mathbb{R}}) = \mathcal{S}_{\mathbb{R}}$

2) *If $L_{|\psi^{\perp}\rangle}$ is the line in the plane $\mathcal{S}_{\mathbb{R}}$ which passes through the origin and which is perpendicular to $|\psi\rangle$, then $I_{|\psi\rangle}$ restricted to $\mathcal{S}_{\mathbb{R}}$ is a reflection in (i.e., a Möbius inversion [1] about) the line $L_{|\psi^{\perp}\rangle}$. A similar statement can be made in regard to $|\chi\rangle$.*
3) *If $|\psi^{\perp}\rangle$ is a unit length vector in $\mathcal{S}_{\mathbb{R}}$ perpendicular to $|\psi\rangle$, then*

$$-I_{|\psi\rangle} = I_{|\psi^{\perp}\rangle} \ .$$

(Hence, $\langle \psi^{\perp} \mid \chi \rangle$ is real.)

Finally we note that, since $I_{|\psi\rangle} = I - 2 |\psi\rangle \langle \psi|$, it follows that

PROPOSITION 2. *If $|\psi\rangle$ is a unit length ket in $\mathcal{H}$, and if U is a unitary transformation on $\mathcal{H}$, then*

$$U I_{|\psi\rangle} U^{-1} = I_{U|\psi\rangle} \ .$$

5. The method in Lov's "madness"

Let $H : \mathcal{H} \longrightarrow \mathcal{H}$ be the Hadamard transform, i.e.,

$$H = \bigotimes_{0}^{n-1} H^{(2)} ,$$

where

$$H^{(2)} = \frac{1}{\sqrt{2}} \begin{pmatrix} 1 & 1 \\ 1 & -1 \end{pmatrix}$$

with respect to the basis $|0\rangle, |1\rangle$.

We begin by using the Hadamard transform H to construct a state $|\psi_0\rangle$ which is an equal superposition of all the standard basis states $|0\rangle, |1\rangle, \ldots, |N-1\rangle$ (including the unknown state $|x_0\rangle$), i.e.,

$$|\psi_0\rangle = H|0\rangle = \frac{1}{\sqrt{N}} \sum_{k=0}^{N-1} |k\rangle .$$

Both $|\psi_0\rangle$ and the unknown state $|x_0\rangle$ lie in the Euclidean plane $\mathcal{S}_\mathbb{R} = Span_\mathbb{R}(|\psi_0\rangle, |x_0\rangle)$. Our strategy is to rotate within the plane $\mathcal{S}_\mathbb{R}$ the state $|\psi_0\rangle$ about the origin until it is as close as possible to $|x_0\rangle$. Then a measurement with respect to the standard basis of the state resulting from rotating $|\psi_0\rangle$, will produce $|x_0\rangle$ with high probability.

To achieve this objective, we use the oracle $I_{|x_0\rangle}$ to construct the unitary transformation

$$Q = -H I_{|0\rangle} H^{-1} I_{|x_0\rangle} ,$$

which by proposition 2 above, can be rewritten as

$$Q = -I_{|\psi_0\rangle} I_{|x_0\rangle} .$$

Let $|x_0^\perp\rangle$ and $|\psi_0^\perp\rangle$ denote unit length vectors in $\mathcal{S}_\mathbb{R}$ perpendicular to $|x_0\rangle$ and $|\psi_0\rangle$, respectively. There are two possible choices for each of $|x_0^\perp\rangle$ and $|\psi_0^\perp\rangle$ respectively. To remove this minor, but nonetheless annoying, ambiguity, we select $|x_0^\perp\rangle$ and $|\psi_0^\perp\rangle$ so that the orientation of the plane $\mathcal{S}_\mathbb{R}$ induced by the ordered spanning vectors $|\psi_0\rangle, |x_0\rangle$ is the same orientation as that induced by each of the ordered bases $|x_0^\perp\rangle, |x_0\rangle$ and $|\psi_0\rangle, |\psi_0^\perp\rangle$. (Please refer to Figure 2.)

REMARK 1. *The removal of the above ambiguities is really not essential. However, it does simplify the exposition given below.*

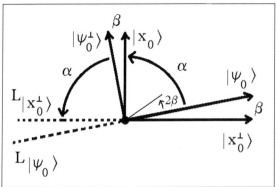

Figure 2. The linear transformation $Q|_{\mathcal{S}_\mathbb{R}}$ is a reflection in the line $L_{|x_0^\perp\rangle}$ followed by reflection in the line $L_{|\psi_0\rangle}$. By Theorem 1, this is the same as rotation by the angle 2β. Thus, $Q|_{\mathcal{S}_\mathbb{R}}$ rotates $|\psi_0\rangle$ by the angle 2β toward $|x_0\rangle$.

We proceed by noting that, by the above proposition 1, the plane $\mathcal{S}_\mathbb{R}$ lying in $\mathcal{H}$ is invariant under the linear transformation Q, and that, when Q is restricted to the plane $\mathcal{S}_\mathbb{R}$, it can be written as the composition of two inversions, i.e.,

$$Q|_{\mathcal{S}_\mathbb{R}} = I_{|\psi_0^\perp\rangle} I_{|x_0\rangle} \ .$$

In particular, $Q|_{\mathcal{S}_\mathbb{R}}$ is the composition of two inversions in $\mathcal{S}_\mathbb{R}$, the first in the line $L_{|x_0^\perp\rangle}$ in $\mathcal{S}_\mathbb{R}$ passing through the origin having $|x_0\rangle$ as normal, the second in the line $L_{|\psi_0\rangle}$ through the origin having $|\psi_0^\perp\rangle$ as normal.[3]

We can now apply the following theorem from plane geometry:

THEOREM 1. *If L_1 and L_2 are lines in the Euclidean plane $\mathbb{R}^2$ intersecting at a point O, and if β is the angle in the plane from L_1 to L_2, then the operation of reflection in L_1 followed by reflection in L_2 is just rotation by angle 2β about the point O.*

Let β denote the angle in $\mathcal{S}_\mathbb{R}$ from $L_{|x_0^\perp\rangle}$ to $L_{|\psi_0\rangle}$, which by plane geometry is the same as the angle from $|x_0^\perp\rangle$ to $|\psi_0\rangle$, which in turn is the same as the angle from $|x_0\rangle$ to $|\psi_0^\perp\rangle$. Then by the above theorem $Q|_{\mathcal{S}_\mathbb{R}} = I_{|\psi_0^\perp\rangle} I_{|x_0\rangle}$ is a rotation about the origin by the angle 2β.

The key idea in Grover's algorithm is to move $|\psi_0\rangle$ toward the unknown state $|x_0\rangle$ by successively applying the rotation Q to $|\psi_0\rangle$ to rotate it around to $|x_0\rangle$.

---

[3]The line $L_{|x_0^\perp\rangle}$ is the intersection of the plane $\mathcal{S}_\mathbb{R}$ with the hyperplane in $\mathcal{H}$ orthogonal to $|x_0\rangle$. A similar statement can be made in regard to $L_{|\psi_0\rangle}$.

This process is called **amplitude amplification**. Once this process is completed, the measurement of the resulting state (with respect to the standard basis) will, with high probability, yield the unknown state $|x_0\rangle$. This is the essence of Grover's algorithm.

But how many times K should we apply the rotation Q to $|\psi_0\rangle$? If we applied Q too many or too few times, we would over- or undershoot our target state $|x_0\rangle$.

We determine the integer K as follows:

Since
$$|\psi_0\rangle = \sin\beta \, |x_0\rangle + \cos\beta \, |x_0^\perp\rangle \ ,$$
the state resulting after k applications of Q is
$$|\psi_k\rangle = Q^k |\psi_0\rangle = \sin\left[(2k+1)\,\beta\right] |x_0\rangle + \cos\left[(2k+1)\,\beta\right] |x_0^\perp\rangle \ .$$
Thus, we seek to find the smallest positive integer $K = k$ such that
$$\sin\left[(2k+1)\,\beta\right]$$
is as close as possible to 1. In other words, we seek to find the smallest positive integer $K = k$ such that
$$(2k+1)\,\beta$$
is as close as possible to $\pi/2$. It follows that
$$K = k = round\left(\frac{\pi}{4\beta} - \frac{1}{2}\right) = \left\lfloor \frac{\pi}{4\beta} \right\rfloor \ ,$$
where "*round*" is the function that rounds to the nearest integer, and where "$\lfloor - \rfloor$" denotes the floor function.

We can determine the angle β by noting that the angle α from $|\psi_0\rangle$ to $|x_0\rangle$ is complementary to β, i.e.,
$$\alpha + \beta = \pi/2 \ ,$$
and hence,
$$\frac{1}{\sqrt{N}} = \langle x_0 \mid \psi_0 \rangle = \cos\alpha = \cos(\frac{\pi}{2} - \beta) = \sin\beta \ .$$
Thus, the angle β is given by
$$\beta = \sin^{-1}\left(\frac{1}{\sqrt{N}}\right) \approx \frac{1}{\sqrt{N}} \quad \text{(for large } N\text{)} \ ,$$
and hence,
$$K = k = \left\lfloor \frac{\pi}{4\sin^{-1}\left(\frac{1}{\sqrt{N}}\right)} \right\rfloor \approx \left\lfloor \frac{\pi}{4}\sqrt{N} \right\rfloor \quad \text{(for large } N\text{)}.$$

6. Summary of Grover's algorithm

In summary, we provide the following outline of Grover's algorithm:

Grover's Algorithm

STEP 0. (Initialization)

$$|\psi\rangle \longleftarrow H|0\rangle = \frac{1}{\sqrt{N}} \sum_{j=0}^{N-1} |j\rangle$$

$$k \longleftarrow 0$$

STEP 1. Loop until $k = \left\lfloor \frac{\pi}{4\sin^{-1}(1/\sqrt{N})} \right\rfloor \approx \left\lfloor \frac{\pi}{4}\sqrt{N} \right\rfloor$

$$|\psi\rangle \longleftarrow Q|\psi\rangle = -HI_{|0\rangle}HI_{|x_0\rangle}|\psi\rangle$$

$$k \longleftarrow k+1$$

STEP 2. Measure $|\psi\rangle$ with respect to the standard basis $|0\rangle, |1\rangle, \ldots, |N-1\rangle$ to obtain the unknown state $|x_0\rangle$ with probability $\geq 1 - \frac{1}{N}$.

We complete our summary with the following theorem:

THEOREM 2. *With a probability of error*

$$Prob_E \leq \frac{1}{N},$$

Grover's algorithm finds the unknown state $|x_0\rangle$ at a computational cost of

$$O\left(\sqrt{N}\lg N\right)$$

PROOF.

Part 1. The probability of error $Prob_E$ of finding the hidden state $|x_0\rangle$ is given by

$$Prob_E = \cos^2\left[(2K+1)\beta\right],$$

where

$$\begin{cases} \beta = \sin^{-1}\left(\frac{1}{\sqrt{N}}\right) \\ K = \left\lfloor \frac{\pi}{4\beta} \right\rfloor \end{cases},$$

and where "$\lfloor - \rfloor$" denotes the floor function. Hence,

$$\frac{\pi}{4\beta} - 1 \leq K \leq \frac{\pi}{4\beta} \implies \frac{\pi}{2} - \beta \leq (2K+1)\beta \leq \frac{\pi}{2} + \beta$$

$$\implies \sin\beta = \cos\left(\frac{\pi}{2} - \beta\right) \geq \cos\left[(2K+1)\beta\right] \geq \cos\left(\frac{\pi}{2} + \beta\right) = -\sin\beta$$

Thus,
$$Prob_E = \cos^2\left[(2K+1)\beta\right] \leq \sin^2\beta = \sin^2\left(\sin^{-1}\left(\frac{1}{\sqrt{N}}\right)\right) = \frac{1}{N}$$

Part 2. The computational cost of the Hadamard transform $H = \bigotimes_0^{n-1} H^{(2)}$ is $O(n) = O(\lg N)$ single qubit operations. The transformations $-I_{|0\rangle}$ and $I_{|x_0\rangle}$ each carry a computational cost of $O(1)$.

STEP 1 is the computationally dominant step. In STEP 1 there are $O\left(\sqrt{N}\right)$ iterations. In each iteration, the Hadamard transform is applied twice. The transformations $-I_{|0\rangle}$ and $I_{|x_0\rangle}$ are each applied once. Hence, each iteration comes with a computational cost of $O(\lg N)$, and so the total cost of STEP 1 is $O(\sqrt{N} \lg N)$.

□

7. An example of Grover's algorithm

As an example, we search a database consisting of $N = 2^n = 8$ records for an unknown record with the unknown label $x_0 = 5$. The calculations for this example were made with OpenQuacks [15], an open source publically available quantum simulator Maple package developed at UMBC.

We are given a blackbox computing device

$$\text{In} \rightarrow \boxed{I_{|?\rangle}} \rightarrow \text{Out}$$

that implements as an oracle the unitary transformation

$$I_{|x_0\rangle} = I_{|5\rangle} = \begin{pmatrix} 1 & 0 & 0 & 0 & 0 & 0 & 0 & 0 \\ 0 & 1 & 0 & 0 & 0 & 0 & 0 & 0 \\ 0 & 0 & 1 & 0 & 0 & 0 & 0 & 0 \\ 0 & 0 & 0 & 1 & 0 & 0 & 0 & 0 \\ 0 & 0 & 0 & 0 & -1 & 0 & 0 & 0 \\ 0 & 0 & 0 & 0 & 0 & 1 & 0 & 0 \\ 0 & 0 & 0 & 0 & 0 & 0 & 1 & 0 \\ 0 & 0 & 0 & 0 & 0 & 0 & 0 & 1 \end{pmatrix}$$

We cannot open up the blackbox $\rightarrow \boxed{I_{|?\rangle}} \rightarrow$ to see what is inside. So we do not know what $I_{|x_0\rangle}$ and x_0 are. The only way that we can glean some information about x_0 is to apply some chosen state $|\psi\rangle$ as input, and then make use of the resulting output.

Using of the blackbox → $\boxed{I_{|?\rangle}}$ → as a component device, we construct a computing device → $\boxed{-HI_{|0\rangle}HI_{|?\rangle}}$ → which implements the unitary operator

$$Q = -HI_{|0\rangle}HI_{|x_0\rangle} = \frac{1}{4}\begin{pmatrix} -3 & 1 & 1 & 1 & -1 & 1 & 1 & 1 \\ 1 & -3 & 1 & 1 & -1 & 1 & 1 & 1 \\ 1 & 1 & -3 & 1 & -1 & 1 & 1 & 1 \\ 1 & 1 & 1 & -3 & -1 & 1 & 1 & 1 \\ 1 & 1 & 1 & 1 & 3 & 1 & 1 & 1 \\ 1 & 1 & 1 & 1 & -1 & -3 & 1 & 1 \\ 1 & 1 & 1 & 1 & -1 & 1 & -3 & 1 \\ 1 & 1 & 1 & 1 & -1 & 1 & 1 & -3 \end{pmatrix}$$

We do not know what unitary transformation Q is implemented by the device → $\boxed{-HI_{|0\rangle}HI_{|?\rangle}}$ → because the blackbox → $\boxed{I_{|?\rangle}}$ → is one of its essential components.

$\boxed{\text{STEP 0.}}$ We begin by preparing the known state

$$\boxed{|\psi_0\rangle = H|0\rangle = \tfrac{1}{\sqrt{8}}(1,1,1,1,1,1,1,1)^{transpose}}$$

$\boxed{\text{STEP 1.}}$ We proceed to loop

$$K = \left\lfloor \frac{\pi}{4\sin^{-1}(1/\sqrt{8})} \right\rfloor = 2$$

times in STEP 1.

ITERATION 1. On the first iteration, we obtain the unknown state

$$\boxed{|\psi_1\rangle = Q|\psi_0\rangle = \tfrac{1}{4\sqrt{2}}(1,1,1,1,5,1,1,1)^{transpose}}$$

ITERATION 2. On the second iteration, we obtain the unknown state

$$\boxed{|\psi_2\rangle = Q|\psi_1\rangle = \tfrac{1}{8\sqrt{2}}(-1,-1,-1,-1,11,-1,-1,-1)^{transpose}}$$

and branch to STEP 2.

$\boxed{\text{STEP 2.}}$ We measure the unknown state $|\psi_2\rangle$ to obtain either

$$|5\rangle$$

with probability

$$Prob_{Success} = \sin^2((2K+1)\beta) = \frac{121}{128} = 0.9453$$

or some other state with probability
$$Prob_{Failure} = \cos^2\left((2K+1)\beta\right) = \frac{7}{128} = 0.0547$$
and then exit.

References

[1] Beardon, Alan F., "The Geometry of Discrete Groups," Springer-Verlag, (1983).
[2] Brassard, Gilles, and Paul Bratley, **"Algorithmics: Theory and Practice,"** Printice-Hall, (1988).
[3] Brassard, Gilles, **Searching a Quantum Phone Book**, Science, 275, (1997), pp 627-628.
[4] Cormen, Thomas H., Charles E. Leiserson, and Ronald L. Rivest, **"Introduction to Algorithms,"** McGraw-Hill, (1990).
[5] Grover, Lov K., **Quantum computer can search arbitrarily large databases by a single querry**, Phys. Rev. Letters (1997), pp 4709-4712.
[6] Grover, Lov K., **A framework for fast quantum mechanical algorithms**, http://xxx.lanl.gov/abs/quant-ph/9711043.
[7] Grover, L., Proc. 28th Annual ACM Symposium on the Theory of Computing, ACM Press, New Yorkm (1996), pp 212 - 219.
[8] Grover, L., Phys. Rev. Lett. 78, (1997), pp 325 - 328.
[9] Gruska, Jozef, **"Quantum Computing,"** McGraw-Hill, (1999).
[10] Hirvensalo, Mika, **"Quantum Computing,"** Springer-Verlag, (2001).
[11] Jozsa, Richard, **Searching in Grover's Algorithm**, http://xxx.lanl.gov/abs/quant-ph/9901021.
[12] Jozsa, Richard, Proc. Roy. Soc. London Soc., Ser. A, 454, (1998), 323 - 337.
[13] Kitaev, A., **Quantum measurement and the abelian stabiliser problem**, (1995), quant-ph preprint archive 9511026.
[14] Lomonaco, Samuel J., Jr., **A Rosetta Stone for quantum mechanics with an introduction to quantum computation**, in "Quantum Computation," edited by S.J. Lomonaco, Jr., this AMS Proceedings of Symposia in Applied Mathematics (PSAPM). (http://xxx.lanl.gov/abs/quant-ph/0007045)
[15] McCubbin, Christopher B., **OpenQuacks**, (masters thesis), http://userpages.umbc.edu/~cmccub1/quacs/quacs.html.
[16] Nielsen, Michael A., and Isaac L. Chuang, **"Quantum Computation and Quantum Information,"** Cambridge University Press, (2000).
[17] Shor, Peter W., **Introduction to quantum algorithms**, n "Quantum Computation," this AMS Proceedings of Symposia in Applied Mathematics (PSAPM). (http://xxx.lanl.gov/abs/quant-ph/0005003)
[18] Vazirani, Umesh, **Quantum complexity theory**, in "Quantum Computation," this AMS Proceedings of Symposia in Applied Mathematics (PSAPM).
[19] Zalka, Christof, **Grover's quantum searching algorithm is optimal**, (2001). (http://xxx.lanl.gov/abs/quant-ph/9711070).

University of Maryland Baltimore County, Baltimore, MD 21250
E-mail address: Lomonaco@UMBC.EDU
URL: http://www.csee.umbc.edu/~lomonaco

A Survey of Quantum Complexity Theory

Umesh V. Vazirani

ABSTRACT. This paper provides an introduction to quantum complexity theory. Two basic models of quantum computers, quantum Turing Machines and quantum circuits, are defined and shown to be polynomially equivalent. The quantum complexity class BQP of is defined, and its relationship to classical complexity classes is established. The basic techniques for proving limits on quantum computers, by showing a lower bound in the black box model are surveyed. Finally the quantum analog of the Cook-Levin theorem is sketched.

Contents

1. Introduction
2. Quantum Turing Machines and Quantum Circuits
3. Quantum Complexity Classes:
4. Recursive Fourier Sampling
5. Quantum Lower Bounds
6. Quantum NP

References

1. Introduction

The main goal of computational complexity theory is to classify computational problems according to the amount of computational resources — typically time or number of steps — required to solve them on a computer. That this quantity is well defined rests upon the modern form of the Church-Turing thesis, which asserts that any "reasonable" model of computation can be *efficiently* simulated on a probabilistic Turing Machine. (An efficient simulation is one whose running time is bounded by some polynomial in the running time of the simulated machine.)

2000 *Mathematics Subject Classification*. [2000]Primary 81P68; Secondary 81-01.

Key words and phrases. Quantum complexity theory, quantum Turing machines, lower bounds, NP, quantum computation.

This work was partially supported by DARPA Grant F30602-00-2-0601 and NSF Grant NSF Grant CCR-9800024.

This thesis may be informally summarized as follows: All physical implementations of computing devices can be simulated with polynomial factor overhead in running time by the probabilistic Turing Machine or the random access machine. Recently, this paradigm has been fundamentally challenged, since there are strong arguments showing that the Church-Turing thesis (in this modern form) does not hold at the level of quantum mechanics.

Early indications of this possibility occurred in the paper by Feynman [14] which pointed out that it was not clear how to simulate quantum mechanical systems of n particles (say n spins) on a computer, without paying an exponential penalty in simulation time. The first formal evidence that quantum computers violate the modified Church-Turing thesis came a decade later, with the result of Bernstein and Vazirani [9], which showed that relative to an oracle, quantum polynomial time properly contains probabilistic polynomial time. This built upon a previous algorithm due to Deutsch and Jozsa [13]. Simon [24] improved this bound by showing that relative to an oracle, quantum polynomial time is not contained in subexponential probabilistic time. Shor [25] followed up with seminal results in quantum algorithms, showing that the problems of prime factorization and discrete logarithms can both be solved in polynomial time on a quantum computer. The computational intractability of these problems for classical computers is the standard computational assumption underlying modern cryptography. Together, these results provide very strong evidence that quantum computers violate the modern Church-Turing thesis.

In view of these developments, one must explore a new complexity theory based on quantum mechanics. The formal models for quantum computers - quantum Turing Machines and quantum circuits - were introduced by Deutsch [11]. These two models were shown to be polynomially equivalent in terms of their computing power by Yao [31]. Creating a universal quantum Turing Machine (a programmable quantum Turing Machine) has proved to be a much more challenging task than the corresponding classical universal constructions. The first part of this paper introduces these models, and sketches the universal quantum Turing Machince construction from [9], and the polynomial equivalence of quantum Turing Machines and quantum circuits from [31]. It also describes universal families of quantum circuits [7].

The class of problems that can be solved in polynomial time (in the length of the input) has a special place in complexity theory — since it is regarded as the class of efficiently solvable problems. The class of problems that can be solved in polynomial time on a probabilistic Turing Machine is denoted by BPP. The corresponding class problems solvable in polynomial time on a quantum Turing Machine is denoted by BQP. The best upper bound known for this class is $BQP \subseteq P^{\#P} \subseteq PSPACE$ [9], i.e., every problem that can be solved in polynomial time on a quantum Turing Machine can also be solved using polynomial amount of memory space on a classical Turing Machine. Thus $P \subseteq BPP \subseteq BQP \subseteq P^{\#P} \subseteq PSPACE$. Since $P =? PSPACE$ is a major open question in computational complexity theory, this implies that any absolute results showing that quantum computers are more powerful than classical computers ($BQP \neq BPP$) will have to await a major breakthrough in complexity theory. Until then, we must be satisfied with evidence such as [9, 24, 25] that quantum computers violate the modern Church-Turing thesis.

Is $NP \subseteq BQP$? In view of the exponential speedups offered by quantum computers for certain computational problems, it is natural to ask whether quantum computers can solve NP-complete problems in polynomial time. An affirmative answer would have astounding consequences, since the NP-complete problems include several thousand of the most important computational problems, and are believed to be classically intractable. It was shown in [10] that relative to a random oracle, a quantum Turing Machine must take exponential time to solve NP-complete problems. This appears to rule out any efficient quantum algorithm for an NP-complete problem, barring a major breakthrough in computational complexity theory. In the same paper an exponential lower bound was established on the problem of inverting a random permutation (thus opening up the possibility of quantum one-way functions). Both results were proved using the hybrid argument. Two other techniques for establishing lower bounds have been introduced. The first is the method of polynomials [8], which was used to give a tight linear lower bound on the quantum complexity of the parity function in the black box model. In the same paper they showed that, in general, the quantum query complexity of any total function in the black box model is bounded by the sixth power of the deterministic query complexity. The second technique is the method of quantum adversaries [2]. This technique appears to be very general, and has been used to obtain tight bounds for a number of problems. In particular, it was used to prove a tight bound on the problem of inverting a random permutation.

Does $BQP \subseteq NP$? Since BQP includes the ability to randomize, the fair way of asking this question is whether BQP is contained in MA — the probabilistic generalization of NP. There are indications that the answer is negative, since the recursive Fourier sampling problem, which has an efficient quantum algorithm [9] is not in MA relative to an oracle. A major question that remains open is whether $BQP \in BPP^{NP}$, i.e., is the ability to do approximate counting sufficient to simulate BQP in polynomial time?

Perhaps the center piece of classical complexity theory is the Cook-Levin theorem, which states that $3 - SAT$ is NP-complete. Recently, Kitaev [18] proved the quantum analogue of this result. He showed that the problem of 'local Hamiltonians', which is a natural generalization of $3 - SAT$ is complete for BQNP. A very non-trivial consequence that follows from this result is that $BQNP \subseteq P^{\#P}$. Our exposition of these results is based on the manuscript [1]. We do not know of any natural examples, other than 'local Hamiltonians', of complete problems for this quantum analog of NP. Developing this theory further is an important open question in quantum complexity theory.

A number of beautiful and deep results in classical complexity theory have emerged out of a study of interactive proof systems (see [23]). Recently two striking results have been proved about QIP, the quantum analog of IP (interactive polynomial time). The first of these shows that for quantum interactive proofs, three rounds are as powerful as polynomially many rounds [28, 19]. The second gives a non-trivial upper bound, showing that $QIP \subseteq EXP$ [19].

In addition to time or number of steps, another complexity measure of interest is space, or number of tape cells used by the computation. In this measure, quantum computation offers at most a polynomial factor advantage over classical computation, since it has been shown that a theorem analogous to Savitch's theorem holds for quantum computation, i.e., $QSPACE(f(n)) \subseteq SPACE(f(n)^2)$ [29].

This survey is based on lecture notes from a course on quantum computation that I taught recently. The lecture notes from that course are posted on my web page www.cs.berkeley.edu/~vazirani.

2. Quantum Turing Machines and Quantum Circuits

Just as a bit (an element of $\{0,1\}$) is a fundamental unit of classical information, a qubit is the fundamental unit of quantum information. A qubit is described by a unit vector in the 2 dimensional Hilbert space $\mathcal{C}^2$. Let $|0\rangle$ and $|1\rangle$ be an orthonormal basis for this space. In general, the state of the qubit is a linear superposition of the form $\alpha|0\rangle + \beta|1\rangle$. The state of n qubits is described by a unit vector in the n-fold tensor product $\mathcal{C}^2 \otimes \mathcal{C}^2 \otimes \cdots \otimes \mathcal{C}^2$. An orthonormal basis for this space is now given by the 2^n vectors $|x\rangle$, where $x \in \{0,1\}^n$. This is often referred to as the computational basis. In general, the state of n qubits is a linear superposition of the 2^n computational basis states. Thus the description of an n qubit system requires 2^n complex numbers. This is the source of the astounding information processing capabilities of quantum computers. An important constraint that quantum mechanics places is that the evolution of the state of the n qubits over time must be unitary. In other words, if $|\phi_0\rangle$ is the initial state of the n qubits, and U is the (2^n dimensional) unitary transformation that describes the evolution of the system, then the state after t steps is $|\phi_t\rangle = U^t |\phi_0\rangle$. For our purposes, it is sufficient to consider the results of measuring the state of a quantum system in the computational basis. If the system is in the superposition $|\Phi\rangle = \sum_x \alpha_x |x\rangle$, then measuring the state of the system yields the outcome $|x\rangle$ with probability $|\alpha_x|^2$. See [22] for an excellent reference for quantum computation.

The two main models of quantum computers: quantum Turing Machines and quantum circuits were first defined by Deutsch [11, 12]. Quantum Turing Machines are much more complex than their classical counterparts, and the construction of universal quantum Turing Machines, as well as implementing programming primitives such as branching and looping requires a considerable amount of work. We section starts with a sketch of these ideas about quantum Turing Machines. Then we turn our attention to quantum circuits, which provide a convenient language in which to describe quantum algorithms. We end with a sketch of Yao's proof showing the equivalence of these two models.

2.1. Quantum Turing Machines.
The definition of a quantum Turing Machines incorporates these concepts into the usual definition of a classical Turing Machine.

Recall that a classical Turing Machine[1] is defined by a triplet (Σ, Q, δ) where Σ is a finite alphabet with an identified blank symbol $\#$, Q is a finite set of states with an identified initial state q_0 and final state q_f, and δ, the *deterministic finite state control*, is a function

$$\delta \ : \ Q \times \Sigma \to \Sigma \times Q \times D$$

with $D = \{L, R\}$. The Turing Machine has a two-way infinite tape of cells indexed by the integers $\mathbf{Z}$ and a single read/write tape head that steps along the tape.

---
[1][23] is an excellent reference.

A configuration of the TM consists of the description of the contents of the tape, the location of the tape head, and the state $q \in Q$. It is standard to restrict attention to configurations where all but a finite number of tape cells contain the blank symbol $\#$. Each configuration of the TM has a successor defined as follows: If $\delta(p, \sigma) = \tau, q, d$, then whenever the TM is in state p with symbol σ under the tape head, it replaces the symbol σ with the symbol τ, enters state q, and steps in direction d, either left one cell or right one cell.

The TM *halts* on input x if it eventually enters the final state q_f. The number of steps a TM takes to halt on input x is its *running time* on input x. If a TM halts then its *output* is a string in Σ^* consisting of the tape contents from the leftmost non-blank symbol to the rightmost non-blank symbol, or the empty string if the entire tape is blank. A TM which halts on all inputs therefore computes a function from $(\Sigma - \#)^*$ to Σ^*.

DEFINITION 1. *A quantum Turing Machine (QTM) is defined by a triplet (Σ, Q, δ), where Σ is a finite alphabet with an identified blank symbol $\#$, Q is a finite set of states with an identified initial state q_0 and final state q_f, and δ, the quantum finite state control is a function*

$$\delta \ : \ Q \times \Sigma \to \widetilde{\mathcal{C}}^{\Sigma \times Q \times D}$$

with $D = \{L, R\}$.[2] *The quantum Turing Machine has a two-way infinite tape of cells indexed by the integers $\mathbf{Z}$, and by a single read/write tape head that steps along the tape.*

The state of a quantum Turing Machine is a linear superposition $\sum_c \alpha_c |a, q, m\rangle$ over classical configurations $c = |a, q, m\rangle$, where c is a classical configuration consisting of a tape configuration a, a state $q \in Q$, and $m \in \mathbf{Z}$. As in the case of classical Turing Machines, we will consider only those tape configurations a in which all but finitely many tape cells contain the blank symbol $\#$. Also, we will restrict ourselves to considering superpositions consisting of finite linear combinations of configurations c. Formally this restricts us to the dense subspace $\mathcal{S}$ of the Hilbert space with a basis element for each configuration $|c\rangle$. The value $\delta(p, \sigma)$ gives the superposition of updates which the machine will take when in state p reading a σ. In this manner, the transition function δ specifies a linear operator on the space $\mathcal{S}$. Not every transition function δ specifies a legal quantum Turing Machine. For the quantum Turing Machine to be legal, the linear operator specified by δ must be unitary. Such transition functions δ are called well-formed. If the state of the quantum Turing Machine is the superposition $\sum_c \alpha_c |c\rangle$, then performing a measurement on the state of the quantum Turing Machine (in the computational basis) yields the result c with probability $|\alpha_c|^2$.

An Efficient Universal QTM

Unlike in the classical case, the construction of an efficient universal QTM is quite non-trivial. Here we will sketch the main steps in the construction of [9].

In order to show that there is an efficient universal quantum Turing Machine, we must show that there is a fixed quantum Turing Machine with a fixed alphabet Σ and fixed set of states Q, that can simulate an arbitrary quantum Turing

---

[2]By $\widetilde{\mathcal{C}}$, we mean all those complex numbers $x + iy$ where the j-th bit of x and y can be computed in time polynomial in j.

Machine, with possibly a much larger tape alphabet and set of states. To do this, we must allow the universal quantum Turing Machine to encode each symbol of M by a block of symbols. Thus we will assume that there is an efficient encoding and decoding scheme to translate between configurations M and those of the universal quantum Turing Machine. We must also allow the universal quantum Turing Machine several computational steps to simulate each step of M. Finally, we will allow the simulation to be ϵ faithful, i.e. for every input x, and every t, the probability distribution that results from measuring the state of M on input x after t steps must be with in ϵ total variation distance (L_1 norm) of the corresponding probability distribution defined by the simulation on the universal QTM.

DEFINITION 2. *We say that a quantum Turing Machine is in* standard form *if all transitions leading to state $q \in Q$ cause the read/write head to move in direction d.*

For quantum Turing Machines in the standard form, the transition function can be simplified to

$$\delta \,:\, Q \times \Sigma \to \widetilde{\mathcal{C}}^{\Sigma \times Q}$$

This is because the direction d is uniquely determined by the new state $q \in Q$. It is not hard to see that a quantum Turing Machine in standard form is well-formed if and only if its transition function δ is unitary.

In this case, the main problem in constructing the universal quantum Turing Machine lies in implementing a close approximation to a specified unitary transformation δ using only the fixed size of unitary transformation allowed to the universal quantum Turing Machine. Suppose that δ is a d dimensional unitary transformation. The universal quantum Turing Machine accomplishes this by implementing a sequence of very elementary operations:

DEFINITION 3. *An $m \times m$ unitary matrix U is* near-trivial *if it satisfies one of the following two conditions:*

- *U is the identity except that one of its diagonal entries is $e^{i\theta}$ for some $\theta \in [0, 2\pi]$.*
- *U is the identity except that the submatrix in one pair of distinct dimensions i and j is the rotation by some angle $\theta \in [0, 2\pi]$.*

The following theorem provides one of the key ingredients in showing how to implement any arbitrary m dimensional unitary transformation δ on a universal quantum Turing Machine:

THEOREM 1. *Any m dimensional unitary transformation U can be ϵ-approximated by a product of near-trivial matrices. Moreover, there is a deterministic algorithm that outputs such a decomposition in time polynomial in m and $\log 1/\epsilon$.*

Now the simulation of a given quantum Turing Machine M proceeds as follows. Each tape cell (symbol) of M is encoded by a block of cells of the universal quantum Turing Machine. Also, the state $q \in Q$ of M is written out explicitly on the tape of the universal quantum Turing Machine. Now, to simulate a step of M, the universal quantum Turing Machine carries out the unitary transformation δ on the block of cells representing the currently scanned symbol of M, and the block of cells representing the state $q \in Q$ of M. It then updates the scanned symbol by simulating the movement of the tape head of M.

Dealing with non-standard form quantum Turing Machines

The states $q \in Q$ of a quantum Turing Machine $M = (\Sigma, Q, \delta)$ may be regarded as an orthonormal basis for the Hilbert space $\mathcal{C}^{|Q|}$. Consider a new quantum Turing Machine M', whose states B correspond to a different orthonormal basis for this Hilbert space. Then each state $b \in B$ is a linear combination of the states in Q. Finally, if we define the transition function δ' of M' as the natural (linear) extension of δ, then it is easy to see that M' has the same time evolution as M. The key property that facilitates the simulation of quantum Turing Machines that are not in standard form is:

THEOREM 2. *For any quantum Turing Machine $M = (\Sigma, Q, \delta)$, there is a unitary change of basis for $\mathcal{C}^{|Q|}$, such that the resulting equivalent quantum Turing Machine M' is in standard form.*

The only remaining problem in carrying out the construction of a universal quantum Turing Machine is that the start state and final state of M under the basis change in the theorem above, might correspond to superpositions of states in M'. To fix this problem, we can simulate each step of M using three steps — change basis from Q to B, simulate a step of M', change basis back from B to Q.

Multi-tape quantum Turing Machines

The construction of the universal quantum Turing Machine easily extends to multi-tape quantum Turing Machines in standard form. However, the construction given above does not work for multi-tape quantum Turing Machines. The difficulty is that, for multi-tape QTMs, there is no change of basis, in general that yields an equivalent QTM in standard form. Nevertheless, Yao's construction [31], showing the polynomial equivalence of quantum Turing Machines and quantum circuits, implies an efficient universal multi-tape QTM. There is a penalty in the simulation overhead, however, that grows as exponentially in the number of tapes of the QTM. (Since the number of tapes is a constant, this still gives a polynomial overhead.) It is an open question whether this simulation overhead is inherent.

2.2. Quantum Circuits.
Quantum circuits [11] are another abstract model for quantum computers. Yao [31] showed that quantum Turing Machines and uniform families of quantum circuits are equivalent in terms of their computing power.

A quantum circuit on m qubits implements a unitary transformation U on the Hilbert space $(\mathcal{C}^2)^{\otimes m}$. A unitary transformation on n qubits is $c-local$ if it operates nontrivially on at most c of the qubits, and preserves the remaining qubits. Quantum computation may be thought of as the study of those unitary transformations that can be realized as a sequence of $c-local$ unitary transformations, i.e., of the form
$$\mathcal{U} = \mathcal{U}_1 \mathcal{U}_2 \cdots \mathcal{U}_k,$$
where the $\mathcal{U}_i$ are $c-local$ and k is bounded by a polynomial in n.

Indeed, it has been shown that we can simplify the picture further, and restrict attention to certain special $2-local$ unitary transformations, also called elementary quantum gates [7]:

Rotation

A rotation or phase shift through an angle θ can be represented by the matrix

$$\mathcal{U} = \begin{pmatrix} \cos\theta & -\sin\theta \\ \sin\theta & \cos\theta \end{pmatrix}.$$

This can be thought of as rotation of the axes (see Figure 1).

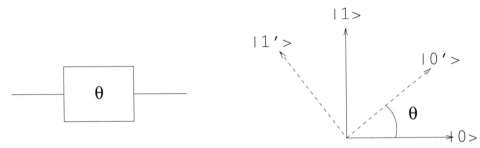

Figure 1. Rotation.

Hadamard transform

If we reflect the axes in the line $\theta = \pi/8$, we get the *Hadamard transform*. (See Figure 2). This can be represented by the matrix

$$\mathcal{H} = \frac{1}{\sqrt{2}} \begin{pmatrix} 1 & 1 \\ 1 & -1 \end{pmatrix}.$$

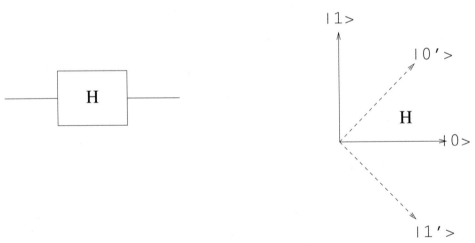

Figure 2. Hadamard transform.

Phase Flip

The phase flip gate operates on a single qubit by the following matrix

$$\begin{pmatrix} 1 & 0 \\ 0 & -1 \end{pmatrix}.$$

Controlled NOT

The controlled NOT gate (Figure 3) operates on 2 qubits and can be represented by the matrix

$$\begin{pmatrix} 1 & 0 & 0 & 0 \\ 0 & 1 & 0 & 0 \\ 0 & 0 & 0 & 1 \\ 0 & 0 & 1 & 0 \end{pmatrix},$$

where the basis elements are (in order) $|00\rangle$, $|01\rangle$, $|10\rangle$, $|11\rangle$. If inputs a and b are basis states, then the outputs are a and $a \oplus b$.

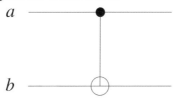

Figure 3. Controlled NOT.

Any unitary transformation can be approximated by using rotation by $\pi/8$, the Hadamard transform, phase flip and controlled NOT gates [7].

Primitives for quantum computation

Let us consider how we can simulate a classical circuit with a quantum circuit. The first observation is that quantum evolution is unitary, and therefore reversible. (The effect of a unitary transformation U can be undone by applying its adjoint). Therefore, if the classical circuit computes the function $f : \{0,1\}^n \to \{0,1\}$, then we must allow the quantum circuit to output $|x\rangle |f(x)\rangle$ on input $|x\rangle$. It follows from the work of Bennett [4] on reversible classical computation that, if $C(f)$ is the size of the smallest classical circuit that computes f, then there exists a quantum circuit of size $O(C(f))$ which, for each input x to f, computes the following unitary transformation U_f on m qubits

$$U_f : |x\rangle |y\rangle \to |x\rangle |y \oplus f(x)\rangle$$

In general though, we don't need to feed U_f a classical state $|x\rangle$. If we feed U_f a superposition

$$\sum_{x \in \{0,1\}^n} \alpha_x |x\rangle |0\rangle$$

then, by linearity,

$$U_f \left(\sum_{x \in \{0,1\}^n} \alpha_x |x\rangle |0\rangle \right) = \sum_{x \in \{0,1\}^n} \alpha_x U_f (|x\rangle |0\rangle) = \sum_{x \in \{0,1\}^n} \alpha_x |x\rangle |f(x)\rangle.$$

At first sight it might seem that we have computed $f(x)$ simultaneously for each basis state $|x\rangle$ in the superposition. However, were we to make a measurement, we would observe $f(x)$ for only one value of x.

Hadamard Transform

The second primitive is the Hadamard transform H_{2^n}, which corresponds to the Fourier transform over the abelian group Z_2^n. The correspondence is based on identifying $\{0,1\}^n$ with Z_2^n, where the group operation is bitwise addition modulo 2. One way to define H_{2^n} is as the $2^n \times 2^n$ matrix in which the (x,y) entry is $2^{-n/2}(-1)^{x \cdot y}$. An equivalent way is as follows. Let H be the unitary transformation on one qubit defined by the matrix

$$\begin{pmatrix} \frac{1}{\sqrt{2}} & \frac{1}{\sqrt{2}} \\ \frac{1}{\sqrt{2}} & -\frac{1}{\sqrt{2}} \end{pmatrix}.$$

If a quantum circuit consists of subcircuits operating on disjoint qubits, then the unitary transformation describing the circuit is the tensor product of the unitary transformations describing the individual subcircuits. Thus, $H_{2^n} = H^{\otimes n}$, or H tensored with itself n times.

Applying the Hadamard transform (or the Fourier transform over Z_2^n) to the state of all zeros gives an equal superposition over all 2^n states

$$\mathcal{H}_{2^n} \left|0 \cdots 0\right\rangle = \frac{1}{\sqrt{2^n}} \sum_{x \in \{0,1\}^n} \left|x\right\rangle.$$

In general, applying the Hadamard transform to the computational basis state $|u\rangle$ yields:

$$\mathcal{H}_{2^n} \left|u\right\rangle = \frac{1}{\sqrt{2^n}} \sum_{x \in \{0,1\}^n} (-1)^{u \cdot x} \left|x\right\rangle$$

2.3. Polynomial Equivalence of Quantum Turing Machines and Quantum Circuits. The central issues to be addressed in constructing an efficient quantum circuit equivalent to a given QTM are the same as those in constructing a universal QTM. In each case, we must decompose one step of the simulated machine (which is a mapping from the computational basis to a new orthonormal basis) into many simple steps, each of which can only map some of the computational basis vectors to their desired destinations. In general, this partial transformation will not be unitary, because the destination vectors will not be orthogonal to the computational basis vectors which have not yet been operated on. The main idea in Yao's construction is to create a second copy of the space, and in each step to map some of the computational basis vectors to their desired destinations, but in the second copy. Once all of the basis vectors have been mapped, then the copies of the space can be interchanged.

THEOREM 3. *A k tape QTM running for T steps can be simulated by a quantum circuit with accuracy ϵ, and size $O(T^2 \log^{O(1)} \epsilon)$.*

PROOF. Let $M = (\Sigma, Q, \delta)$ be a well-formed generalized QTM, and fix $T, \epsilon > 0$.

The simulation uses an extra qubit b, to create two copies of the space of configurations of M. Now if U denotes the unitary transformation carried out by M, then the goal of the simulation is to carry out the transformation $|1\rangle\langle 0| U + |0\rangle\langle 1| U^{-1}$.

The main principle that enables the simulation to be carried out is the following. If U is a unitary transformation acting on the space V, and if $\{P_i\}$ are mutually orthogonal subspaces of V, then the transformation $|1\rangle\langle 0| U + |0\rangle\langle 1| U^{-1}$ on $\mathcal{C}^{\epsilon} \otimes \mathcal{V}$

maps $|0\rangle \otimes P_i$ to $|1\rangle \otimes U(P_i)$ and vice versa. For the simulation of M, the subspaces P_i is spanned by all those configurations of M in which the tape head is in tape cell i. Since the action of the QTM is local, we may write the transformation $|1\rangle\langle 0| U + |0\rangle\langle 1| U^{-1}$ acting on P_i, as a unitary transformation X_i on the smaller set of qubits describing the contents of the three tape cells $i-1, i, i+1$, the finite state control Σ, and the extra qubit b. Since the tape head must be confined between tape cells $-(T-1)$ and $T-1$ in the first T steps, to simulate a step of M, it suffices to carry out the sequence of unitary transformations X_i for i ranging from $-(T-1)$ to $T-1$. Finally, the qubit b must be flipped to restore the state to the original half of the space.

To finish the proof, we observe that each of the transformations X_i is defined on a bounded number of qubits. To simulate them to within a constant ϵ, we must invoke a theorem of Kitaev and Solovay (see [22] for an exposition), which states that if a set of transformations is dense in $SU(2)$ and closed under Hermitian conjugation, then any single qubit gate can be approximated to within ϵ by a product of $\log^{O(1)} \epsilon$ transformations from the set. $\square$

Solovay and Yao [26] point out that the same technique can be used to simulate a multi-tape QTM on a single-tape QTM. Instead of building a quantum gate that operates on three adjacent cells to update any configuration with the tape head in the middle cell, we build a quantum gate that operates on sets of three adjacent cells from each of the tapes to update any configuration with each of the tape heads in the middle of the three cells. Since we have $\Theta(T)$ cells for each tape, this means we need to apply the quantum gate $\Theta(T^k)$ times to handle every possible combination of tape head positions. This proof technique therefore only allows a QTM with a constant number of tapes to be simulated with a polynomial slowdown.

Bernstein [5] showed how to improve the simulation overhead from T^k to 3^k. It is an open question whether this exponential dependence on k is necessary.

3. Quantum Complexity Classes:

Recall that a language $L \subseteq \Sigma^*$ is in the class P (polynomial time), if there is a polynomial time deterministic Turing Machine which on input $x \in \Sigma^*$ decides whether or not x is in L. The class of efficiently solvable problems is identified with the class BPP (bounded-error probabilistic polynomial time).

Definition: A language $L \subseteq \Sigma^*$ is in the class BPP if there is a probabilistic polynomial time Turing Machine which on input $x \in \Sigma^*$ accepts with probability at least $2/3$ when $x \in L$, and rejects with probability at least $2/3$ when $x \notin L$. The probability of correct answer $2/3$ can be boosted to $1-\epsilon$ by running the probabilistic Turing Machine $O(\log 1/\epsilon)$ times and taking the majority answer.

The class of efficiently solvable problems on a quantum computer is BQP (bounded-error quantum polynomial time).

Definition: A language $L \subseteq \Sigma^*$ is in the class BQP if there is a polynomial time quantum Turing Machine which on input $x \in \Sigma^*$ accepts with probability at least $2/3$ when $x \in L$, and rejects with probability at least $2/3$ when $x \notin L$. Once again the probability of correct answer, $2/3$, can be boosted to $1 - \epsilon$ by running the probabilistic Turing Machine $O(\log 1/\epsilon)$ times and taking the majority answer.

BQP may be equivalently defined as the class of languages accepted by a polynomial time uniform family of polynomial size quantum circuits.

Since well-formed quantum Turing Machines must have unitary time evolution, it is not a priori clear that BQP contains P. To show this, one must appeal to work by Bennett [4] showing that the class P can be recognized by reversible polynomial time Turing Machines.

Furthermore, a probabilistic Turing Machine can be simulated by a quantum Turing Machine as follows. If the probabilistic Turing Machine M flips k coins, then the quantum Turing Machine starts with a block of k qubits initially in the state $|0^k\rangle$. Now it performs a $\pi/4$ rotation on each qubit to end up in the state $(\frac{|0\rangle+|1\rangle}{\sqrt{2}})^k = 2^{-k/2}\sum_{x\in\{0,1\}^k}|x\rangle$. Now, if the quantum Turing Machine simulates M, using these k bits as the outcomes of the coin flips (but otherwise leaving these k bits unchanged), then the state at the end of this computation is $2^{-k/2}\sum_{x\in\{0,1\}^k}|x\rangle|M(x)\rangle$. Measuring the resulting state yields $M(x)$ for a uniformly random value of x as desired.

Thus we have the containments $P \subseteq BPP \subseteq BQP$.

Next, we show that $BQP \subseteq P^{\#P}$. Recall that, given a predicate $f(x,y)$ that can be computed in time polynomial in $|x|$, the corresponding counting function $g(x) = |\{y : f(x,y)\}|$ is in $P^{\#P}$.

THEOREM 4. $P \subseteq BPP \subseteq BQP \subseteq P^{\#P} \subseteq PSPACE$.

We give a sketch of the proof that $BQP \subseteq P^{\#P}$. It uses the fact from [9] that we can assume without loss of generality that all the transition amplitudes specified in the transition function δ are real. The action of the transition function δ of QTM M may be described by a tree. The root of the tree corresponds to the initial configuration, and applying δ to the configuration corresponding to any node yields a superposition of configurations represented by the children of that node. The leaves of the tree correspond to accepting and rejecting configurations. Let p be a root-leaf path in this tree. The amplitude of this path β_p is just the product of the branching amplitudes along the path, and is computable to within $1/2^j$ in time polynomial in j. Several paths may lead to the same configuration c. Thus the amplitude of c after T steps of computation is the following sum over all T length paths p: $\alpha_c = \sum_{p \text{ to } c} \beta_p$. The probability that QTM M accepts is $\sum_{accepting\ c} |\alpha_c|^2$. Let $a_p = max(\beta_p, 0)$ and $b_p = max(-\beta_p, 0)$. Then $|\alpha_c|^2$ can be written as $|\alpha_c|^2 = \sum_{p \text{ to } c}(a_p - b_p)^2 = \sum_{p \text{ to } c} a_p^2 + b_p^2 - \sum_{p,p' \text{ to } c} 2a_p b_p$. It follows that the acceptance probability of M can be written as the difference between the two quantities $\sum_{accepting\ c}\sum_{p \text{ to } c} a_p^2 + b_p^2$, and $\sum_{accepting\ c}\sum_{p,p' \text{ to } c} 2a_p b_{p'}$. Since each of these quantities is easily seen to be in $P^{\#P}$, it follows that $BQP \subseteq P^{\#P}$.

In view of this theorem, we cannot expect to prove that BQP strictly contains BPP without resolving the long standing open question in computational complexity theory, namely, whether or not $P = PSPACE$.

4. Recursive Fourier Sampling

In this section, we shall introduce the recursive Fourier sampling problem, show that it can be solved in polynomial time on a quantum computer. However, we shall also show that it does not even lie in the class MA (the probabilistic generalization of NP, see section 6 for a definition) relative to some oracle.

The recursive Fourier sampling problem was defined by Bernstein and Vazirani [9], and was inspired by earlier work by Deutsch and Jozsa [13]. It not only provided the first example of a problem that demonstrates a superpolynomial speedup of quantum algorithms over probabilistic algorithms, but it also shows that $BQP \nsubseteq MA$ relative to an oracle. The proof presented here is a modification of the proof from [9], followed by an elegant argument from [17].

First we introduce the basic primitive, Fourier sampling.

Fourier Sampling

In general, if we start with a state $|\phi\rangle = \sum_x \alpha_x |x\rangle$, then, after applying the Fourier transform over Z_2^n, we obtain the new state $|\widehat{\phi}\rangle = \sum_x \alpha_x |\widehat{x}\rangle$. Notice that this transform can be computed by applying only n single qubit gates, whereas it is computing the Fourier transform on a 2^n dimensional vector. However, the output of the Fourier transform is not accessible to us. To read out the answer, we must make a measurement, and now we obtain x with probability $|\widehat{\alpha}_x|^2$. This process of computing the Fourier transform and then performing a measurement is called Fourier sampling, and is one of the basic primitives in quantum computation.

To see the power of Fourier sampling, suppose we are given a function $f : \{0,1\}^n \to \{1,-1\}$ such that there is an $s \in \{0,1\}^n$ such that for all x, $f(x) = s \cdot x$, where $s \cdot x$ denotes the dot product $s_1 x_1 + \cdots + s_n x_n \bmod 2$, i.e., f is one of the Fourier basis functions. The task is to determine which one it is. The following quantum algorithm carries out this task using two quantum registers, the first consisting of n qubits, and the second consisting of a single qubit.

- Start with the registers in the state $|0^n\rangle |0\rangle$
- Compute the Fourier transform on the first register to get $\sum_{x \in \{0,1\}^n} |x\rangle \otimes |0\rangle$.
- Compute f to get $\sum_x |x\rangle |f(x)\rangle$.
- Apply a conditional phase based on $f(x)$ to get $\sum_x (-1)^{f(x)} |x\rangle |f(x)\rangle$.
- Uncompute f to get $\sum_x (-1)^{f(x)} |x\rangle \otimes |0\rangle$.
- Compute the Fourier transform on the first register to get $|s\rangle \otimes |0\rangle$.

Measuring the first register now yields the string s. Notice that the quantum algorithm queried the boolean function f only twice, but obtained n bits of information about it in the process! An easy information theoretic argument shows that any classical (probabilistic) algorithm must make at least n queries to obtain this information. This single iteration of Fourier sampling is carried out by exactly the quantum circuit devised by Deutsch and Jozsa [13]. The only difference here is that, unlike the Deutsch-Jozsa problem, here all the n output bits are measured (and the promise that f must satisfy is different).

One way to use this difference in the number of queries in order to demonstrate a gap between quantum and probabilistic algorithms is to make the queries very expensive. Then the quantum algorithm would be $n/2$ times faster than any probabilistic algorithm for the given task. The idea behind proving a superpolynomial gap is to make each query itself be the answer to a Fourier sampling problem. Now each query itself is much easier for the quantum algorithm than for any probabilistic algorithm. Carrying this out recursively for $\log n$ levels leads to the superpolynomial speedup for quantum algorithms.

Recursive Fourier sampling

We say that a boolean function f defines a Fourier sampling tree if $f(x_1, \ldots, x_k) = s(x_1, \ldots, x_{k-1}) \cdot x_k$, where $\cdot$ denotes the dot product as defined above. We say that f is *derived from* g (and g *specifies* the Fourier sampling tree f) if $f(x_1, \ldots, x_{k-1}) = g(x_1, \ldots, x_{k-1}, s)$. In the recursive Fourier sampling problem, the x_j's form a telescoping series, so that $|x_{j+1}| = |x_j|/2 = 2^{l-j}$, down to $|x_{l+1}| = 1$. Now, given an oracle for g, and the values of f for the leaf nodes (of the form $(x_1, x_2, \ldots, x_{l+1})$), the challenge is to determine $f(x_1)$.

A recursive implementation of the Fourier sampling primitive introduced above can be used to give a polynomial time quantum algorithm to solve this problem as follows. For the base case, since f is known at the leaf nodes, to determine $f(x_1, x_2, \ldots, x_l)$, we simply perform Fourier sampling with respect to the two leaves in this subtree, to determine $s(x_1, x_2, \ldots, x_l)$ and then output $g(x_1, x_2, \ldots, x_l, s)$. In general, to determine $f(x_1, \ldots, x_{k-1})$, we perform Fourier sampling (recursively using the values of $f(x_1, \ldots, x_{k-1}, x_k)$) to determine $s(x_1, \ldots, x_{k-1})$, and then output $g(x_1, \ldots, x_{k-1}, s)$. Since the Fourier sampling requires only two queries to $f(x_1, \ldots, x_{k-1}, x_k)$, the running time of this quantum algorithm satisfies the recurrence relation $T(n) \leq 2T(\frac{n}{2}) + O(n)$, whose solution is $T(n) = O(n \log n)$.

On the other hand, if a classical algorithm were to take the same kind of recursive approach, it must make n queries at the highest level, each requiring $n/2$ queries at the next level, etc. The recurrence relation is now the following — $T(n) \leq nT(\frac{n}{2}) + \Omega(n)$, which implies $T(n) = \Omega(n^{\log n})$. Proving such a lower bound for any probabilistic algorithm (and generalizing it to any Merlin-Arthur protocol) requires more work. We start by proving that if g is chosen uniformly at random from among functions specifying Fourier sampling trees f, and we condition on the values of g at any choice of $o(n^{\log n})$ points, then the value of $f(x_1)$ is almost unbiased.

Consider a set S of query-answer pairs that are consistent with some legal Fourier sampling tree. Consider a node $y = (x_1, \ldots, x_{k-1})$ in the tree. Consider a legal Fourier sampling tree f which agrees with S below y in the tree. Then f determines string s such that $f(x_1, \ldots, x_k) = x_k \cdot s$. Say that y is a hit if $(x_1, \ldots, x_{k-1}, s)$ is a query in S. Let $P(y)$ denote the probability that y is a hit (with respect to S) when g is chosen uniformly among all functions such that the derived Fourier sampling tree f agrees with S below y in the tree. Similarly, if z is an ancestor of y, denote by $P_z(y)$ the probability that y is a hit when g is chosen uniformly among all functions such that the derived Fourier sampling tree f agrees with S below z in the tree.

LEMMA 1. $P_z(y) \leq 2P(y)$.

Proof Sketch We observe that the queries outside of the subtree rooted at y can only determine the value of $f(y)$, but not the string s. If y is not a hit, then the only constraint this places is on $g(y, s)$. The lemma follows.

LEMMA 2. *If there are q queries of S in the subtree rooted at y, then $P_x(y) \leq q/\gamma(n/4)$, where $n = 2^\ell$, and ℓ is the height of y in the tree. Here $\gamma(n)$ is the product $n(n/2) \cdots 1 = 2^{\ell(\ell+1)/2}$.*

Proof Sketch The proof is by induction on ℓ. For the inductive step, if q' of the q queries are at y, and there are exactly c hits among the 2^n children of y, then the probability that y is a hit is at most $\frac{q'}{2^{n-c}}$. Thus the probability that y is a hit is bounded above by the sum of $\frac{q'}{2^{n/2}}$ and the probability that at least $n/2$ of the children of y are hits. By the induction hypothesis and an easy manipulation the latter probability, it can be bounded by $\frac{q-q'}{2\gamma(n/4)}$. It follows that $P(y) \leq \frac{q-q'}{2\gamma(n/4)} + \frac{q'}{2^{n/2}} \leq \frac{q}{2\gamma(n/4)}$.

By the above lemma, it follows that any deterministic algorithm that makes $o(n^{\log n})$ queries gives the wrong answer on at least $1/2 - o(1)$ fraction of inputs g chosen uniformly at random from among functions specifying Fourier sampling trees f. Now by Yao's lemma [**30**], it follows that any probabilistic algorithm that makes $o(n^{\log n})$ queries must have error probability at least $1/2 - o(1)$ on some input. This completes the proof that relative to some oracle, BQP properly contains BPP.

To show that relative to some oracle BQP contains a problem that is not in MA, we start with the following observation. The above lemma actually shows that any non-deterministic algorithm that makes $o(n^{\log n})$ queries gives the wrong answer on at least $1/2 - o(1)$ fraction of inputs g chosen uniformly at random from among functions specifying Fourier sampling trees f.

Now we show that there is an oracle relative to which the recursive Fourier sampling problem is not in MA. The proof is by contradiction.

Assume to the contrary that there is a Merlin-Arthur protocol that runs in time $poly(n)$. Then if the input g is in the language, there is a proof of length at most $poly(n)$ such that the verifier accepts with probability at least $2/3$ (and running time at most $poly(n)$). And if the input g is not in the language, then for every proof the verifier rejects with probability at least $2/3$. Since the protocol is Merlin-Arthur, we can boost the success probability by running the verifier's algorithm several times using independent random strings (without increasing the length of the proof that prover sends). Thus by increasing the running time of the verifier to $O(poly(n) \times poly'(n))$, we can replace the error probability $1/3$ above by $\frac{1}{2^{poly'(n)}}$. Now let us pick g uniformly at random from among functions specifying Fourier sampling trees f. Then if g is in the language, clearly the verifier rejects with probability at most $\frac{1}{2^{poly'(n)}}$. If g is not in the language, then the verifier accepts each possible proof with probability at most $\frac{1}{2^{poly'(n)}}$. But since there are at most $2^{poly(n)}$ proofs of length $poly(n)$, it follows that the verifier accepts with probability at most $\frac{2^{poly(n)}}{2^{poly'(n)}}$. Thus in each case the verifier errs with probability at most $\frac{2^{poly(n)}}{2^{poly'(n)}}$. It follows that there is a fixed random tape for the verifier such that the verifier errs on at most $\frac{2^{poly(n)}}{2^{poly'(n)}}$ fraction of the g's. But with a fixed random tape, the verifier is deterministic, and this contradicts the consequence of the lemma above that any non-deterministic algorithm that makes $o(n^{\log n})$ queries must give the wrong answer on at least $1/2 - o(1)$ fraction of inputs g. This completes the proof of the following theorem:

THEOREM 5. *There is an oracle relative to which* $BQP \not\subseteq MA$.

5. Quantum Lower Bounds

In view of the exponential advantage offered by quantum algorithms for certain computational problems such as prime factorization, it is tempting to ask whether quantum computers can solve NP-complete problems in polynomial time.

Consider the NP-complete problem — 3-SAT. Given a $3CNF$ Boolean formula $f(x_1, x_2, \ldots, x_n)$, is there an boolean assignment to $x_1, \ldots, x_n$ that satisfies f? Brute force search would take $O(N)$ steps, where $N = 2^n$. Is it possible that the exponential advantage of quantum computers could be used to determine such a truth assignment in polynomial time?

We will abstract this problem in a black box or oracle model as follows. Assume that the input to the problem is a table with N boolean entries, and the task is to find whether any entry in the table is 1. Classical algorithms are allowed random access to the table entries, and quantum algorithms can query the entries of the table in superposition, i.e., a query is of the form $\sum_x \alpha_x \ket{x} \ket{0}$, and the answer to the query is $\sum_x \alpha_x \ket{x} \ket{f(x)}$. It was shown in [10] that any quantum algorithm for this problem must make $\Omega(\sqrt{N})$ queries to the table. Their proof is based on a technique called the hybrid argument. In view of this result, resolving the question whether $NP \subseteq BQP$ will require a non-relativizing proof technique, and is not likely barring a major advance in computational complexity theory. There is a striking algorithm due to Grover [15] that provides a matching upper bound.

A second technique for proving quantum lower bounds — the method of polynomials — was introduced in [8]. It is based on the fact that the acceptance probability of a quantum algorithm after T queries to the input can be described by a polynomial of degree $2T$. The lower bound on a specific function f is proved by showing that it cannot be approximated by a polynomial of degree $2T$ unless T is very large. This method was used, for example, to give a tight linear lower bound for the quantum complexity of the parity function. They also proved that for any total function, the deterministic query complexity $D(h) = O(Q(h)^6)$, where $Q(h)$ is the quantum query complexity. In this section, we sketch a simpler proof of this result using the hybrid argument.

Recently, a third technique for proving lower bounds, the method of quantum adversaries [2] has been introduced. We illustrate the quantum adversary technique by re-proving the $\Omega(2^{n/2})$ lower bound for the search problem using this technique. This technique appears very general, and indeed has been used to obtain tight bounds on the query complexity of a number of problems.

A notable example is the following problem. Let $f : \{0,1\}^n \to \{0,1\}^n$ be a random permutation. Find $f^{-1}(0^n)$. The best previous bound on this problem was $\Omega(2^{n/3})$, via the hybrid argument [10]. However, [2] gives a tight $\Omega(2^{n/2})$ lower bound using this quantum adversary technique.

There are still important open questions in this area of quantum lower bounds. Perhaps the most striking is the following: Let $f : \{0,1\}^n \to \{0,1\}^n$ be a $2-1$ function. Find x and y such that $f(x) = f(y)$. No non-trivial lower bound is known for this problem. An efficient algorithm for this problem would imply that quantum computation precludes the possibility of collision-intractible hashing — an important cryptographic primitive.

5.1. The hybrid argument.

THEOREM 6. *Any quantum algorithm, in the black box model, for determining whether there exist $x_1, \ldots, x_n$ such that $f(x_1, \ldots, x_n) = 1$ must make $\Omega(\sqrt{N})$ queries to f.*

PROOF. Consider any quantum algorithm A for solving this search problem. First do a test run of A on function $f \equiv 0$. Define the query magnitude of x to be $\sum_t |\alpha_{x,t}|^2$, where $\alpha_{x,t}$ is the amplitude with which A queries x at time t. The expectation value of the query magnitudes $E_x \left(\sum_t |\alpha_{x,t}|^2 \right) = T/N$. Thus $\min_x \left(\sum_t |\alpha_{x,t}|^2 \right) \leq T/N$. If the minimum occurs at z, then by the Cauchy-Schwarz inequality $\sum_t |\alpha_{z,t}| \leq T/\sqrt{N}$.

Let $|\phi_t\rangle$ be the states of A_f after the t-th step. Now run the algorithm A on the function $g : g(z) = 1, g(y) = 0 \forall y \neq z$. Suppose the final state of A_g is $|\psi_T\rangle$. By the claim that follows, $||\,|\phi_T\rangle - |\psi_T\rangle\,|| \leq \sum_t |\alpha_{z,t}| \leq T/\sqrt{N}$. This implies that the two states can be distinguished with probability at most $O(T/\sqrt{N})$ by any measurement. Thus any quantum algorithm that distinguishes f from g with constant probability of success must make $T = \Omega(\sqrt{N})$ queries. □

CLAIM 1. $|\psi_T\rangle = |\phi_T\rangle + |E_0\rangle + |E_1\rangle + \ldots + |E_{T-1}\rangle$, where $||\,|E_t\rangle\,|| \leq |\alpha_{z,t}|$.

PROOF. Consider two runs of the algorithm A, which differ only on the t-th step. The first run queries the function f on the first t steps and queries g for the remaining $T - t$ steps; the second run queries f on the first $t - 1$ steps and g for the remaining $T - t + 1$ steps. Then at the end of the t-th step, the state of the first run is $|\phi_t\rangle$, whereas the state of the second run is $|\phi_t\rangle + |F_t\rangle$, where $||\,|F_t\rangle\,|| \leq |\alpha_{x,t}|$. Now if U is the unitary transform describing the remaining $T - t$ steps (of both runs), then the final state after T steps for the two runs are $U|\phi_t\rangle$ and $U(|\phi_t\rangle + |F_t\rangle)$, respectively. The latter state can be written as $U|\phi_t\rangle + |E_t\rangle$, where $|E_t\rangle = U|F_t\rangle$. Thus the effect of switching the queried function only on the t-th step can be described by an "error" $|E_t\rangle$ in the final state of the algorithm, where $||\,|E_t\rangle\,|| \leq |\alpha_{z,t}|$.

We can transform the run A_f to A_g by a succession of T changes of the kind described above. Therefore, by the linearity of quantum mechanics, the difference between the final states of A_f and A_g is $|E_0\rangle + |E_1\rangle + \ldots + |E_{T-1}\rangle$, where $||\,|E_t\rangle\,|| \leq |\alpha_{z,t}|$. □

5.2. Block Sensitivity and the Black Box Model.

In this section we will work more explicitly in the black box model, where the algorithm has to explicitly query the input $w \in \{0, 1\}^n$, and the complexity of the algorithm is the number of queries that the algorithm makes. We give a simple proof via the hybrid argument of the result of [8] showing that $D(f) = O(Q(f)^6)$, where $D(f)$ is the deterministic query complexity of f and $Q(f)$ is its quantum query complexity:

For a boolean function $f : \{0,1\}^n \to \{0,1\}$, $D(f)$ is the minimum number of bits of the input that a deterministic algorithm must query to compute f. $Q(f)$ is the minimum number of queries to the input w that a quantum algorithm must make to compute $f(w)$ with error probability at most $1/3$.

$C(f)$, the certificate complexity or the nondeterministic complexity of f, is the minimum number of bits of the input that must be revealed (by someone who knows all the input bits) to convince a deterministic algorithm about the value of $f(w)$.

A key result, that was first discovered by Blum [6], shows that in the black box model the deterministic and nondeterministic (certificate) complexity of a function are polynomially related. Recall that in the black box model we only count the number of queries made by the algorithm, not the number of steps of computation performed by the algorithm between queries. In fact, the $C(f)^2$ upper bound on the deterministic complexity is established by giving an algorithm that requires $2^{C(f)}$ steps of computation, but only $C(f)^2$ queries.

LEMMA 3. $C(f) \leq D(f) \leq C(f)^2$.

Another closely related property is a structural property of f called its block sensitivity:

Notation: For a string $w \in \{0,1\}^n$ and a set $S \subseteq \{1,2,\ldots,n\}$, define $w^{(S)}$ to be the boolean string y that differs from w on exactly the bit positions in the set S.

DEFINITION 4. *For a boolean function $f : \{0,1\}^n \to \{0,1\}$ the block sensitivity of f, $bs(f)$ is defined to be the maximum number t such that there exists an input $w \in \{0,1\}^n$ and t disjoint subsets $S_1, \ldots S_t \subseteq \{1,2,\ldots,n\}$ such that for all $1 \leq i \leq t$, $f(w) \neq f(w^{(S_i)})$.*

Nisan [21] proved the following fundamental lemma:

LEMMA 4. $\sqrt{C(f)} \leq bs(f) \leq C(f)$.

COROLLARY 1. $D(f) \leq bs(f)^4$.

[8] improve this bound by showing:

LEMMA 5. $D(f) \leq C(f)bs(f) \leq bs(f)^3$.

We are now ready to prove the relationship between $D(f)$ and $Q(f)$:

THEOREM 7. $Q(f) = \Omega(\sqrt{bs(f)})$.

PROOF. The proof mirrors the $\sqrt{N}$ lower bound from the last section. Let $w \in \{0,1\}^n$ be the input defining the block sensitivity of f. Given a quantum algorithm A define the query magnitude of set $S \subseteq \{1,2,\ldots,n\}$ on input w to be $q_S(w) = sum_{j \in S} q_j(w)$. Now if A runs for T steps, then since the sets S_i are disjoint, the expected query magnitude for a random set S_i is $q_{S_i}(w) \leq T/bs(f)$. Let $z = w^{(S_i)}$, where $q_{S_i}(w)$ is the minimum among these query magnitudes. Let ϕ_T and ψ_T denote the final states on inputs w and z, respectively. Then, as in the $\sqrt{N}$ lower bound from the previous subsection, $|||\psi_T\rangle - |\phi_T\rangle|| \leq T/\sqrt{bs(f)}$. Since A must distinguish these states with constant probability, it follows that $T = \Omega(\sqrt{bs(f)})$. □

COROLLARY 2. $D(f) = O(Q(f)^6)$.

5.3. The Method of Quantum Adversaries. In this section we illustrate the method of quantum adversaries, by proving the lowerbound for unstructured search: given a boolean function $f : \{0,1\}^n \to \{0,1\}$, is there an $x \in \{0,1\}^n$ such that $f(x) = 1$?

We assume that the quantum computer is partitioned into two registers, the input register (specifying the function f) and the work space register. Now, for the search problem, the input function f may be identified for convenience by x, the unique n bit string on which $f(x) = 1$. Thus, the initial state of the registers may be written as $|x\rangle |0\rangle$. If the quantum algorithm correctly solves the search problem, when it is done the work space must look like $|x\rangle |x\rangle |junk_x\rangle$. But now if the algorithm works correctly on every input, it must work correctly on a superposition of inputs as well. In the case of the search problem, if the input is a uniform superposition over $|x\rangle$, then the initial state is $(\sum_x |x\rangle) \otimes |0\rangle$, and the final state is $\sum_x |x\rangle \otimes (|x\rangle |junk_x\rangle)$. The main point of the quantum adversary argument is that the two registers are initially unentangled, whereas they are maximally entangled at the end of the algorithm. Now establishing an upper bound on the increase in entanglement per step of the quantum algorithm implies a corresponding lower bound on the number of steps that the quantum algorithm must take.

The first step towards fleshing out the above argument is to quantify how entangled the two registers are. Consider the density matrix that describes the state of the input register when the work register is measured. For the search problem, this density matrix is initially (at the beginning of the algorithm)

$$\rho_0 = \left(\sum_{i=1}^N |i\rangle\right)\left(\sum_{i=1}^N \langle i|\right) = \begin{pmatrix} \frac{1}{N} & \cdots & \frac{1}{N} \\ \vdots & & \vdots \\ \frac{1}{N} & \cdots & \frac{1}{N} \end{pmatrix}$$

corresponding to the state $1/\sqrt{N} \sum_i |i\rangle$. The final density matrix at the end of the algorithm is

$$\rho_F = \begin{pmatrix} \frac{1}{N} & & 0 \\ & \ddots & \\ 0 & & \frac{1}{N} \end{pmatrix}$$

corresponding to the uniform distribution over all $|i\rangle$.

As a measure of entanglement, consider the sum of absolute values of the off-diagonal entries. Initially, this quantity is 0 and finally it is $(N^2 - N) * \frac{1}{N} = \theta(N)$.

How fast do the off-diagonal entries of the density matrix increase? Each step of a quantum algorithm may be regarded as performing a query on the input tape, followed by a unitary transformation on the work register. Clearly the unitary transformation on the work register does not affect the reduced density matrix describing the input register.

To understand the effect of a query, recall how a query is carried out. The input to be queried is written out on a part of the work register designated to be the query register. (Thus the contents of the register might be $\sum_j \alpha_j |j\rangle$.) The reduced density matrix of the input register at this step (just before the query is actually performed) may be written in the form $\rho = \sum_j p_j \rho_j$, where $p_j = |\alpha_j|^2$.

Each ρ_i can be expanded as $\sum_{j,k} \rho_{j,k}^{(i)} = \sum_j \rho_{i,j}^{(i)} + \sum_k \rho_{k,i}^{(i)} = 2\sum_j \rho_{i,j}^{(i)}$, since all the other entries in the sum cannot occur. If the query register contains $\sum \alpha_i |i\rangle$, the query modifies $\rho_{i,j}$ with probability $\alpha_i \alpha_j$.

To bound the effect of the query on ρ, it suffices to bound its effect on each ρ_j. But since this part of the query accesses the input only on location j, the only entries of ρ_j that change are those in the j-th row and in the j-th column. There are only $O(N)$ such entries, and the maximum is achieved when they are equal, and is therefore bounded by $N * O(\frac{1}{\sqrt{N}}) = O(\sqrt{N})$. Since ρ is a convex combination of the ρ_j, it follows that each query can change the off-diagonal entries of ρ by $O(\sqrt{N})$. Since the total change in the off-diagonal entries is $\Theta(N)$, it follows that any quantum algorithm must take $\Omega(\sqrt{N})$ steps to solve the search problem.

Thus the changes to the density matrix are only to those summands, and they change at most $|\alpha_i| \sum |\alpha_j| = O\left(\sqrt{N}\right)$. Since the starting and ending density matrices differ by $O(\sqrt{N})$, the number of queries required is $O(\sqrt{N})$.

So far we assumed that the quantum algorithm gives the correct answer with certainty. To prove a meaningful lower bound, we must allow the quantum algorithm to give an incorrect answer with a small constant probability ϵ. But it is easy to see that the above argument still works, since it can be easily verified that the total change in the off-diagonal entries of the initial density matrix and the final density matrix is still $\Theta(N)$, since the final density matrix satisfies $|\rho_{ij}| \leq \frac{2\sqrt{\epsilon(1-\epsilon)}}{N}$.

6. Quantum NP

The class NP (non-deterministic polynomial time) contains many thousand of the most important computational problems. Of these problems, the vast majority are NP-complete. This means that these are the hardest problems in NP. By this we mean that, if anyone of them can be solved by a polynomial time algorithm, then every problem in NP can be solved by a polynomial time algorithm. The cornerstone of this theory of NP-completeness is the Cook-Levin theorem, which states that 3-SAT is NP-complete.

A language L is in NP if there is a polynomial time proof checker C and a polynomial $poly$, with the following property: if $x \in L$ then there is a string y with $|y| \leq poly|x|$, such that $C(x, y) = 1$. If $x \notin L$, then for every y such that $|y| \leq poly(|x|)$, $C(x, y) = 0$.

Recently Kitaev [18] gave the quantum analogue of the Cook-Levin theorem by showing that QSAT the quantum analogue of 3-SAT is complete for BQNP or QMA. Our exposition of this result is based upon the manuscript by Aharonov and Nave [1].

BQNP or QMA is the quantum generalization of MA — the probabilistic analogue of NP. To define MA, we simply replace the deterministic polynomial time proof checker with a probabilistic polynomial time proof checker C. Now if $x \in L$, then there is a string y with $|y| \leq poly|x|$, such that $C(x, y) = 1$ with probability at least 2/3. If $x \notin L$, then for every y such that $|y| \leq poly(|x|)$, $C(x, y) = 0$ with probability at least 2/3.

To define BQNP, the quantum analogue of MA, we replace the probabilistic polynomial time proof checker by a quantum polynomial time proof checker. Equally important, the witness string y is now allowed to be a quantum witness, i.e., it can be a superposition over strings of length at most $poly(|x|)$.

BQP is trivially contained in BQNP since it can be simulated by the verifier alone. MA is also contained in BQNP since quantum machines can perform the classical computations of their classical counterparts. Kitaev's proof that QSAT is BQNP-complete implies a non-trivial upper bound, showing that $BQNP \subseteq P^{\#P}$.

A BQNP-Complete Problem

Recall that a Hamiltonian acting on n qubits is a 2^n dimensional Hermitian matrix. Say that a Hamiltonian is c-local if it acts as the identity on all except c of the qubits. Consider the following problem:

DEFINITION 5. **Local Hamiltonians or Q5SAT:** Let H_j (for $j = 1,\ldots r$) be 5-local Hamiltonians on n qubits (each specified by complex $2^5 \times 2^5$ matrices.). Assume that each H_j is scaled so that all eigenvalues λ of H_j satisfy $0 \leq \lambda \leq 1$. Let $H = \sum_{j=1}^{r} H_j$. There is a promise about H that either all eigenvalues of H are $\geq b$ or there is an eigenvalue of H that is $\leq a$, where $0 \leq a < b \leq 1$ and the difference $b - a$ is at least inverse polynomial in n, i.e., $b - a \geq \frac{1}{poly(n)}$. The problem asks whether H has an eigenvalue $\leq a$.

The Connection with 3-SAT

In 3-SAT, we are given a formula f on n variables in 3-CNF (conjunctive normal form.) That is, f is a conjunction of many clauses c_i:

$$f(x_1, x_2, \ldots, x_n) = c_1 \wedge c_2 \wedge \ldots \wedge c_m ,$$

where each clause c_j is a disjunction of three variables or their negations. For example, c_j may be $(x_a \vee \overline{x_b} \vee x_c)$.

We would like to make a corresponding Hamiltonian H_i for each clause c_i. H_i should penalize an assignment which does not satisfy the clause c_i. In the example where $c_j = (x_a \vee \overline{x_b} \vee x_c)$, we want to penalize the assignment state $|010\rangle$. If our notion of *penalize* is to have a positive eigenvalue, then we can let $H_j = |010\rangle\langle 010|$, and define the other H_i's similarly, i.e., each H_i has a 1 eigenvalue with a corresponding eigenvector that causes clause c_i to be false.

Finally, we let

$$H = \sum_{i=1}^{m} H_i,$$

so that H is a sum of 3-local Hamiltonians. It is not hard to see that the smallest eigenvalue of H is the minimum (over all assignments) number of unsatisfied clauses. In particular, H has a 0 eigenvalue exactly when there is a satisfying assignment for f.

For general QSAT instances, the Hamiltonians H_j cannot be simultaneously diagonalized in general, and the problem appears much harder.

Membership in BQNP

We can assume without loss of generality that each H_j is just a projection matrix $|\phi_j\rangle\langle\phi_j| \otimes I$. The prover would like to provide convincing and easily verifiable evidence that $H = \sum H_j = \sum (A_j \otimes I)$ has a small eigenvalue $\lambda \leq a$. The proof consists of (a tensor product of) polynomial in n copies of the corresponding eigenvector η.

$\lambda = \sum_j \langle\eta| H_j |\eta\rangle$. Given a single copy of $|\eta\rangle$, the verifier can flip a coin with bias $\frac{\lambda}{r}$ as follows:

(1) Pick $H_j = |\phi_j\rangle\langle\phi_j|$ at random
(2) Measure $|\eta\rangle$ by projecting onto $|\phi_j\rangle$.

This succeeds with probability $\frac{\lambda}{r}$. Given the promise that $\lambda \leq a$ or $\lambda \geq b$, it suffices for the verifier to repeat this test $\frac{r^2}{(b-a)^2}$ times to conclude with high confidence that $\lambda \leq a$. Thus polynomial in n copies of $|\eta\rangle$ are sufficient. Note that since the verifier is performing each test randomly and independently, the prover gains no advantage by sending an entangled state to the verifier.

BQNP-Completeness

To show that QSAT is complete in BQNP, we need to show that the universal BQNP problem reduces to it. That is, given a quantum circuit $U = U_L U_{L-1} \ldots U_1$ and a promise that exactly one of the following holds:

(1) $\exists |\eta\rangle$, U accepts on input $|\eta\rangle$ with probability $\geq p_1 = 1 - \epsilon$
(2) $\forall |\eta\rangle$, U accepts on input $|\eta\rangle$ with probability $\leq p_0 = \epsilon$,

The challenge is to design an instance of QSAT which allows us to distinguish the above two cases. i.e. we wish to specify a sum of local Hamiltonians that has an eigenvector with small eigenvalue if and only if $\exists |\eta\rangle$ that causes U to accept with high ($\geq p_1$) probability.

The construction of the local Hamiltonian is analogous to Cook's theorem. The quantum analogue of the accepting tableau in Cook's theorem will be the computational history of the quantum circuit:

$$|T\rangle = \sum_{t=0}^{L} |\phi_t\rangle \otimes |t\rangle$$

where $|\phi_0\rangle$ is a valid initial state and $|\phi_i\rangle = U_i |\phi_{i-1}\rangle$. Thus the computation history $|T\rangle$ is an element of $(\mathcal{C}^2)^{\otimes n} \otimes \mathcal{C}^{L+1}$. It is a superposition over time steps of the state of the quantum bits as the quantum circuit operates on them.

Now the idea of the BQNP-completeness proof is to design the Hamiltonian H such that:

(1) if there exists $|\eta\rangle$ where $U|\eta\rangle$ accepts with probability at least $1 - \epsilon$, then the computational history $|T\rangle$ of the quantum circuit U on input η is an eigenvector with eigenvalue at most $\frac{\epsilon}{L+1}$
(2) if U rejects every input with probability at least $1 - \epsilon$, then all the eigenvalues of H are at least $\frac{c(1-\epsilon)}{L^3}$

H will be the sum $H_{initial} + H_{final} + H_{propagate}$. The first two terms are simple and express the condition that the computational history starts with a valid input state, and ends in an accepting state.

We consider the first m bits of U's state the input bits and the remaining $n - m$ bits to be the clean work bits. The design of the $H_{initial}$ component should then reflect that at time 0, all of the work bits are clear:

$$H_{initial} = \sum_{s=m+1}^{n} \Pi_s^{(1)} \otimes |0\rangle \langle 0|$$

where $\Pi_s^{(1)}$ denotes projection onto the s-th qubit with value 1.

Assume that the state of the first qubit at the output determines whether or not the input is accepted. Then H_{final} needs to indicate that at time L the first qubit is a 1:

$$H_{final} = \Pi_1^{(0)} \otimes |L\rangle \langle L|.$$

The most complicated component of H is $H_{propagate}$, which captures transitions between time steps. $H_{propagate} = \sum_{j=1}^{L} H_j$, where

$$H_j = -\frac{1}{2} U_j \otimes |j\rangle \langle j-1|$$
$$- \frac{1}{2} U_j^\dagger \otimes |j-1\rangle \langle j|$$
$$+ \frac{1}{2} I \otimes (|j\rangle \langle j| + |j-1\rangle \langle j-1|)$$

The fact that the computational history is a superposition over time steps is quite crucial here. To check that the correct operation has been applied in step j, it suffices to restrict attention to the $j-1$-st and j-th bit of the clock (assuming that the clock is represented in unary). Now the quantum register is in a superposition over its state at time $j-1$ and at time j. Locally checking this superposition is sufficient to determine whether its clock j component is the result of applying the quantum gate U_j to the clock $j-1$ component. This is precisely what the Hamiltonian H_j above is designed to do.

Next we show that an accepting history of computation is an eigenvector of H with eigenvalue 0.

Let $|T\rangle = \sum_{t=0}^{L} |\phi_t\rangle \otimes |t\rangle$. We analyze the contribution from each component of H. If $|T\rangle$ starts with qubits $m+1$ through n clear, $H_{initial}$ does not contribute to $H|T\rangle$. If $|T\rangle$ is a computation of U, that is, $|\phi_t\rangle = U_t |\phi_{t-1}\rangle$ for all t, then from $H_{propagate}$ we get:

$$H_j T = -\frac{1}{2} U_j |\phi_{j-1}\rangle |j\rangle - \frac{1}{2} U_j^\dagger |\phi_j\rangle |j-1\rangle$$
$$+ \frac{1}{2} |\phi_j\rangle |j\rangle + \frac{1}{2} |\phi_{j-1}\rangle |j-1\rangle$$
$$= -\frac{1}{2} |\phi_j\rangle |j\rangle - \frac{1}{2} |\phi_{j-1}\rangle |j-1\rangle$$
$$+ \frac{1}{2} |\phi_j\rangle |j\rangle + \frac{1}{2} |\phi_{j-1}\rangle |j-1\rangle$$
$$= 0,$$

for no contribution from $H_{propagate}$.

Finally, if U accepts with probability at least $1-\epsilon$, only H_{final} contributes a penalty to the sum, for an eigenvalue of at most $\frac{\epsilon}{L+1}$.

The hard part of the proof lies in showing the converse. That if there is no $|\eta\rangle$ which U accepts with high probability, then all eigenvalues of H are large. We refer the interested reader to [1] for the proof of this.

Upper bound on BQNP

One consequence of this proof of BQNP-completeness is the following:

Theorem: $BQNP \subseteq P^{\#P}$.

Consider the trace of H^k. This is either at least b^k or at most Na^k. We can make sure that $b^k >> Na^k$, by choosing $k >> n^d \log N$. So we just need to estimate $Tr(H^k)$ in $P^{\#P}$.

To see this, write $Tr(H^k) = Tr((\sum_j H_j)^k) = Tr(\sum_{j_1,...j_k} H_{j_1} \cdots H_{j_k}) = \sum_{j_1,...j_k} Tr(H_{j_1} \cdots H_{j_k})$. Each trace in this sum is itself just a sum of exponentially many easy to compute contributions, and thus the entire sum is easily seen to be estimated in $P^{\#P}$.

Kitaev's results may well be the first steps towards a rich new theory of BQNP-completeness. Perhaps the most important open question in this area is to find other examples of natural BQNP-complete problems.

References

[1] Aharonov, D, and Nave, T., **Quantum NP**, manuscript.
[2] Ambainis, A., **Quantum lower bounds by quantum arguments**, *Proceedings of the 32nd Annual ACM Symposium on Theory of Computing*, 2000.
[3] Arora, S., Impagliazzo, R., and Vazirani, U., **On the Role of the Cook-Levin Theorem in Complexity Theory**, manuscript (1994).
[4] Bennett, C. H., **Logical reversibility of computation**, *IBM J. Res. Develop.*, Vol. 17, 1973, pp. 525-532.
[5] BERNSTEIN, E., **Quantum complexity theory**, PhD dissertation, U.C. Berkeley, 1996.
[6] Blum, M. and Impagliazzo, R., **Generic oracles and oracle classes**, 28th Annual Symposium on Foundations of Computer Science, IEEE computer society press, 1987.
[7] Barenco, A., Bennett, C., Cleve, R., DiVincenzo, D., Margolus, N., Shor, P., Sleator, T., Smolin, J., and Weinfurter, H., **Elementary gates for quantum computation**, Phys. Rev. A **52**, 3457 (1995).
[8] Beals, R., Buhrman, H., Cleve, R., Mosca, M., de Wolf, R., **Quantum Lower Bounds by Polynomials**, *Proceedings of the 39th Annual IEEE Symposium on Foundations of Computer Science*, 1998.
[9] Bernstein, E. and Vazirani, U., **Quantum complexity theory**, *Proceedings of the 25th Annual ACM Symposium on Theory of Computing*, 1993, pp, 11-20 . SIAM J. Computing, 26, pp. 1411-1473 (1997).
[10] Bennett, C., Bernstein, E., Brassard, G., and Vazirani, U., **Strengths and weaknesses of quantum computation**, ISI Torino tech report (1994), SIAM J. Computing, 26, pp. 1510-1523 (1997).
[11] Deutsch, D., **Quantum theory, the Church-Turing principle and the universal quantum computer**, *Proc. R. Soc. Lond.*, Vol. A400, 1985, pp. 97-117.
[12] Deutsch, D., "Quantum computational networks," *Proc. R. Soc. Lond.*, Vol. A425, 1989, pp. 73-90.
[13] Deutsch, D. and Jozsa, R., **Rapid solution of problems by quantum computation**, *Proceedings of the Royal Society*, London, vol. A439, 1992, pp. 553-558.
[14] Feynman, R., **Simulating physics with computers**, *International Journal of Theoretical Physics*, Vol. 21, nos. 6/7, 1982, pp. 467-488.
[15] Grover, L., **Quantum mechanics helps in searching for a needle in a haystack**, Phys. Rev. Letters, 78, pp. 325-328 (1997).
[16] Hales,L. and Hallgren, L., **An Improved Quantum Fourier Sampling Algorithm and Applications**, *Proceedings of the 41st Annual IEEE Symposium on Foundations of Computer Science*, 2000.
[17] Hales, L. and Hallgren, S., manuscript, 1998.
[18] Kitaev, A., **Quantum NP**, AQIP '99.
[19] Kitaev, A., and Watrous, J., **Parallelization, amplification, and exponential time simulation of quantum interactive proof sy stems**, *Proceedings of the 32nd ACM Symposium on Theory of Computing*, pages 608-617, 2000.
[20] Manin, Y., **Computable and uncomputable**, Moscow, Sovetskoye Radio, 1980.
[21] Nisan, N., **CREW PRAM's and decision trees**, STOC 1989, pages 327-335.
[22] Nielson, M., and Chuang, I., "**Quantum Computation and Quantum Information,**" Cambridge University Press, 2000.
[23] Papadimitriou, C., "**Computational Complexity,**" Addison-Wesley Publishing Company.
[24] Simon, D., **On the power of quantum computation**, SIAM J. Computing, 26, pp. 1474-1483 (1997).

[25] Shor, P., **Polynomial-time algorithms for prime factorization and discrete logarithms on a quantum computer**, SIAM J. Computing, 26, pp. 1484-1509 (1997).
[26] Solovay, B., and A. Yao, manuscript 1997.
[27] Vazirani, U., **On the power of quantum computation**, *Philosophical Transactions of the Royal Society of London, Series A*, 356:1759-1768, August 1998.
[28] Watrous, J., **PSPACE has constant-round quantum interactive proof systems**, *Proceedings of the 40th Annual Symposium on Foundations of Computer Science*, pp. 112-119, 1999.
[29] Watrous, J., **On quantum and classical space-bounded processes with algebraic transition amplitudes**, quant-ph 9911008.
[30] Yao, A., **Probabilistic Computations: Towards a Unified Measure of Complexity**, *Proceedings of the 18th Annual IEEE Symposium on Foundations of Computer Science*, 1977, pp. 222-227.
[31] Yao, A., **Quantum circuit complexity**, *Proceedings of the 34th Annual IEEE Symposium on Foundations of Computer Science*, 1993.

UNIVERSITY OF CALIFORNIA, BERKELEY, BERKELEY, CA 94720
E-mail address: vazirani@cs.berkeley.edu
URL: http://www.cs.berkeley.edu/~vazirani

Chapter III

Quantum Error Correcting Codes and Quantum Cryptography

An Introduction to Quantum Error Correction

Daniel Gottesman

ABSTRACT. Quantum states are very delicate, so it is likely some sort of quantum error correction will be necessary to build reliable quantum computers. The theory of quantum error-correcting codes has some close ties to and some striking differences from the theory of classical error-correcting codes. Many quantum codes can be described in terms of the stabilizer of the codewords. The stabilizer is a finite Abelian group, and allows a straightforward characterization of the error-correcting properties of the code. The stabilizer formalism for quantum codes also illustrates the relationships to classical coding theory, particularly classical codes over GF(4), the finite field with four elements.

Contents

1. Background: the need for error correction
2. The nine-qubit code
3. General properties of quantum error-correcting codes
4. Stabilizer codes
5. Some other important codes
6. Codes over GF(4)
7. Fault-Tolerant Quantum Computation
8. Summary (Quantum Error Correction Sonnet)

References

1. Background: the need for error correction

Quantum computers have a great deal of potential, but to realize that potential, they need some sort of protection from noise.

Classical computers don't use error correction. One reason for this is that classical computers use a large number of electrons, so when one goes wrong, it is not too serious. A single qubit in a quantum computer will probably be just one, or a small number, of particles, which already creates a need for some sort of error correction.

2000 *Mathematics Subject Classification.* Primary 81P68; Secondary 94B60.

© 2002 American Mathematical Society

Another reason is that classical computers are digital: after each step, they correct themselves to the closer of 0 or 1. Quantum computers have a continuum of states, so it would seem, at first glance, that they cannot do this. For instance, a likely source of error is over-rotation: a state $\alpha|0\rangle + \beta|1\rangle$ might be supposed to become $\alpha|0\rangle + \beta e^{i\phi}|1\rangle$, but instead becomes $\alpha|0\rangle + \beta e^{i(\phi+\delta)}|1\rangle$. The actual state is very close to the correct state, but it is still wrong. If we don't do something about this, the small errors will build up over the course of the computation, and eventually will become a big error.

Furthermore, quantum states are intrinsically delicate: looking at one collapses it. $\alpha|0\rangle + \beta|1\rangle$ becomes $|0\rangle$ with probability $|\alpha|^2$ and $|1\rangle$ with probability $|\beta|^2$. The environment is constantly trying to look at the state, a process called *decoherence*. One goal of quantum error correction will be to prevent the environment from looking at the data.

There is a well-developed theory of classical error-correcting codes, but it doesn't apply here, at least not directly. For one thing, we need to keep the phase correct as well as correcting bit flips. There is another problem, too. Consider the simplest classical code, the repetition code:

(1.1) $$0 \to 000$$
(1.2) $$1 \to 111$$

It will correct a state such as 010 to the majority value (becoming 000 in this case).[1]

We might try a quantum repetition code:

(1.3) $$|\psi\rangle \to |\psi\rangle \otimes |\psi\rangle \otimes |\psi\rangle$$

However, no such code exists because of the No-Cloning theorem [6, 20]:

THEOREM 1 (No-Cloning). *There is no quantum operation that takes a state $|\psi\rangle$ to $|\psi\rangle \otimes |\psi\rangle$ for all states $|\psi\rangle$.*

PROOF. This fact is a simple consequence of the linearity of quantum mechanics. Suppose we had such an operation and $|\psi\rangle$ and $|\phi\rangle$ are distinct. Then, by the definition of the operation,

(1.4) $$|\psi\rangle \to |\psi\rangle |\psi\rangle$$
(1.5) $$|\phi\rangle \to |\phi\rangle |\phi\rangle$$
(1.6) $$|\psi\rangle + |\phi\rangle \to (|\psi\rangle + |\phi\rangle)(|\psi\rangle + |\phi\rangle).$$

(Here, and frequently below, I omit normalization, which is generally unimportant.)

But by linearity,

(1.7) $$|\psi\rangle + |\phi\rangle \to |\psi\rangle |\psi\rangle + |\phi\rangle |\phi\rangle.$$

This differs from (1.6) by the crossterm

(1.8) $$|\psi\rangle |\phi\rangle + |\phi\rangle |\psi\rangle.$$

□

---

[1] Actually, a classical digital computer is using a repetition code – each bit is encoded in many electrons (the repetition), and after each time step, it is returned to the value held by the majority of the electrons (the error correction).

Identity	$I = \begin{pmatrix} 1 & 0 \\ 0 & 1 \end{pmatrix}$		$I\|a\rangle = \|a\rangle$
Bit Flip	$X = \begin{pmatrix} 0 & 1 \\ 1 & 0 \end{pmatrix}$		$X\|a\rangle = \|a \oplus 1\rangle$
Phase Flip	$Z = \begin{pmatrix} 1 & 0 \\ 0 & -1 \end{pmatrix}$		$Z\|a\rangle = (-1)^a\|a\rangle$
Bit & Phase	$Y = \begin{pmatrix} 0 & -i \\ i & 0 \end{pmatrix}$	$= iXZ$	$Y\|a\rangle = i(-1)^a\|a \oplus 1\rangle$

TABLE 1. The Pauli matrices

2. The nine-qubit code

To solve these problems, we will try a variant of the repetition code [16].

(2.1) $\quad |0\rangle \to |\bar{0}\rangle = (|000\rangle + |111\rangle)(|000\rangle + |111\rangle)(|000\rangle + |111\rangle)$

(2.2) $\quad |1\rangle \to |\bar{1}\rangle = (|000\rangle - |111\rangle)(|000\rangle - |111\rangle)(|000\rangle - |111\rangle)$

Note that this does not violate the No-Cloning theorem, since an arbitrary codeword will be a linear superposition of these two states

(2.3) $\quad \alpha|\bar{0}\rangle + \beta|\bar{1}\rangle \neq [\alpha(|000\rangle + |111\rangle) + \beta(|000\rangle - |111\rangle)]^{\otimes 3}.$

The superposition is linear in α and β. The complete set of codewords for this (or any other) quantum code form a linear subspace of the Hilbert space, the *coding space*.

The inner layer of this code corrects bit flip errors: We take the majority within each set of three, so

(2.4) $\quad |010\rangle \pm |101\rangle \to |000\rangle \pm |111\rangle.$

The outer layer corrects phase flip errors: We take the majority of the three signs, so

(2.5) $\quad (|\cdot\rangle + |\cdot\rangle)(|\cdot\rangle - |\cdot\rangle)(|\cdot\rangle + |\cdot\rangle) \to (|\cdot\rangle + |\cdot\rangle)(|\cdot\rangle + |\cdot\rangle)(|\cdot\rangle + |\cdot\rangle).$

Since these two error correction steps are independent, the code also works if there is both a bit flip error *and* a phase flip error.

Note that in both cases, we must be careful to measure just what we want to know and no more, or we would collapse the superposition used in the code. I'll discuss this in more detail in section 4.

The bit flip, phase flip, and combined bit and phase flip errors are important, so let's take a short digression to discuss them. We'll also throw in the identity matrix, which is what we get if no error occurs. The definitions of these four operators are given in table 1. The factor of i in the definition of Y has little practical significance — overall phases in quantum mechanics are physically meaningless — but it makes some manipulations easier later. It also makes some manipulations harder, so either is a potentially reasonable convention.

The group generated by tensor products of these 4 operators is called the Pauli group. X, Y, and Z anticommute: $XZ = -ZX$ (also written $\{X, Z\} = 0$). Similarly, $\{X, Y\} = 0$ and $\{Y, Z\} = 0$. Thus, the n-qubit Pauli group $\mathcal{P}_n$ consists of the 4^n tensor products of I, X, Y, and Z, and an overall phase of ± 1 or $\pm i$, for a total of 4^{n+1} elements. The phase of the operators used is not generally very important, but we can't discard it completely. For one thing, the fact that this is not an Abelian group is quite important, and we would lose that if we dropped the phase!

$\mathcal{P}_n$ is useful because of its nice algebraic properties. Any pair of elements of $\mathcal{P}_n$ either commute or anticommute. Also, the square of any element of $\mathcal{P}_n$ is ± 1. We shall only need to work with the elements with square $+1$, which are tensor products of I, X, Y, and Z with an overall sign ± 1; the phase i is only necessary to make $\mathcal{P}_n$ a group. Define the *weight* of an operator in $\mathcal{P}_n$ to be the number of tensor factors which are not I. Thus, $X \otimes Y \otimes I$ has weight 2.

Another reason the Pauli matrices are important is that they span the space of 2×2 matrices, and the n-qubit Pauli group spans the space of $2^n \times 2^n$ matrices. For instance, if we have a general phase error

$$(2.6) \qquad R_{\theta/2} = \begin{pmatrix} 1 & 0 \\ 0 & e^{i\theta} \end{pmatrix} = e^{i\theta/2} \begin{pmatrix} e^{-i\theta/2} & 0 \\ 0 & e^{i\theta/2} \end{pmatrix}$$

(again, the overall phase does not matter), we can write it as

$$(2.7) \qquad R_{\theta/2} = \cos\frac{\theta}{2} I - i \sin\frac{\theta}{2} Z.$$

It turns out that our earlier error correction procedure will also correct this error, without any additional effort. For instance, the earlier procedure might use some extra qubits (*ancilla* qubits) that are initialized to $|0\rangle$ and record what type of error occurred. Then we look at the ancilla and invert the error it tells us:

$$(2.8) \qquad Z\left(\alpha|\overline{0}\rangle + \beta|\overline{1}\rangle\right) \otimes |0\rangle_{\text{anc}} \to Z\left(\alpha|\overline{0}\rangle + \beta|\overline{1}\rangle\right) \otimes |Z\rangle_{\text{anc}}$$

$$(2.9) \qquad \to \left(\alpha|\overline{0}\rangle + \beta|\overline{1}\rangle\right) \otimes |Z\rangle_{\text{anc}}$$

$$(2.10) \qquad I\left(\alpha|\overline{0}\rangle + \beta|\overline{1}\rangle\right) \otimes |0\rangle_{\text{anc}} \to I\left(\alpha|\overline{0}\rangle + \beta|\overline{1}\rangle\right) \otimes |\text{no error}\rangle_{\text{anc}}$$

$$(2.11) \qquad \to \left(\alpha|\overline{0}\rangle + \beta|\overline{1}\rangle\right) \otimes |\text{no error}\rangle_{\text{anc}}$$

When the actual error is $R_{\theta/2}$, recording the error in the ancilla gives us a superposition:

$$(2.12) \quad \cos\frac{\theta}{2} I\left(\alpha|\overline{0}\rangle + \beta|\overline{1}\rangle\right) \otimes |\text{no error}\rangle_{\text{anc}} - i\sin\frac{\theta}{2} Z\left(\alpha|\overline{0}\rangle + \beta|\overline{1}\rangle\right) \otimes |Z\rangle_{\text{anc}}$$

Then we measure the ancilla, which with probability $\sin^2 \theta/2$ gives us

$$(2.13) \qquad Z\left(\alpha|\overline{0}\rangle + \beta|\overline{1}\rangle\right) \otimes |Z\rangle_{\text{anc}},$$

and with probability $\cos^2 \theta/2$ gives us

$$(2.14) \qquad I\left(\alpha|\overline{0}\rangle + \beta|\overline{1}\rangle\right) \otimes |\text{no error}\rangle_{\text{anc}}.$$

In each case, inverting the error indicated in the ancilla restores the original state.

It is easy to see this argument works for any linear combination of errors [16, 18]:

THEOREM 2. *If a quantum code corrects errors A and B, it also corrects any linear combination of A and B. In particular, if it corrects all weight t Pauli errors, then the code corrects all t-qubit errors.*

So far, we have only considered individual unitary errors that occur on the code. But we can easily add in all possible quantum errors. The most general quantum operation, including decoherence, interacts the quantum state with some extra qubits via a unitary operation, then discards some qubits. This process can turn pure quantum states into mixed quantum states, which are normally described using density matrices. We can write the most general operation as a transformation on density matrices

$$\rho \to \sum_i E_i \rho E_i^\dagger, \tag{2.15}$$

where the E_is are normalized so $\sum E_i^\dagger E_i = I$. The density matrix ρ can be considered to represent an ensemble of pure quantum states $|\psi\rangle$, each of which, in this case, should be in the coding space of the code. Then this operation simply performs the following operation on each $|\psi\rangle$:

$$|\psi\rangle \to E_i|\psi\rangle \text{ with probability } |E_i|\psi\rangle|^2. \tag{2.16}$$

If we can correct each of the individual errors E_i, then we can correct this general error as well. For instance, for quantum operations that only affect a single qubit of the code, E_i will necessarily be in the linear span of I, X, Y, and Z, so we can correct it. Thus, in the statement of theorem 2, "all t-qubit errors" really does apply to *all* t-qubit errors, not just unitary ones.

We can go even further. It is not unreasonable to expect that every qubit in our nine-qubit code will be undergoing some small error. For instance, qubit i experiences the error $I + \epsilon E_i$, where E_i is some single-qubit error. Then the overall error is

$$\bigotimes (I + \epsilon E_i) = I + \epsilon \left(E_1 \otimes I^{\otimes 8} + I \otimes E_2 \otimes I^{\otimes 7} + \ldots \right) + O(\epsilon^2) \tag{2.17}$$

That is, to order ϵ, the actual error is the sum of single-qubit errors, which we know the nine-qubit code can correct. That means that after the error correction procedure, the state will be correct to $O(\epsilon^2)$ (when the two-qubit error terms begin to become important). While the code cannot completely correct this error, it still produces a significant improvement over not doing error correction when ϵ is small. A code correcting more errors would do even better.

3. General properties of quantum error-correcting codes

Let us try to understand what properties are essential to the success of the nine-qubit code, and derive conditions for a subspace to form a quantum error-correcting code.

One useful feature was *linearity*, which will be true of any quantum code. We only need to correct a basis of errors (I, X, Y, and Z in the one-qubit case), and all other errors will follow, as per theorem 2.

In any code, we must never confuse $|\bar{0}\rangle$ with $|\bar{1}\rangle$, even in the presence of errors. That is, $E|\bar{0}\rangle$ is orthogonal to $F|\bar{1}\rangle$:

$$\langle \bar{0}|E^\dagger F|\bar{1}\rangle = 0. \tag{3.1}$$

It is *sufficient* to distinguish error E from error F when they act on $|\bar{0}\rangle$ and $|\bar{1}\rangle$. Then a measurement will tell us exactly what the error is and we can correct it:

$$\langle \bar{0}|E^\dagger F|\bar{0}\rangle = \langle \bar{1}|E^\dagger F|\bar{1}\rangle = 0 \tag{3.2}$$

for $E \ne F$.

But (3.2) is not *necessary*: in the nine-qubit code, we cannot distinguish between Z_1 and Z_2, but that is OK, since we can correct either one with a single operation. To understand the necessary condition, it is helpful to look at the operators $F_1 = (Z_1 + Z_2)/2$ and $F_2 = (Z_1 - Z_2)/2$ instead of Z_1 and Z_2. F_1 and F_2 span the same space as Z_1 and Z_2, so Shor's code certainly corrects them; let us try to understand how. When we use the Fs as the basis errors, now equation (3.2) *is* satisfied. That means we can make a measurement and learn what the error is. We also have to invert it, and this is a potential problem, since F_1 and F_2 are not unitary. However, F_1 acts the same way as Z_1 on the coding space, so $Z_1^\dagger$ suffices to invert F_1 on the states of interest. F_2 acts the same way as the 0 operator on the coding space. We can't invert this, but we don't need to — since F_2 annihilates codewords, it can never contribute a component to the actual state of the system.

The requirement to invert the errors produces a third condition:

$$(3.3) \qquad \langle \bar{0} | E^\dagger E | \bar{0} \rangle = \langle \bar{1} | E^\dagger E | \bar{1} \rangle.$$

Either this value is nonzero, as for F_1, in which case some unitary operator will act the same way as E on the coding space, or it will be zero, as for F_2, in which case E annihilates codewords and never arises.

These arguments show that if there is some basis for the space of errors for which equations (3.1), (3.2), and (3.3) hold, then the states $|\bar{0}\rangle$ and $|\bar{1}\rangle$ span a quantum error-correcting code. Massaging these three equations together and generalizing to multiple encoded qubits, we get the following theorem [2, 11]:

THEOREM 3. *Suppose $\mathcal{E}$ is a linear space of errors acting on the Hilbert space $\mathcal{H}$. Then a subspace C of $\mathcal{H}$ forms a quantum error-correcting code correcting the errors $\mathcal{E}$ iff*

$$(3.4) \qquad \langle \psi | E^\dagger E | \psi \rangle = C(E)$$

for all $E \in \mathcal{E}$. The function $C(E)$ does not depend on the state $|\psi\rangle$.

PROOF. Suppose $\{E_a\}$ is a basis for $\mathcal{E}$ and $\{|\psi_i\rangle\}$ is a basis for C. By setting E and $|\psi\rangle$ equal to the basis elements and to the sum and difference of two basis elements (with or without a phase factor i), we can see that (3.4) is equivalent to

$$(3.5) \qquad \langle \psi_i | E_a^\dagger E_b | \psi_j \rangle = C_{ab} \delta_{ij},$$

where C_{ab} is a Hermitian matrix independent of i and j.

Suppose equation (3.5) holds. We can diagonalize C_{ab}. This involves choosing a new basis $\{F_a\}$ for $\mathcal{E}$, and the result is equations (3.1), (3.2), and (3.3). The arguments before the theorem show that we can measure the error, determine it uniquely (in the new basis), and invert it (on the coding space). Thus, we have a quantum error-correcting code.

Now suppose we have a quantum error-correcting code, and let $|\psi\rangle$ and $|\phi\rangle$ be two distinct codewords. Then we must have

$$(3.6) \qquad \langle \psi | E^\dagger E | \psi \rangle = \langle \phi | E^\dagger E | \phi \rangle$$

for all E. That is, (3.4) must hold. If not, E changes the relative size of $|\psi\rangle$ and $|\phi\rangle$. Both $|\psi\rangle + |\phi\rangle$ and $|\psi\rangle + c|\phi\rangle$ are valid codewords, and

$$(3.7) \qquad E(|\psi\rangle + |\phi\rangle) = N(|\psi\rangle + c|\phi\rangle),$$

where N is a normalization factor and

$$(3.8) \qquad c = \langle \psi | E^\dagger E | \psi \rangle / \langle \phi | E^\dagger E | \phi \rangle.$$

The error E will actually change the encoded state, which is a failure of the code, unless $c = 1$. □

There is a slight subtlety to the phrasing of equation (3.4). We require $\mathcal{E}$ to be a linear space of errors, which means that it must be closed under sums of errors which may act on different qubits. In contrast, for a code that corrects t errors, in (3.5), it is safe to consider only E_a and E_b acting on just t qubits. We can restrict even further, and only use Pauli operators as E_a and E_b, since they will span the space of t-qubit errors. This leads us to a third variation of the condition:

$$\langle \psi | E | \psi \rangle = C'(E), \tag{3.9}$$

where E is now any operator acting on $2t$ qubits (that is, it replaces $E_a^\dagger E_b$ in (3.5)). This can be easily interpreted as saying that no measurement on $2t$ qubits can learn information about the codeword. Alternatively, it says we can *detect* up to $2t$ errors on the code without necessarily being able to say what those errors are. That is, we can distinguish those errors from the identity.

If the matrix C_{ab} in (3.5) has maximum rank, the code is called *nondegenerate*. If not, as for the nine-qubit code, the code is *degenerate*. In a degenerate code, different errors look the same when acting on the coding subspace.

For a nondegenerate code, we can set a simple bound on the parameters of the code simply by counting states. Each error E acting on each basis codeword $|\psi_i\rangle$ produces a linearly independent state. All of these states must fit in the full Hilbert space of n qubits, which has dimension 2^n. If the code encodes k qubits, and corrects errors on up to t qubits, then

$$\left(\sum_{j=0}^{t} 3^j \binom{n}{j} \right) 2^k \leq 2^n. \tag{3.10}$$

The quantity in parentheses is the number of errors of *weight* t or less: that is, the number of tensor products of I, X, Y, and Z that are the identity in all but t or fewer places. This inequality is called the *quantum Hamming bound*. While the quantum Hamming bound only applies to nondegenerate codes, we do not know of any codes that beat it.

For $t = 1$, $k = 1$, the quantum Hamming bound tells us $n \geq 5$. In fact, there is a code with $n = 5$, which you will see later. A code that corrects t errors is said to have *distance* $2t + 1$, because it takes $2t + 1$ single-qubit changes to get from one codeword to another. We can also define distance as the minimum weight of an operator E that violates equation (3.9) (a definition which also allows codes of even distance). A quantum code using n qubits to encode k qubits with distance d is written as an $[[n, k, d]]$ code (the double brackets distinguish it from a classical code). Thus, the nine-qubit code is a $[[9, 1, 3]]$ code, and the five-qubit code is a $[[5, 1, 3]]$ code.

We can also set a lower bound telling us when codes exist. I will not prove this here, but an $[[n, k, d]]$ code exists when

$$\left(\sum_{j=0}^{d-1} 3^j \binom{n}{j} \right) 2^k \leq 2^n \tag{3.11}$$

(known as the quantum Gilbert-Varshamov bound [3]). This differs from the quantum Hamming bound in that the sum goes up to $d-1$ (which is equal to $2t$) rather than stopping at t.

THEOREM 4. *A quantum $[[n,k,d]]$ code exists when (3.11) holds. Any nondegenerate $[[n,k,d]]$ code must satisfy (3.10). For large n, $R = k/n$ and $p = d/2n$ fixed, the best nondegenerate quantum codes satisfy*

$$1 - 2p\log_2 3 - H(2p) \leq R \leq 1 - p\log_2 3 - H(p),$$

where $H(x) = -x\log_2 x - (1-x)\log_2(1-x)$.

One further bound, known as the Knill-Laflamme bound [11] or the quantum Singleton bound, applies even to degenerate quantum codes. For an $[[n,k,d]]$ quantum code,

(3.12) $$n - k \geq 2d - 2.$$

This shows that the $[[5,1,3]]$ code really is optimal — a $[[4,1,3]]$ code would violate this bound.

I will not prove the general case of this bound, but the case of $k = 1$ can be easily understood as a consequence of the No-Cloning theorem. Suppose r qubits of the code are missing. We can substitute $|0\rangle$ states for the missing qubits, but there are r errors on the resulting codeword. The errors are of unknown type, but all the possibilities are on the same set of r qubits. Thus, all products $E_a^\dagger E_b$ in condition (3.5) have weight r or less, so this sort of error (an "erasure" error [9]) can be corrected by a code of distance $r+1$. Now suppose we had an $[[n,1,d]]$ code with $n \leq 2d - 2$. Then we could split the qubits in the code into two groups of size at most $d-1$. Each group would have been subject to at most $d-1$ erasure errors, and could therefore be corrected without access to the other group. This would produce two copies of the encoded state, which we know is impossible.

4. Stabilizer codes

Now let us return to the nine-qubit code, and examine precisely what we need to do to correct errors.

First, we must determine if the first three qubits are all the same, and if not, which is different. We can do this by measuring the parity of the first two qubits and the parity of the second and third qubits. That is, we measure

(4.1) $$Z \otimes Z \otimes I \text{ and } I \otimes Z \otimes Z.$$

The first tells us if an X error has occurred on qubits one or two, and the second tells us if an X error has occurred on qubits two or three. Note that the error detected in both cases anticommutes with the error measured. Combining the two pieces of information tells us precisely where the error is.

We do the same thing for the other two sets of three. That gives us four more operators to measure. Note that measuring $Z \otimes Z$ gives us just the information we want and no more. This is crucial so that we do not collapse the superpositions used in the code. We can do this by bringing in an ancilla qubit. We start it in the state $|0\rangle + |1\rangle$ and perform controlled-Z operations to the first and second qubits

$$\begin{array}{ccccccccc}
Z & Z & I & I & I & I & I & I & I \\
I & Z & Z & I & I & I & I & I & I \\
I & I & I & Z & Z & I & I & I & I \\
I & I & I & I & Z & Z & I & I & I \\
I & I & I & I & I & I & Z & Z & I \\
I & I & I & I & I & I & I & Z & Z \\
X & X & X & X & X & X & I & I & I \\
I & I & I & X & X & X & X & X & X
\end{array}$$

TABLE 2. The stabilizer for the nine-qubit code. Each column represents a different qubit.

of the code:

$$(4.2) \quad (|0\rangle + |1\rangle) \sum_{abc} c_{abc} |abc\rangle \to \sum_{abc} c_{abc} \left(|0\rangle |abc\rangle + (-1)^{a \oplus b} |1\rangle |abc\rangle \right)$$

$$(4.3) \qquad\qquad = \sum_{abc} c_{abc} \left(|0\rangle + (-1)^{\text{parity}(a,b)} |1\rangle \right) |abc\rangle.$$

At this point, measuring the ancilla in the basis $|0\rangle \pm |1\rangle$ will tell us the eigenvalue of $Z \otimes Z \otimes I$, but nothing else about the data.

Second, we must check if the three signs are the same or different. We do this by measuring

$$(4.4) \qquad X \otimes X \otimes X \otimes X \otimes X \otimes X \otimes I \otimes I \otimes I$$

and

$$(4.5) \qquad I \otimes I \otimes I \otimes X \otimes X \otimes X \otimes X \otimes X \otimes X.$$

This gives us a total of 8 operators to measure. These two measurements detect Z errors on the first six and last six qubits, correspondingly. Again note that the error detected anticommutes with the operator measured.

This is no coincidence: in each case, we are measuring an operator M which should have eigenvalue $+1$ for any codeword:

$$(4.6) \qquad\qquad\qquad M|\psi\rangle = |\psi\rangle.$$

If an error E which anticommutes with M has occurred, then the true state is $E|\psi\rangle$, and

$$(4.7) \qquad\qquad M(E|\psi\rangle) = -EM|\psi\rangle = -E|\psi\rangle.$$

That is, the new state has eigenvalue -1 instead of $+1$. We use this fact to correct errors: each single-qubit error E anticommutes with a particular set of operators $\{M\}$; which set, exactly, tells us what E is.

In the case of the nine-qubit code, we cannot tell exactly what E is, but it does not matter. For instance, we cannot distinguish Z_1 and Z_2 because

$$(4.8) \qquad\qquad Z_1 Z_2 |\psi\rangle = |\psi\rangle \iff Z_1 |\psi\rangle = Z_2 |\psi\rangle.$$

This is an example of the fact that the nine-qubit code is degenerate.

Table 2 summarizes the operators we measured. These 8 operators generate an Abelian group called the *stabilizer* of the nine-qubit code. The stabilizer contains all operators M in the Pauli group for which $M|\psi\rangle = |\psi\rangle$ for all $|\psi\rangle$ in the code.

Conversely, given an Abelian subgroup S of the Pauli group $\mathcal{P}_n$ (which, if you recall, consists of tensor products of I, X, Y, and Z with an overall phase of $\pm 1, \pm i$), we can define a quantum code $T(S)$ as the set of states $|\psi\rangle$ for which $M|\psi\rangle = |\psi\rangle$ for all $M \in S$. S must be Abelian and cannot contain -1, or the code is trivial: If $M, N \in S$,

$$MN|\psi\rangle = M|\psi\rangle = |\psi\rangle \tag{4.9}$$
$$NM|\psi\rangle = N|\psi\rangle = |\psi\rangle \tag{4.10}$$

so

$$[M, N]|\psi\rangle = MN|\psi\rangle - NM|\psi\rangle = 0. \tag{4.11}$$

Since elements of the Pauli group either commute or anticommute, $[M, N] = 0$. Clearly, if $M = -1 \in S$, there is no nontrivial $|\psi\rangle$ for which $M|\psi\rangle = |\psi\rangle$.

If these conditions are satisfied, there will be a nontrivial subspace consisting of states fixed by all elements of the stabilizer. We can tell how many errors the code corrects by looking at operators that commute with the stabilizer. We can correct errors E and F if either $E^\dagger F \in S$ (so E and F act the same on codewords), or if $\exists M \in S$ s.t. $\{M, E^\dagger F\} = 0$, in which case measuring the operator M distinguishes between E and F. If the first condition is ever true, the stabilizer code is degenerate; otherwise it is nondegenerate.

We can codify this by looking at the normalizer $N(S)$ of S in the Pauli group (which is in this case equal to the centralizer, composed of Pauli operators which commute with S). The distance d of the code is the minimum weight of any operator in $N(S) \setminus S$ [**3, 7**].

THEOREM 5. *Let S be an Abelian subgroup of order 2^a of the n-qubit Pauli group, and suppose $-1 \notin S$. Let d be the minimum weight of an operator in $N(S) \setminus S$. Then the space of states $T(S)$ stabilized by all elements of S is an $[[n, n-a, d]]$ quantum code.*

To correct errors of weight $(d-1)/2$ or below, we simply measure the generators of S. This will give us a list of eigenvalues, the *error syndrome*, which tells us whether the error E commutes or anticommutes with each of the generators. The error syndromes of E and F are equal iff the error syndrome of $E^\dagger F$ is trivial. For a nondegenerate code, the error syndrome uniquely determines the error E (up to a trivial overall phase) — the generator that anticommutes with $E^\dagger F$ distinguishes E from F. For a degenerate code, the error syndrome is not unique, but error syndromes are only repeated when $E^\dagger F \in S$, implying E and F act the same way on the codewords.

If the stabilizer has a generators, then the code encodes $n - a$ qubits. Each generator divides the allowed Hilbert space into $+1$ and -1 eigenspaces of equal sizes. To prove the statement, note that we can find an element G of the Pauli group that has any given error syndrome (though G may have weight greater than $(d-1)/2$, or even greater than d). Each G maps $T(S)$ into an orthogonal but isomorphic subspace, and there are 2^a possible error syndromes, so $T(S)$ has dimension at most $2^n/2^a$. In addition, the Pauli group spans $U(2^n)$, so its orbit acting on any single state contains a basis for $\mathcal{H}$. Every Pauli operator has *some* error syndrome, so $T(S)$ has dimension exactly 2^{n-a}.

$$\begin{array}{ccccccc}
Z & Z & Z & Z & I & I & I \\
Z & Z & I & I & Z & Z & I \\
Z & I & Z & I & Z & I & Z \\
X & X & X & X & I & I & I \\
X & X & I & I & X & X & I \\
X & I & X & I & X & I & X
\end{array}$$

TABLE 3. Stabilizer for the seven-qubit code.

5. Some other important codes

Stabilizers make it easy to describe new codes. For instance, we can start from classical coding theory, which describes a linear code by a generator matrix or its dual, the parity check matrix. Each row of the generator matrix is a codeword, and the other codewords are all linear combinations of the rows of the generator matrix. The rows of the parity check matrix specify parity checks all the classical codewords must satisfy. (In quantum codes, the stabilizer is closely analogous to the classical parity check matrix.) One well-known code is the seven-bit Hamming code correcting one error, with parity check matrix

(5.1)
$$\begin{pmatrix} 1 & 1 & 1 & 1 & 0 & 0 & 0 \\ 1 & 1 & 0 & 0 & 1 & 1 & 0 \\ 1 & 0 & 1 & 0 & 1 & 0 & 1 \end{pmatrix}.$$

If we replace each 1 in this matrix by the operator Z, and 0 by I, we are really changing nothing, just specifying three operators that implement the parity check measurements. The statement that the classical Hamming code corrects one error is the statement that each bit flip error of weight one or two anticommutes with one of these three operators.

Now suppose we replace each 1 by X instead of Z. We again get three operators, and they will anticommute with any weight one or two Z error. Thus, if we make a stabilizer out of the three Z operators and the three X operators, as in table 3, we get a code that can correct any single qubit error [18]. X errors are picked up by the first three generators, Z errors by the last three, and Y errors are distinguished by showing up in both halves. Of course, there is one thing to check: the stabilizer must be Abelian; but that is easily verified. The stabilizer has 6 generators on 7 qubits, so it encodes 1 qubit — it is a $[[7, 1, 3]]$ code.

In this example, we used the same classical code for both the X and Z generators, but there was no reason we had to do so. We could have used any two classical codes C_1 and C_2 [5, 19]. The only requirement is that the X and Z generators commute. This corresponds to the statement that $C_2^\perp \subseteq C_1$ ($C_2^\perp$ is the dual code to C_2, consisting of those words which are orthogonal to the codewords of C_2). If C_1 is an $[n, k_1, d_1]$ code, and C_2 is an $[n, k_2, d_2]$ code (recall single brackets means a classical code), then the corresponding quantum code is an $[[n, k_1+k_2-n, \min(d_1, d_2)]]$ code.[2] This construction is known as the CSS construction after its inventors Calderbank, Shor, and Steane.

The codewords of a CSS code have a particularly nice form. They all must satisfy the same parity checks as the classical code C_1, so all codewords will be

---

[2]In fact, the true distance of the code could be larger than expected because of the possibility of degeneracy, which would not have been a factor for the classical codes.

$$\begin{array}{ccccc} X & Z & Z & X & I \\ I & X & Z & Z & X \\ X & I & X & Z & Z \\ Z & X & I & X & Z \end{array}$$

TABLE 4. The stabilizer for the five-qubit code.

superpositions of words of C_1. The parity check matrix of C_2 is the generator matrix of $C_2^\perp$, so the X generators of the stabilizer add a word of $C_2^\perp$ to the state. Thus, the codewords of a CSS code are of the form

$$(5.2) \qquad \sum_{w \in C_2^\perp} |u+w\rangle,$$

where $u \in C_1$ ($C_2^\perp \subseteq C_1$, so $u + w \in C_1$). If we perform a Hadamard transform

$$(5.3) \qquad |0\rangle \longleftrightarrow |0\rangle + |1\rangle$$
$$(5.4) \qquad |1\rangle \longleftrightarrow |0\rangle - |1\rangle$$

on each qubit of the code, we switch the Z basis with the X basis, and C_1 with C_2, so the codewords are now

$$(5.5) \qquad \sum_{w \in C_1^\perp} |u+w\rangle \quad (u \in C_2).$$

Thus, to correct errors for a CSS code, we can measure the parities of C_1 in the Z basis, and the parities of C_2 in the X basis.

Another even smaller quantum code is the $[[5,1,3]]$ code I promised earlier [**2, 13**]. Its stabilizer is given in table 4. I leave it to you to verify that it commutes and actually does have distance 3. You can also work out the codewords. Since multiplication by $M \in S$ merely rearranges elements of the group S, the sum

$$(5.6) \qquad \left(\sum_{M \in S} M\right) |\phi\rangle$$

is in the code for any state $|\phi\rangle$. You only need find two states $|\phi\rangle$ for which (5.6) is nonzero. Note that as well as telling us about the error-correcting properties of the code, the stabilizer provides a more compact notation for the coding subspace than listing the basis codewords.

A representation of stabilizers that is often useful is as a pair of binary matrices, frequently written adjacent with a line between them [**3**]. The first matrix has a 1 everywhere the stabilizer has an X or a Y, and a 0 elsewhere; the second matrix has a 1 where the stabilizer has a Y or a Z. Multiplying together Pauli operators corresponds to adding the two rows for both matrices. Two operators M and N commute iff their binary vector representations $(a_1|b_1)$, (a_2,b_2) are orthogonal under a symplectic inner product: $a_1 b_2 + b_1 a_2 = 0$. For instance, the stabilizer for the five-qubit code becomes the matrix

$$(5.7) \qquad \left(\begin{array}{ccccc|ccccc} 1 & 0 & 0 & 1 & 0 & 0 & 1 & 1 & 0 & 0 \\ 0 & 1 & 0 & 0 & 1 & 0 & 0 & 1 & 1 & 0 \\ 1 & 0 & 1 & 0 & 0 & 0 & 0 & 0 & 1 & 1 \\ 0 & 1 & 0 & 1 & 0 & 1 & 0 & 0 & 0 & 1 \end{array}\right)$$

Stabilizers	GF(4)
I	0
Z	1
X	ω
Y	ω^2
tensor products	vectors
multiplication	addition
$[M, N] = 0$	$tr\,(M \cdot \overline{N}) = 0$
$N(S)$	dual

TABLE 5. Connections between stabilizer codes and codes over GF(4).

6. Codes over GF(4)

I will finish by describing another connection to classical coding theory. Frequently, classical coding theorists consider not just binary codes, but codes over larger finite fields. One of the simplest is GF(4), the finite field with four elements. It is a field of characteristic 2, containing the elements $\{0, 1, \omega, \omega^2\}$.

(6.1) $$\omega^3 = 1,\ \omega + \omega^2 = 1$$

It is also useful to consider two operations on GF(4). One is conjugation, which switches the two roots of the characteristic polynomial $x^2 + x + 1$:

(6.2) $\qquad\qquad\overline{1} = 1 \qquad\qquad \overline{\omega} = \omega^2$

(6.3) $\qquad\qquad\overline{0} = 0 \qquad\qquad \overline{\omega^2} = \omega$

The other is trace. $\mathbf{tr}\,x$ is the trace of the linear operator "multiplication by x" when GF(4) is considered as a vector space over $\mathbb{Z}_2$:

(6.4) $\qquad\qquad\mathbf{tr}\,0 = \mathbf{tr}\,1 = 0$

(6.5) $\qquad\qquad\mathbf{tr}\,\omega = \mathbf{tr}\,\omega^2 = 1$

Stabilizer codes make extensive use of the Pauli group $\mathcal{P}_n$. We can make a connection between stabilizer codes and codes over GF(4) by identifying the four operators I, X, Y, and Z with the four elements of GF(4), as in table 5 [4].

The commutativity constraint in the Pauli group becomes a symplectic inner product between vectors in GF(4). The fact that the stabilizer is Abelian can be phrased in the language of GF(4) as the fact that the code must be contained in its dual with respect to this inner product. To determine the number of errors corrected by the code, we must examine vectors which are in the dual (corresponding to $N(S)$) but not in the code (corresponding to S).

The advantage of making this correspondence is that a great deal of classical coding theory instantly becomes available. Many classical codes over GF(4) are known, and many of them are self-dual with respect to the symplectic inner product, so they define quantum codes. For instance, the five-qubit code is one such — in fact, it is just a Hamming code over GF(4)! Of course, mostly classical coding theorists consider *linear* codes (which are closed under addition and scalar multiplication), whereas in the quantum case we wish to consider the slightly more general class of *additive* GF(4) codes (that is, codes which are closed under addition of elements, but not necessarily scalar multiplication).

7. Fault-Tolerant Quantum Computation

Hopefully, this paper has given you an understanding of quantum error-correcting codes, but there is still a major hurdle before the goal of making quantum computers resistant to errors. You must also understand how to perform operations on a state encoded in a quantum code without losing the code's protection against errors, and how to safely perform error correction when the gates used are themselves noisy. For a full discussion of this problem and its resolutions, see [**14**] or [**15**].

Shor presented the first protocols for fault-tolerant quantum computation [**17**]. While those protocols can be extended to work for arbitrary stabilizer codes, including those with multiple encoded qubits per block [**8**], the gates which can be performed easily on the code arise from symmetries of the stabilizer. The stabilizer of the seven-qubit code has a particularly large symmetry group and therefore is particularly good for fault-tolerant computation.

When the error rate per gate is low enough, encoding a state in a quantum code and performing fault-tolerant operations will reduce the effective error rate. By concatenating the seven-qubit code or another code (i.e., encoding each qubit of the code with another copy of the seven-qubit code, and possibly repeating the procedure multiple times), we can compound this improvement, giving a threshold result [**1, 10, 12**]: if the error rate is below some threshold value, concatenating a code allows us to perform arbitrarily long fault-tolerant quantum computations, with overhead that is polylogarithmic in the length of the computation.

8. Summary (Quantum Error Correction Sonnet)

We cannot clone, perforce; instead, we split
    Coherence to protect it from that wrong
    That would destroy our valued quantum bit
    And make our computation take too long.
Correct a flip and phase - that will suffice.
    If in our code another error's bred,
    We simply measure it, then God plays dice,
    Collapsing it to X or Y or Zed.
We start with noisy seven, nine, or five
    And end with perfect one. To better spot
    Those flaws we must avoid, we first must strive
    To find which ones commute and which do not.
With group and eigenstate, we've learned to fix
    Your quantum errors with our quantum tricks.

References

[1] D. Aharonov and M. Ben-Or, *Fault-tolerant quantum computation with constant error*, Proc. 29th Ann. ACM Symp. on Theory of Computation (ACM, New York, 1998), pp. 176–188, http://xxx.lanl.gov/abs/quant-ph/9611025; D. Aharonov and M. Ben-Or, *Fault-tolerant quantum computation with constant error rate*, http://xxx.lanl.gov/abs/quant-ph/9906129.

[2] C. Bennett, D. DiVincenzo, J. Smolin, and W. Wootters, *Mixed state entanglement and quantum error correction*, Phys. Rev. A **54** (1996), 3824–3851; http://xxx.lanl.gov/abs/quant-ph/9604024.
[3] A. R. Calderbank, E. M. Rains, P. W. Shor, and N. J. A. Sloane, *Quantum error correction and orthogonal geometry*, Phys. Rev. Lett. **78** (1997), 405–408; http://xxx.lanl.gov/abs/quant-ph/9605005.
[4] ———, *Quantum error correction via codes over* GF(4), IEEE Trans. Inform. Theory **44** (1998), 1369–1387; http://xxx.lanl.gov/abs/quant-ph/9605005.
[5] A. R. Calderbank and P. W. Shor, *Good quantum error-correcting codes exist*, Phys. Rev. A **54** (1996), 1098–1105; http://xxx.lanl.gov/abs/quant-ph/9512032.
[6] D. Dieks, *Communication by EPR devices*, Phys. Lett. A **92** (1982), 271–272.
[7] D. Gottesman, *Class of quantum error-correcting codes saturating the quantum Hamming bound*, Phys. Rev. A **54** (1996), 1862–1868; http://xxx.lanl.gov/abs/quant-ph/9604038.
[8] ———, *Theory of fault-tolerant quantum computation*, Phys. Rev. A **57** (1998), 127–137; http://xxx.lanl.gov/abs/quant-ph/9702029.
[9] M. Grassl, T. Beth, and T. Pellizzari, *Codes for the quantum erasure channel*, Phys. Rev. A **56** (1997), 33–38; http://xxx.lanl.gov/abs/quant-ph/9610042.
[10] A. Y. Kitaev, *Quantum error correction with imperfect gates*, Quantum Communication, Computing, and Measurement (Proc. 3rd Int. Conf. of Quantum Communication and Measurement) (Plenum Press, New York, 1997), p. 181–188.
[11] E. Knill and R. Laflamme, *A theory of quantum error-correcting codes*, Phys. Rev. A **55** (1997), 900–911; http://xxx.lanl.gov/abs/quant-ph/9604034.
[12] E. Knill, R. Laflamme, and W. H. Zurek, *Threshold accuracy for quantum computation*, quant-ph/9610011; E. Knill, R. Laflamme, and W. H. Zurek, *Resilient quantum computation*, Science **279** (1998), 342–345; E. Knill, R. Laflamme, and W. H. Zurek, *Resilient quantum computation: error models and thresholds*, Proc. Royal Soc. London A **454** (1998), 365–384, http://xxx.lanl.gov/abs/quant-ph/9702058.
[13] R. Laflamme, C. Miquel, J. P. Paz, and W. Zurek, *Perfect quantum error correction code*, Phys. Rev. Lett. **77** (1996), 198–201; http://xxx.lanl.gov/abs/quant-ph/9602019.
[14] J. Preskill, *Reliable quantum computers*, Proc. Roy. Soc. A **454**, 385–410 (1998); quant-ph/9705031.
[15] ———, "Fault-tolerant quantum computation," ch. 8 in *Introduction to Quantum Computation and Information*, eds. H.-K. Lo, S. Popescu, and T. Spiller (World Scientific, New Jersey, 1998), pp. 213–269; http://xxx.lanl.gov/abs/quant-ph/9712048.
[16] P. W. Shor, *Scheme for reducing decoherence in quantum memory*, Phys. Rev. A **52** (1995), 2493–2496.
[17] ———, *Fault-tolerant quantum computation*, Proc. 35th Ann. Symp. on Fundamentals of Computer Science (IEEE Press, Los Alamitos, 1996), pp. 56–65; http://xxx.lanl.gov/abs/quant-ph/9605011.
[18] A. M. Steane, *Error correcting codes in quantum theory*, Phys. Rev. Lett. **77** (1996), 793–797.
[19] ———, *Multiple particle interference and quantum error correction*, Proc. Roy. Soc. London A **452** (1996), 2551–2577; http://xxx.lanl.gov/abs/quant-ph/9601029.
[20] W. K. Wooters and W. H. Zurek, *A single quantum cannot be cloned*, Nature **299** (1982), 802–803.

SODA HALL 585, EECS: COMPUTER SCIENCE DIVISION, UNIVERSITY OF CALIFORNIA BERKELEY, BERKELEY, CA 94720

E-mail address: gottesma@eecs.berkeley.edu

URL: http://www.cs.berkeley.edu/~gottesma/

A Talk on Quantum Cryptography
or
How Alice Outwits Eve

Samuel J. Lomonaco, Jr.

ABSTRACT. Alice and Bob wish to communicate without the archvillainess Eve eavesdropping on their conversation. Alice, decides to take two college courses, one in cryptography, the other in quantum mechanics. During the courses, she discovers she can use what she has just learned to devise a cryptographic communication system that automatically detects whether or not Eve is up to her villainous eavesdropping. Some of the topics discussed are Heisenberg's Uncertainty Principle, the Vernam cipher, the BB84 and B92 cryptographic protocols. The talk ends with a discussion of some of Eve's possible eavesdropping strategies, opaque eavesdropping, translucent eavesdropping, and translucent eavesdropping with entanglement.

Contents

1. **Preface**
1.1. The Unique Contribution of Quantum Cryptography
1.2. A Note to the Reader

2. **Introduction**

3. **A Course on Classical Cryptography**
3.1. Alice's enthusiastic decision
3.2. Plaintext, ciphertext, key, and ... Catch 22

2000 *Mathematics Subject Classification.* Primary 81P68, 94A60,81-02; Secondary 81V80, 68P25 .

Key words and phrases. Quantum cryptography, cryptography, quantum secrecy, quantum communication, quantum information, qubit, eavesdropping, intrusion detection, authentication, entanglement, quantum mechanics.

Partially supported by Army Research Laboratory (ARL) Contract #DAAL01-95-P-1884, by Army Research Office (ARO) Grant #P-38804-PH-QC, by the National Institute of Standards and Technology (NIST), by the Defense Advanced Research Projects Agency (DARPA) and Air Force Materiel Command USAF under agreement number F30602-01-0522, and by the L-O-O-P Fund. This paper is a revised version of a paper published in "Coding Theory, and Cryptography: From Geheimscheimschreiber and Enigma to Quantum Theory," (edited by David Joyner), Lecture Notes in Computer Science and Engineering, Springer-Verlag, 1999 (pp. 144-174). It has been reproduced with the permission of Springer-Verlag.

© 2002 American Mathematical Society

3.3. Practical Secrecy
3.4. Perfect Secrecy
3.5. Computational Security

4. **A Course on Quantum Mechanics**
4.1. Alice's Reluctant Decision
4.2. The Classical World – Introducing the Shannon Bit
4.3. The Quantum World – Introducing the Qubit
4.4. Where do qubits live?
4.5. Some Dirac notation – Introducing kets
4.6. Finally, a definition of a qubit
4.7. More Dirac notation – Introducing bras and bra-c-kets
4.8. Activities in the quantum world – Unitary transformations
4.9. Observables in quantum mechanics – Hermitian operators
4.10. The Heisenberg uncertainty principle – A limitation on what we can actually observe
4.11. Young's two slit experiment – An example of Heisenberg's uncertainty principle

5. **The Beginnings of Quantum Cryptography**
5.1. Alice has an idea
5.2. Quantum secrecy – The BB84 protocol without noise
5.3. Quantum secrecy – The BB84 protocol with noise

6. **The B92 quantum cryptographic protocol**

7. **There are many other quantum cryptographic protocols**

8. **A comparison of quantum cryptography with classical and public key cryptography**

9. **Eavesdropping strategies and counter measures**
9.1. Opaque eavesdropping
9.2. Translucent eavesdropping without entanglement
9.3. Translucent eavesdropping with entanglement
9.4. Eavesdropping based on implementation weaknesses

10. **Implementations**

11. **Conclusion**

12. **Acknowledgement**

References

1. Preface

1.1. The Unique Contribution of Quantum Cryptography. Before beginning our story, I'd like to state precisely what is the unique contribution of quantum cryptography.

> Quantum cryptography provides a new mechanism enabling the parties communicating with one another to:
>
> $\boxed{\text{Automatically Detect Eavesdropping}}$

Consequently, it provides a means for determining when an encrypted communication has been compromised.

1.2. A Note to the Reader.

This paper is based on an invited talk given at the Conference on Coding theory, Cryptology, and Number Theory held at the US Naval Academy in Annapolis, Maryland in October of 1998. It was also given as an invited talk at the Quantum Computational Science Workshop held in conjunction with the Frontiers in Computing Conference in Annapolis, Maryland in February of 1999, at a Bell Labs Colloquium in Murray Hill, New Jersey in April of 1999, at the Security and Technology Division Colloquium of NIST in Gaithersburg, Maryland, and at the Quantum Computation Seminar at the U.S. Naval Research Labs in Washington, DC.

My objective in creating this paper was to write it exactly as I had given the talk. But ... Shortly after starting this manuscript, I succumbed to the temptation of greatly embellishing the story that had been woven into the original talk. I leave it to the reader to decide whether or not this detracts from or enhances the paper.

2. Introduction

We begin our crypto drama with the introduction of two of the main characters, **Alice** ♡and♡ **Bob**, representing respectively the **sender** and the **receiver**. As in every drama, there is a triangle. The triangle is completed with the introduction of the third main character, the archvillainess **Eve**, representing the **eavesdropper**.

Our story begins with Alice and Bob attending two different universities which are unfortunately separated by a great distance. Alice would like to communicate with Bob without the ever vigilant Eve eavesdropping on their conversation. In other words, how can Alice talk with Bob while at the same time preventing the evil Eve from listening in on their conversation?

3. A Course on Classical Cryptography

3.1. Alice's enthusiastic decision. Hoping to find some way out of her dilemma, Alice elects to take a course on cryptography, Crypto 351 taught by Professor Shannon with guest lecturers Diffie, Rivest, Shamir, and Adleman. Alice thinks to herself, "Certainly this is a wise choice. It is a very applied course, and surely relevant to the real world. Maybe I will learn enough to outwit Eve?"

3.2. Plaintext, ciphertext, key, and ... Catch 22. Professor Shannon begins the course with a description of classical cryptographic communication systems, as illustrated in Fig. 1. Alice, the sender, encrypts her **plaintext** P into **ciphertext** C using a **secret key** K which she shares only with Bob, and sends the ciphertext C over an **insecure channel** on which the evil Eve is ever vigilantly eavesdropping. Bob, the receiver, receives the ciphertext C, and uses the secret key K, shared by him and Alice only, to decrypt the ciphertext C into plaintext P.

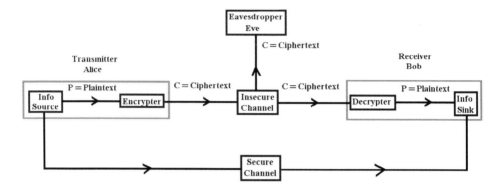

Figure 1. A classical cryptographic communication system.

What is usually not mentioned in the description of a classical cryptographic communication system is that Alice and Bob must first communicate over a **secure channel** to establish a secret key K shared only by Alice and Bob before they can communicate in secret over the insecure channel. Such a channel could consist, for example, of a trusted courier, wearing a trench coat and dark sunglasses, transporting from Alice to Bob a locked briefcase chained to his wrist. In other words, we have the famous Catch 22 of classical cryptography, namely:

> **Catch 22.** There are perfectly good ways to communicate in secret, provided we can communicate in secret ...

Professor Shannon then goes on to discuss the different types of classical communication security.

3.3. Practical Secrecy. A cryptographic communication system is **practically secure** if the encryption scheme can be broken after X years, where X is determined by one's security needs and by existing technology. Practically secure cryptographic systems have existed since antiquity. One example would be the Caesar cipher used by Julius Caesar during the during the Gallic wars, a cipher that was difficult for his opponents to break at that time, but easily breakable by today's standards. A modern day example of a practically secure classical cryptographic system is the digital encryption standard (DES) which has just recently been broken[1]. For this and many other reasons, DES is to be replaced by a more

---

[1] Tim O'Reilly and the Electronic Frontier Foundation have constructed a computing device for $250,000 which does an exhaustive key search on DES in 4.5 days[24]. See also [3] and [18]. As far as I know, triple DES has not been broken.

practically secure classical encryption system, the Advanced Encryption Standard (AES). In turn, AES will be replaced by an even more secure cryptographic system should the advances in technology ever challenge its security.

3.4. Perfect Secrecy. A cryptographic communication is said to be **perfectly secure** if the ciphertext C gives no information whatsoever about the plaintext P, even when the design of the cryptographic system is known. In mathematical terms, this can be stated succinctly with the equation:

$$PROB(P \mid C) = PROB(P).$$

In other words, the probability of plaintext P given ciphertext C, written $PROB(P|C)$, is equal to the probability of the plaintext P.

An example of a perfectly secure classical cryptographic system is the **Vernam Cipher**, better known as the **One-Time-Pad**. The plaintext P is a binary sequence of zeroes and ones, i.e.,

$$P = P_1, P_2, P_3, \ldots P_n, \ldots$$

The secret key K consists of a totally random binary sequence of the same length, i.e.,

$$K = K_1, K_2, K_3, \ldots, K_n, \ldots$$

The ciphertext C is the binary sequence

$$C = C_1, C_2, C_3, \ldots C_n, \ldots$$

obtained by adding the sequences P and K bitwise modulo 2, i.e.,

$$C_i = P_i + K_i \bmod 2 \quad \text{for} \quad i = 1, 2, 3, \ldots$$

For example,

$$
\begin{array}{rcl}
P & = & 0110\ \ 0101\ \ 1101 \\
K & = & 1010\ \ 1110\ \ 0100 \\
\hline
C = P \oplus K & = & 1100\ \ 1011\ \ 1001
\end{array}
$$

This cipher is perfectly secure if key K is totally random and shared only by Alice and Bob. It is easy to encode with the key K. If, however, one succumbs to the temptation of using the same key K to encode two different plaintexts $P^{(1)}$ and $P^{(2)}$ into ciphertexts $C^{(1)}$ and $C^{(2)}$, then the cipher system immediately changes from a perfectly secure cipher to one that is easily broken by even the most amateur cryptanalyst. For, $C^{(1)} \oplus C^{(2)} = P^{(1)} \oplus P^{(2)}$ is easily breakable because of the redundancy that is usually present in plaintext.

The only problem with the one-time-pad is that long bit sequences must be sent over a secure channel before it can be used. This once again leads us to the Catch 22 of classical cryptography, i.e.,

> **Catch 22.** There are perfectly good ways to communicate in secret, provided we can communicate in secret ...

... and to the:

- **Key Problem 1.** *Catch 22*: A secure means of communicating key is needed.[2]

Finally, there are two other key problems in classical cryptography in need of a solution, namely:

- **Key Problem 2.** *Authentication*: Alice needs to determine with certainty that she is actually talking to Bob, and not to an impostor such as Eve.
- **Key Problem 3.** *Intrusion Detection*: Alice needs a means of determining whether or not Eve is eavesdropping.

In summary, we have the following checklist for classical cryptographic systems:

Check List for Classical Crypto Systems	
■ Catch 22 Solved?	**NO**
■ Authentication?	**NO**
■ Intrusion Detection?	**NO**

3.5. Computational Security.

Relatively recently in the history of cryptography, Diffie and Hellman [8], [9] suggested a new type cryptographic secrecy. A cipher is said to be **computationally secure** if the computational resources required to break it exceed anything possible now and into the future. For example, a cipher would be computationally secure if the number of bits of computer memory required to break it were greater than the number of atoms in the universe, or if the computational time required to break it exceeded the age of the universe. Cryptographic systems can be created in such a way that it is computationally infeasible to find the decryption key D even when the encryption key E is known. To create such a cryptographic system, all one would need is a trap-door function f.

DEFINITION 1. *A function f is a **trap-door function** if*

1) *f is easy to compute, i.e., polynomial time computable, and*
2) *Given the function f, the inverse function f^{-1} can not be computed from f in polynomial time, i.e., such a computation is superpolynomial time, intractable, or worse.*

---

[2]Hired trench coats are exorbitantly expensive and time consuming.

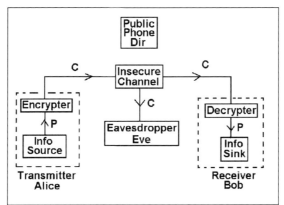

Figure 2. A public key cryptographic communication system

A trap-door function E can be used to create a **public key cryptographic system** as illustrated in Fig.2. All parties who wish to communicate in secret should choose their own trap-door function E and place it in a **public directory**, the "**yellow pages**," for all the world to see. But they should keep their decryption key $D = E^{-1}$ secret. Since E is a trap-door function, it is computationally infeasible for anyone to use the publicly known E to find the decryption key D. So D is secure in spite of the fact that its inverse E is publicly known.

If Alice wishes to send a secret communication to Bob, she first looks up in the yellow pages Bob's encryption key E_B, encrypts her plaintext P with Bob's encryption key E_B to produce ciphertext $C = E_B(P)$, and then sends the ciphertext C over a public channel. Bob receives the ciphertext C, and decrypts it back into plaintext $P = D_B(C)$ using his secret decryption key D_B.

Alice can even do more than this. She can authenticate, i.e., sign her encrypted communication to Bob so that Bob knows with certitude that the message he received actually came from Alice and not from an Eve masquerading as Alice. Alice can do this by encrypting her signature $\mathcal{ALICE}$ using her secret decryption key D_A into $D_A(\mathcal{ALICE})$. She then encrypts plaintext P plus her signature $D_A(\mathcal{ALICE})$ using Bob's publicly known encryption key E_B to produce the signed ciphertext $C_S = E_B(P + D_A(\mathcal{ALICE}))$, and then sends her signed ciphertext C_S over the public channel to Bob. Bob can then decrypt the message as he did before to produce the signed plaintext $P + D_A(\mathcal{ALICE})$. Bob can verify Alice's digital signature $D_A(\mathcal{ALICE})$ by looking up Alice's encryption key E_A in the "yellow pages," and using it to find her signature $E_A(D_A(\mathcal{ALICE})) = \mathcal{ALICE}$. In this way, he authenticates that Alice actually sent the message because only she knows her secret decryption key. Hence, only she could have signed the plaintext.[3]

The RSA cryptographic system is believed to be one example of a public key cryptographic system. There are many public software implementations of RSA, e.g., PGS (Pretty Good Security).

---

[3]Because of the need for brevity, we have not discussed all the subtleties involved with digital signatures. For example, for more security, Alice should add a time stamp and some random symbols to her signature. For more information on digital signatures, please refer to one of the standard references such as [22].

Thus, besides solving the authentication problem for cryptography, public key cryptographic systems appear also to solve the Catch 22 of cryptography. However, frequently the encryption and decryption keys of a public key cryptographic system are managed by a central key bank. In this case, the Catch 22 problem is still there. For that reason, we have entered 'MAYBE' in the summary given below.

Check List for PKS

- Catch 22 Solved?   MAYBE
- Authentication?   YES
- Intrusion Detection? NO

---

4. A Course on Quantum Mechanics

4.1. Alice's Reluctant Decision. In spite of Alice's many intense efforts to avoid taking a course in quantum mechanics, she was finally forced by her university's General Education Requirements (GERs) to register for the course Quantum 317, taught by Professor Dirac with guest lecturers Feynman, Bennett, and Brassard. She did so reluctantly. "After all," she thought, "Certainly this is an insane requirement. Quantum mechanics is not applied. It's too theoretical to be relevant to the real world. Ugh! But I do want to graduate."

4.2. The Classical World – Introducing the Shannon Bit. Professor Dirac began the course with a brief introduction to the classical world of information. In particular, Alice was introduced to the classical Shannon Bit, and shown that he/she/it is a very decisive individual. The Shannon Bit is either 0 or 1, but by no means both at the same time.

"Hmm ... ," she thought, "I bet that almost everyone I know is gainfully employed because of the Shannon Bit."

The professor ended his brief discussion of the Shannon Bit by mentioning that there is one of its properties that we take for granted. I.e., it can be copied.

4.3. The Quantum World – Introducing the Qubit. Next Professor Dirac switched to the mysterious world of the quantum. He began by introducing the runt of the Bit clan, i.e., the Quantum Bit, nicknamed **Qubit**. He began by showing the class a small dot, i.e., a quantum dot. In fact it was so small that Alice couldn't see it at all. He promptly pulled out a microscope[4], and projected a large image on a screen for the entire class to view.

Professor Dirac went on to say, "In contrast to the decisive classical Shannon Bit, the Qubit is a very indecisive individual. It is both 0 and 1 at the same time! Moreover, unlike the Shannon Bit, the Qubit cannot be copied because of the no cloning theorem of Dieks, Wootters, and Zurek[**7**][**33**]. Qubits are very slippery characters, exceedingly difficult to deal with."

---

[4]This is a most unusual microscope!

"One example of a qubit is a spin $\frac{1}{2}$ particle which can be in a spin-up state $|1\rangle$ which we label as 1, in a spin-down state $|0\rangle$ which we label as 0, or in a **superposition** of these states, which we interpret as being both 0 and 1 at the same time." (The term "superposition" will be explained shortly.)

"Another example of a qubit is the polarization state of a photon. A photon can be in a vertically polarized state $|\updownarrow\rangle$. We assign a label of 1 to this state. It can be in a horizontally polarized state $|\leftrightarrow\rangle$. We assign a label of 0 to this state. Or, it can be in a superposition of these states. In this case, we interpret its state as representing both 0 and 1 at the same time."

"Anyone who has worn polarized sunglasses should be familiar with the polarization states of the photon. Polarized sunglasses eliminate glare because they let through only vertically polarized light while filtering out the horizontally polarized light that is reflected from the road."

4.4. Where do qubits live? But where do qubits live? They live in a Hilbert space $\mathcal{H}$. By a Hilbert space, we mean:

DEFINITION 2. *A **Hilbert Space** is a vector space over the complex numbers $\mathbb{C}$ together with an inner product*

$$\langle\ ,\ \rangle : \mathcal{H} \times \mathcal{H} \longrightarrow \mathbb{C}$$

such that
1) $\langle u_1 + u_2, v \rangle = \langle u_1, v \rangle + \langle u_2, v \rangle$ *for all* $u_1, u_2, v \in \mathcal{H}$
2) $\langle u, \lambda v \rangle = \langle \lambda u, v \rangle$ *for all* $u, v \in \mathcal{H}$ *and* $\lambda \in \mathbb{C}$
3) $\langle u, v \rangle^* = \langle v, u \rangle$ *for all* $u, v \in \mathcal{H}$, *where the superscript '*' denotes complex conjugation.*
4) *For every Cauchy sequence* $u_1, u_2, u_3, \ldots$ *in* $\mathcal{H}$,

$$\lim_{n \to \infty} u_n \text{ exists and lies in } \mathcal{H}$$

In other words, a Hilbert space is a vector space over the complex numbers $\mathbb{C}$ with a sequilinear inner product in which sequences that should converge actually do converge to points in the space.

4.5. Some Dirac notation – Introducing kets. The elements of $\mathcal{H}$ are called **kets**, and will be denoted by

$$|label\rangle\ ,$$

where '|' and '>' are left and right delimiters, and '*label*' denotes any label, i.e., name, we wish to assign to a ket.

4.6. Finally, a definition of a qubit.
So finally, we can define what is meant by a qubit.

DEFINITION 3. *A **qubit** is a ket (state) in a two dimensional Hilbert space $\mathcal{H}$.*

Thus, if we let $|0\rangle$ and $|1\rangle$ denote an arbitrary orthonormal basis of a two dimensional Hilbert space $\mathcal{H}$, then each qubit in $\mathcal{H}$ can be written in the form

$$|qubit\rangle = \alpha_0 |0\rangle + \alpha_1 |0\rangle$$

where $\alpha_0, \alpha_1 \in \mathbb{C}$. Since any scalar multiple of a ket represents the same state of an isolated quantum system, we can assume, without loss of generality, that $|qubit\rangle$ is a ket of unit length, i.e., that

$$|\alpha_0|^2 + |\alpha_1|^2 = 1$$

The above qubit is said to be in a **superposition** of the states $|0\rangle$ and $|1\rangle$. This is what we mean when we say that a qubit can be simultaneously both 0 and 1. However, if the qubit is observed it immediately "makes a decision." It "decides" to be 0 with probability $|\alpha_0|^2$ and 1 with probability $|\alpha_1|^2$. Some physicists call this the **"collapse"** of the wave function[5].

4.7. More Dirac notation – Introducing bras and bra-c-kets.
Given a Hilbert space $\mathcal{H}$, let

$$\mathcal{H}^* = Hom(\mathcal{H}, \mathbb{C})$$

denote the set of all linear maps from $\mathcal{H}$ to $\mathbb{C}$. Then $\mathcal{H}^*$ is actually a Hilbert space, called the **dual** Hilbert space of $\mathcal{H}$, with scalar product and vector sum defined by:

$$\begin{cases} (\lambda \cdot f)(|\Psi\rangle) = \lambda(f(|\Psi\rangle)), & \text{for all } \lambda \in \mathbb{C} \text{ and for all } f \in \mathcal{H}^* \\ (f_1 + f_2)(|\Psi\rangle) = f_1(|\Psi\rangle) + f_2(|\Psi\rangle), & \text{for all } f_1, f_2 \in \mathcal{H}^* \end{cases}$$

We call the elements of $\mathcal{H}^*$ **bra**'s, and denote them as:

$$\langle label |$$

We can now define a bilinear map

$$\mathcal{H}^* \times \mathcal{H} \longrightarrow \mathbb{C}$$

by

$$(\langle \Psi_1 |)(|\Psi_2\rangle) \in \mathbb{C}$$

since bra $\langle \Psi_1 |$ is a complex valued function of kets. We denote this product more simply as

$$\langle \Psi_1 | \Psi_2 \rangle$$

and call it the **Bra-c-Ket** (or **bracket**) of bra $\langle \Psi_1 |$ and ket $|\Psi_2\rangle$.

Finally, the bracket induces a dual correspondence[6] between $\mathcal{H}$ and $\mathcal{H}^*$, i.e.,

$$|\Psi_2\rangle \stackrel{D.C.}{\longleftrightarrow} \langle \Psi_1 |$$

---

[5]It is very difficult, if not impossible, to find two physicists who agree on the subject of quantum measurement. The phrase "collapse of the wave function" immediately engenders a "war cry" in most physicists. For that reason, "collapse" is enclosed in quotes.

[6]This is true for finite dimensional Hilbert spaces. It is more subtle for infinite dimensional Hilbert spaces.

4.8. Activities in the quantum world – Unitary transformations.

All "activities" in the quantum world are linear transformations

$$U : \mathcal{H} \longrightarrow \mathcal{H}$$

from the Hilbert space $\mathcal{H}$ into itself, called **unitary transformations** (or, **unitary operators**). If we think of linear transformations as matrices, then a **unitary transformation** U is a square matrix of complex numbers such that

$$\overline{U}^T U = I = U\overline{U}^T$$

where $\overline{U}^T$ denotes the matrix obtained from U by conjugating all its entries and then transposing the matrix. We denote $\overline{U}^T$ by $U^\dagger$, and refer to it as the **adjoint** of U.

Thus, an "activity" in the quantum world would be, for example, a unitary transformation U that carries a state ket $|\Psi_0\rangle$ at time $t = 0$ to a state ket $|\Psi_1\rangle$ at time $t = 1$, i.e.,

$$U : |\Psi_0\rangle \longmapsto |\Psi_1\rangle$$

4.9. Observables in quantum mechanics – Hermitian operators.

In quantum mechanics, what does an observer observe?

All **observables** in the quantum world are linear transformations

$$\mathcal{O} : \mathcal{H} \longrightarrow \mathcal{H}$$

from the Hilbert space $\mathcal{H}$ into itself, called **Hermitian operators** (or, **self-adjoint operators**). If we think of linear transformations as matrices, then a **Hermitian operator** $\mathcal{O}$ is a square matrix of complex numbers such that

$$\overline{\mathcal{O}}^T = \mathcal{O}$$

where $\overline{\mathcal{O}}^T$ again denotes the matrix obtained from $\mathcal{O}$ by conjugating all its entries, and then transposing the matrix. As before, we denote $\overline{\mathcal{O}}^T$ by $\mathcal{O}^\dagger$, and refer to it as the **adjoint** of $\mathcal{O}$.

Let $|\varphi_i\rangle$ denote the eigenvectors, called **eigenkets**, of an observable $\mathcal{O}$, and let a_i denote the corresponding eigenvalue, i.e.,

$$\mathcal{O} : |\varphi_i\rangle = a_i |\varphi_i\rangle$$

In the cases we consider in this talk, the eigenkets form an orthonormal basis of the underlying Hilbert space $\mathcal{H}$.

Finally, we can answer our original question, i.e.,

$$\boxed{\text{What does an observer observe?}}$$

Let us suppose that we have a physical device M that is so constructed that it measures an observable $\mathcal{O}$, and that we wish to use M to measure a quantum system which just happens to be in a quantum state $|\Psi\rangle$. We assume $|\Psi\rangle$ is a ket of unit length. The quantum state $|\Psi\rangle$ can be written as a linear combination of the eigenkets of $\mathcal{O}$, i.e.,

$$|\Psi\rangle = \sum \alpha_i |\varphi_i\rangle$$

When we use the device M to measure $|\Psi\rangle$, we observe the eigenvalue a_i with probability $p_i = |\alpha_i|^2$, and in addition, after the measurement the quantum system has "collapsed" into the state $|\varphi_i\rangle$. Thus, the outcome of a measurement is usually random, and usually has a lasting impact on the state of the quantum system.

We can use Dirac notation to write down an expression for the average observed value. Namely, the **averaged observed value** is given by the expression $\langle\Psi|\,(\mathcal{O}\,|\Psi\rangle)$, which is written more succinctly as $\langle\Psi|\,\mathcal{O}\,|\Psi\rangle$, or simply as $\langle\mathcal{O}\rangle$.

4.10. The Heisenberg uncertainty principle – A limitation on what we can actually observe.

There is, surprisingly enough, a limitation of what can be observed in quantum mechanics.

Two observables A and B are said to be **compatible** if they commute, i.e., if

$$AB = BA.$$

Otherwise, they are said to be **incompatible**.

Let $[A, B]$, called the **commutator** of A and B, denote the expression

$$[A, B] = AB - BA$$

In this notation, two operators A and B are compatible if and only if $[A, B] = 0$. Finally, let

$$\triangle A = A - \langle A \rangle$$

The following principle is one expression of how quantum mechanics places limits on what can be observed:

Heisenberg's Uncertainty Principle[7]

$$\left\langle (\triangle A)^2 \right\rangle \left\langle (\triangle B)^2 \right\rangle \geq \frac{1}{4} \|\langle [A, B]\rangle\|^2$$

where $\left\langle (\triangle A)^2 \right\rangle = \langle\Psi|\,(\triangle A)^2\,|\Psi\rangle$ is the **mean squared standard deviation** of the observed eigenvalue, written in Dirac notation. It is a measure of the uncertainty in A.

This if A and B are incompatible, i.e., do not commute, then, by measuring A more precisely, we are forced to measure B less precisely, and vice versa. We can not simultaneously measure both A and B to to unlimited precision. Measurement of A somehow has an impact on the measurement of B.

---

[7] We have assumed units have been chosen such that $\hbar = 1$.

4.11. Young's two slit experiment − An example of Heisenberg's uncertainty principle.

For the purpose of illustrating Heisenberg's Uncertainty Principle, Professor Dirac wheeled out into the classroom a device to demonstrate Young's two slit experiment. The device consisted of an electron gun which spewed out electrons[8] in the direction of a wall with two slits. The electrons that managed to pass through the two slits then impacted on a backstop which consisted of a 1600×1200 rectangular lattice of extremely small counters. From the back of the backstop, all of the 1,920,000 tiny counters on the backstop were individually connected to a PC by a cable consisting of a dense bundle of filaments.

Professor Dirac pointed to the PC, and explained that the PC was setup to display on the CRT's 1600×1200 pixel screen the individual running total counts of all the backstop counters. He went on to say that the intensity $P(i,j)$ of pixel (i,j) on the screen was proportional to the total number of electrons counted so far by the counter at position (i,j) in the backstop.

Professor Dirac proceeded to demonstrate what the device could actually do. He turned on the electron gun, and turned down its intensity so low that the probability of more than two electrons being emitted at the same time was negligibly small.

His first experiment with the device was to cover slit 2, allowing the incoming electrons to pass only through slit 1. He reset all the counters on the backstop to zero, and then stepped back to let the students in his class view the screen.

Initially nothing could be seen on the screen but blackness. However, gradually an intensity pattern began to form on the screen. At first the displayed pattern was indiscernible. But eventually it began to look like the intensity pattern of a classical two dimensional Gaussian distribution. He then pressed a key on his computer to show a three dimensional plot of the intensity $P(i,j)$ as a surface in 3-space. Then with the click of a mouse, he displayed a plot of the intensity $P(i,j)$ along the vertical line $j = 800$ going down the center of the screen. The plot was that of the bell shaped classical one dimensional Gaussian distribution curve P_1, as shown in Fig. 3a. This was a clear indication that the random impacts on the backstop were obeying the classical Gaussian distribution.

When he repeated the experiment with the slit 1 instead of slit 2 covered, exactly the same pattern of a classical two dimensional Gaussian distribution pattern was seen, but only this time shifted vertically down a short distance on the screen. A plot of the intensity $P(i,j)$ of along the vertical line $j = 800$ going down the center of the screen is indicated by curve P_2 shown in Fig. 3a.

---

[8]The original Young's two slit experiment used photons rather than electrons.

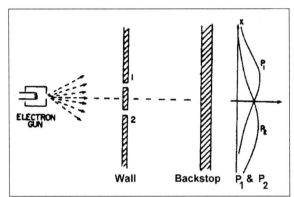

Figure 3a. Young's two slit experiment with one slit closed.

Professor Dirac then asked the students in the class what pattern they thought would appear if he left both of the slits uncovered. Most of the class responded by saying that the resulting light pattern would simply be the sum of the two patterns, i.e., the bell shaped curve $P_1 + P_2$, as illustrated in Fig. 3c by the curve labeled P'_{12}. Most of the class was convinced that the two classical probability distributions would simply add, as many of them had learned in the probability course Prob 323.

The remainder of the class stated quite emphatically that they did not care what happened. What was being illustrated was far from an applied area, and hence not relevant to their real world. Or so they thought ...

Professor Dirac smiled, and then proceeded to uncover both slits. What appeared on the screen to almost everyone's surprise was not the pattern with the bell shape $P_1 + P_2$. It was instead a light pattern with a wavy bell shaped curve, as illustrated by the curve P_{12} in Fig. 3b.

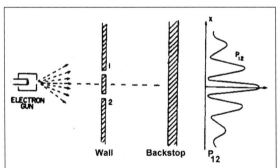

Figure 3b. Young's two slit experiment with both slits open.

Professor Dirac explained, "Something non-classical had occurred. Unlike classical probabilities, the quantum probabilities (or more correctly stated, the quantum amplitudes) had interfered with one another to produce an interference pattern. In the dark areas, one finds destructive interference. In the bright areas, one finds constructive interference."

"Indeed, something non-classical is happening here."

"Strangely enough, quantum mechanics is telling us that each electron is actually passing through both slits simultaneously! It is as if each electron were a wave and not a particle."

"But what happens when we actually try to observe through which slit each electron passes?"

Professor Dirac pulled out his trusty microscope[9] to observe which of the two slits each electron passed through. He reset all the backstop counters to zero, turned on the device, and began observing through which slit each electron passed through. The class was much surprised to find that the wavy interference pattern did not appear on the screen this time. Instead, what appeared was the classical intensity pattern all had initially expected to see in the first place, i.e., the intensity pattern of the bell shaped curve $P_{12} = P_1 + P_2$, as shown in Fig. 3c.

"So we see that, when observed, the electrons act as particles and not as waves!"

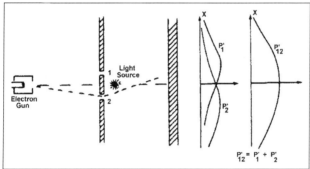

Figure 3c. Young's 2 slit experiment when the slit through which the electron passes is determined by observation.

After a brief pause, Professor Dirac said, "This is actually an example of the Heisenberg Uncertainty Principle. We can see this as follows:"

"In the experiment, we are effectively observing two incompatible observables, the position operator X (i.e., which slit each electron passes through) and the momentum operator P (i.e., the momentum with which each electron leaves the slitted wall.) When we observe the momentum P, the interference pattern is present. But when we observe the position X, the interference pattern vanishes. We can not observe position without disturbing momentum, and vice versa."

5. The Beginnings of Quantum Cryptography

5.1. Alice has an idea. After class on her way back to her dorm room, Alice began once again to ruminate over her dilemma in regard to Bob and Eve.

"If only her message to Bob were like the interference pattern in Young's two slit experiment. Then, if the prying Eve were to observe which of the two slits each of the electrons emerged from (i.e., 'listen in'), Bob would know of her presence. For, if Eve were observing the individual electrons as they left the slits, the pattern on the screen would be distorted from the beautiful wavy interference pattern in a

---
[9]This is a most unusual microscope.

direction toward the dull ugly Gaussian distribution pattern. Bob would see this distortion, and thereby be able to surmise that Eve was eavesdropping."

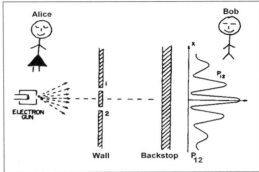

Figure 4a. Bob sees an interference pattern when Eve is not eavesdropping.

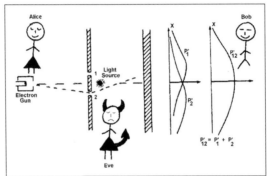

Figure 4b. Bob sees no interference pattern when Eve is eavesdropping.

"This idea has possibilities. Maybe quantum mechanics is relevant after all!"

Her mind began to race. "Perhaps something like Young's two slit experiment could be used to communicate random key K? Then Bob could tell which key had been compromised by an intruder such as Eve. But most importantly, he could also surmise which key had not been compromised. Bob could then communicate to me over the phone (or even over any public channel available also to Eve) whether or not the key had been compromised, without, of course, revealing the key itself. Any uncompromised key could then be employed to send Bob a message by using the one-time-pad that was mentioned yesterday in Crypto 351."

"The beauty of this approach is that the one-time-pad is perfectly secure. There is no way whatsoever that Eve could get any information about our conversation. This would be true even if I used the campus radio station to send my encrypted message."

"The evil Eve is foiled! Eureka! Contrary to student conventional wisdom, both cryptography and quantum mechanics are relevant to the real world!"

"I have discovered a new kind of secrecy, i.e., quantum secrecy, which has built-in detection of eavesdropping based on the principles of quantum mechanics. I can hardly wait to tell Professor Dirac. She ran immediately to his office."

After listening to Alice's excited impromptu, and at times disjointed, explanation, Professor Dirac suggested that she present her newly found discoveries in his next class. Alice happily agreed to do so.

5.2. Quantum secrecy – The BB84 protocol without noise. Two days later, after two sleepless but productive nights of work, Alice was prepared for her presentation. She walked in the classroom for Quantum 317 carrying an overhead projector and a sizable bundle of transparencies.

After Professor Dirac had turned the large lecture hall over to her, she began as follows:

"Let us suppose that I (Alice) would like to transmit a secret key K to Bob. Let us also suppose that someone by the name of Eve intends to make every effort to eavesdrop on the transmission and learn the secret key."

Wouldn't you know it. Eve just so happens to be sitting in the classroom!

"My objective today is to show you how the principles of quantum mechanics can be used to build a cryptographic communication system in such a way that the system detects if Eve is eavesdropping, and which also gives a guarantee of no intrusion if Eve is not eavesdropping."

"A diagrammatic outline of the system I'm about to describe is shown on the screen. (Please refer to Fig. 5.) Please note that the system consists of two communication channels. One is a non-classical one-way quantum communication channel, which I will soon describe. The other is an ordinary run-of the-mill classical two-way public channel, such as a two-way radio communication system. I emphasize that this classical two-way channel is public, and open to whomever would like to listen in. For the time being, I will assume that the two-way public channel is noise free."

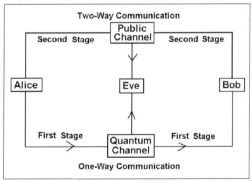

Figure 5. A quantum cryptographic communication system.

"I will now describe how the polarization states of the photon can be used to construct a quantum one-way communication channel[10]."

---

[10] Any two dimensional quantum system such as a spin $\frac{1}{2}$ particle could be used.

"From Professor Dirac's last lecture, we know that the polarization states of a photon lie in a two dimensional Hilbert space $\mathcal{H}$. For this space, there are many orthonormal bases. We will use only two for our quantum channel."

"The first is the basis consisting of the vertical and horizontal polarization states, i.e., the kets $|\updownarrow\rangle$ and $|\leftrightarrow\rangle$, respectively. We will refer to this orthonormal basis as the **vertical/horizontal (V/H) basis**, and denote this basis with the symbol '⊞.' "

"The second orthonormal basis consists of the polarization states $|\nearrow\rangle$ and $|\nwarrow\rangle$, which correspond to polarizations directions formed respectively by 45% clockwise and counter-clockwise rotations off from the vertical. We call this the **oblique basis**, and denote this basis with the symbol '⊠.' "

"If I (Alice) decide to use the VH basis ⊞ on the quantum channel, then I will use the following **quantum alphabet**:

$$\begin{cases} \text{"1"} = |\updownarrow\rangle \\ \text{"0"} = |\leftrightarrow\rangle \end{cases}$$

In other words, if I use this quantum alphabet on the quantum channel, I will transmit a "1" to Bob simply by sending a photon in the polarization state $|\updownarrow\rangle$., and I will transmit a "0" by sending a photon in the polarization state $|\leftrightarrow\rangle$."

"On the other hand, if I (Alice) decide to use the oblique basis ⊠, then I will use the following **quantum alphabet**:

$$\begin{cases} \text{"1"} = |\nearrow\rangle \\ \text{"0"} = |\nwarrow\rangle \end{cases},$$

sending a "1" as a photon in the polarization state $|\nearrow\rangle$, and sending a "0" as a photon in the polarization state $|\nwarrow\rangle$."

"I have chosen these two bases because the Heisenberg Uncertainty Principle implies that observations with respect to the ⊞ basis are incompatible with observations with respect to the ⊠ basis. We will soon see how this incompatibility can be translated into intrusion detection."

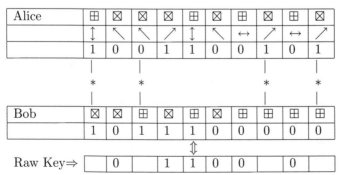

Fig. 6a. The BB84 protocol without Eve present (No noise)

Alice and Bob now communicate with one another using a two stage protocol, called the **BB84 protocol**[?]. (Please refer to Figs. 6a and 6b.)

In stage 1, Alice creates a random sequence of bits, which she sends to Bob over the quantum channel using the following protocol:

Stage 1 protocol: Communication over a quantum channel

Step 1. Alice flips a fair coin to generate a random sequence S_{Alice} of zeroes and ones. This sequence will be used to construct a secret key shared only by Alice and Bob.

Step 2. For each bit of the random sequence, Alice flips a fair coin again to choose at random one of the two quantum alphabets. She then transmits the bit as a polarized photon according to the chosen alphabet.

Step 3. Each time Bob receives a photon sent by Alice, he has no way of knowing which quantum alphabet was chosen by Alice. So he simply uses the flip of a fair coin to select one of the two alphabets and makes his measurement accordingly. Half of the time he will be lucky and choose the same quantum alphabet as Eve. In this case, the bit resulting from his measurement will agree with the bit sent by Alice. However, the other half of the time he will be unlucky and choose the alphabet not used by Alice. In this case, the bit resulting from his measurement will agree with the bit sent by Alice only 50% of the time. After all these measurements, Bob now has in hand a binary sequence S_{Bob}.

Alice and Bob now proceed to communicate over the public two-way channel using the following stage 2 protocol:

Stage 2 protocol: Communication over a public channel

Phase 1. *Raw key extraction*

Step 1. Over the public channel, Bob communicates to Alice which quantum alphabet he used for each of his measurements.

Step 2. In response, Alice communicates to Bob over the public channel which of his measurements were made with the correct alphabet.

Step 3. Alice and Bob then delete all bits for which they used incompatible quantum alphabets to produce their resulting **raw keys**. If Eve has not eavesdropped, then their resulting raw keys will be the same. If Eve has eavesdropped, their resulting raw keys will not be in total agreement.

Phase 2. *Error estimation*

Step 1. Over the public channel, Alice and Bob compare small portions of their raw keys to estimate the error-rate R, and then delete the disclosed bits from their raw keys to produce their **tentative final keys**. If through their public disclosures, Alice and Bob find no errors (i.e., $R = 0$), then they know that Eve was not eavesdropping and that their tentative keys must be the same **final key**. If they discover at least one error during their public disclosures (i.e., $R > 0$), then they know that Eve has been eavesdropping. In this case, they discard their tentative final keys and start all over again[11].

Alice	⊞	⊠	⊠	⊠	⊞	⊠	⊞	⊠	⊞	⊠
	↕	↖	↖	↗	↕	↖	↔	↗	↔	↗
	1	0	0	1	1	0	0	1	0	1

Eve	⊠	⊞	⊞	⊠	⊞	⊞	⊠	⊠	⊞	⊞
	1	0	1	1	1	1	0	1	0	0

Bob	⊠	⊠	⊞	⊠	⊞	⊠	⊞	⊞	⊞	⊞
	1	0	1	1	1	1	1	0	0	0
	*	0	*	1	1	1	1	*	0	*
					E	E				

Fig 6b. The BB84 with Eve present (No noise)

5.3. Quantum secrecy – The BB84 protocol with noise.

Alice continues her presentation by addressing the issue of noise.

"So far we have assumed that our cryptographic communication system is noise free. But every realistic communication system has noise present. Consequently, we now need to modify our quantum protocol to allow for the presence of noise."

"We must assume that Bob's raw key is noisy. Since Bob can not distinguish between errors caused by noise and by those caused by Eve's intrusion, the only practical working assumption he can adopt is that all errors are caused by Eve's eavesdropping. Under this working assumption, Eve is always assumed to have some information about bits transmitted from Alice to Bob. Thus, raw key is always only **partially secret**."

"What is needed is a method to distill a smaller secret key from a larger partially secret key. We call this **privacy amplification**. We will now create from the old protocol a new protocol that allows for the presence of noise, a protocol that includes privacy amplification."

Stage 1 protocol: Communication over a quantum channel

This stage is exactly the same as before, except that errors are now also induced by noise.

Stage 2 protocol: Communication over a public channel

Phase 1 protocol: Raw key extraction.

---

[11]If Eve were to intercept each qubit received from Alice, to measure it, and then to masqurade as Alice by sending on to Bob a qubit in the state she measured, then Eve would be introducing a 25% error rate in Bob's raw key. This method of eavesdropping is called **opaque eavesdropping**. We will discuss this eavesdropping strategy as well as others at a later time.

This phase is exactly the same as in the noise-free protocol, except that Alice and Bob also delete those bit locations at which Bob should have received but did not receive a bit. Such "non-receptions" could be caused by Eve's intrusion or by **dark counts** in Bob's detection device. The location of dark counts are communicated by Bob to Alice over the public channel.

Phase 2 protocol: Error estimation.

Over the public channel, Alice and Bob compare small portions of their raw keys to estimate the error-rate R, and then delete the disclosed bits from their raw key to produce their **tentative final keys**. If R exceeds a certain threshold R_{Max}, then privacy amplification is not possible If so, Alice and Bob return to stage 1 to start over. On the other hand, if $R \leq R_{Max}$, then Alice and Bob proceed to phase 3.

Phase 3 protocol: Extraction of reconciled key[12].

In this phase[13], Alice and Bob remove all errors from what remains of raw key to produce a common error-free key, called **reconciled key**.

Step 1. Alice and Bob publicly agree upon a random permutation, and apply it to what remains of their respective raw keys. Next Alice and Bob partition the remnant raw key into blocks of length ℓ, where the length ℓ is chosen so that blocks of that length are unlikely to have more than one error. For each of these blocks, Alice and Bob publicly compare overall parity checks, making sure each time to discard the last bit of each compared block. Each time an overall parity check does not agree, Alice and Bob initiate a binary search for the error, i.e., bisecting the block into two subblocks, publicly comparing the parities for each of these subblocks, discarding the right most bit of each subblock. They continue their bisective search on the subblock for which their parities are not in agreement. This bisective search continues until the erroneous bit is located and deleted. They then continue to the next ℓ-block.

This step is repeated, i.e., a random permutation is chosen, a remnant raw key is partitioned into blocks of length ℓ, parities are compared, etc.. This is done until it becomes inefficient to continue in this fashion.

Step 2. Alice and Bob publicly select randomly chosen subsets of remnant raw key, publicly compare parities, each time discarding an agreed upon bit from their chosen key sample. If a parity should not agree, they employ the binary search strategy of Step 1 to locate and delete the error.

---

[12]There are more efficient and elegant procedures than the procedure descibed in Stage 2 Phase 3. See [**16**] for references.

[13]The procedure given in Stage 2 Phase 3 is only one of many different possible procedures. In fact, there are much more efficient and elegant procedures than the one described herein.

- Finally, when, for some fixed number N of consecutive repetitions of Step 2, no error is found, Alice and Bob assume that to a high probability, the remnant raw key is without error. Alice and Bob now rename the remnant raw key **reconciled key**, and proceed to the next phase.

Phase 4: Privacy amplification

Alice and Bob now have a common reconciled key which they know is only partially secret from Eve. They now begin the process of **privacy amplification**, which is the extraction of a secret key from a partially secret one.

Step 1. Alice and Bob compute from the error-rate R obtained in Phase 2 of Stage 2 an upper bound k of the number of bits of reconciled key known by Eve.

Let n denote the number of bits in reconciled key, and let s be a **security parameter** to be adjusted as required.

Step 2. Alice and Bob publicly select $n - k - s$ random subsets of reconciled key, without revealing their contents. The undisclosed parities of these subsets become the final secret key.

It can be shown that Eve's average information about the final secret key is less than $2^{-s}/\ln 2$ bits.

The bell rang, indicating the end of the period. The entire class with two exceptions, immediately raced out of the lecture hall, almost knocking Alice down as they passed by. Professor Dirac thanked Alice for an excellent presentation.

As Alice left, she saw Eve in one of the dark recesses of the large lecture hall with her head resting on the palm of her hand as if in deep thought. She had a frown on her face. Alice left with a broad smile on her face.

6. The B92 quantum cryptographic protocol

In the next class, Alice continued her last presentation.

In thinking about the BB84 protocol this weekend, I was surprised to find that it actually is possible to build a different quantum protocol that uses only one quantum alphabet instead of two. I'll call this new quantum protocol **B92**."

"As before, we will describe the protocol in terms of the polarization states of the photon[14]."

"As our quantum alphabet, we choose

$$\begin{cases} \text{``1''} &= |\theta_+\rangle \\ \text{``0''} &= |\theta_-\rangle \end{cases},$$

where $|\theta_+\rangle$ and $|\theta_-\rangle$ denote respectively the polarization states of a photon linearly polarized at angles θ and $-\theta$ with respect to the vertical, where $0 < \theta < \frac{\pi}{4}$."

---

[14] Any two dimensional quantum system such as a spin $\frac{1}{2}$ particle could be used.

"We assume that Bob's quantum receiver, called a **POVM receiver**[4], is base on the following observables[15]:

$$\begin{cases} A_{\theta_+} = \dfrac{1 - |\theta_-\rangle\langle\theta_-|}{1 + \langle\theta_+|\theta_-\rangle} \\ \\ A_{\theta_-} = \dfrac{1 - |\theta_+\rangle\langle\theta_+|}{1 + \langle\theta_+|\theta_-\rangle} \\ \\ A_? = 1 - A_{\theta_+} - A_{\theta_-} \end{cases},$$

where A_{θ_+} is the observable for $|\theta_+\rangle$, A_{θ_-} the observable for $|\theta_-\rangle$ and $A_?$ is the observables for inconclusive receptions."

The **B92** quantum protocol is as follows:

Stage 1 protocol. Communication over a quantum channel.

Step 1. The same as in the BB84 protocol. Alice flips a fair coin to generate a random sequence S_{Alice} of zeroes and ones. This sequence will be used to construct a secret key shared only by Alice and Bob.

Step 2. The same as in the previous protocol, except this time Alice uses only one alphabet, the one above. So she does not have to flip a coin to choose an alphabet.

Step 3. Bob uses his POVM receiver to measure photons received from Alice.

Stage 2. Communication in four phases over a public channel.

This stage is the same as in the BB84 protocol, except that in phase 1, Bob publicly informs Alice as to which time slots he received non-erasures. The bits in these time slots become Alice's and Bob's raw keys.

Alice completed her discussion of the B92 protocol with,

"Eve's presence is again detected by an unusual error rate in Bob's raw key. Moreover, for some but not all eavesdropping strategies, Eve can also be detected by an unusual erasure rate for Bob."

Alice then stepped down from the lecture hall podium and returned to her seat.

7. There are many other quantum cryptographic protocols

Before continuing our story about Alice, Bob, and Eve, there are a few points that need to be made:

There are many other quantum cryptographic protocols. Quantum protocols showing the greatest promise for security are those based on EPR pairs. Unfortunately, the technology for implementing such protocols is not yet available. For references on various protocols, please refer to [**16**].

---

[15]The observables A_{θ_+}, A_{θ_-}, and $A_?$ form a postive operator value measure (POVM).

8. A comparison of quantum cryptography with classical and public key cryptography

Quantum cryptography's unique contribution is that it provides a mechanism for eavesdropping detection. This is an entirely new contribution to cryptography. On the other hand, one of the main drawbacks of quantum cryptography is that it provides no mechanism for authentication, i.e., for detecting whether or not Alice and Bob are actually communicating with each other, and not with an intermediate Eve masquerading as each of them. Thus, the Catch 22 problem is not solved by quantum cryptography. Before Alice and Bob can begin their quantum protocol, they first need to send an authentication key over a secure channel.

Thus, quantum cryptography's unique contribution is to provide a means of expanding existing secure key. Quantum protocols are secure key expanders. First a small authentication key is exchanged over a secure channel. Then that key can be amplified to an arbitrary length through quantum cryptography.

Check List for Q. Crypto. Sys.	
■ Catch 22 Solved?	**YES & NO**
■ Authentication?	**NO**
■ Intrusion Detection?	**YES**

9. Eavesdropping strategies and counter measures

Now let us resume our story:

Not a split second after Alice had seated herself, Eve raised her hand and asked for permission to make her own presentation to the class. Professor Dirac yielded the podium, not knowing exactly what to expect, but nonetheless elated that his usually phlegmatic class was beginning to show signs of something he had not seen for some time, class participation and initiative.

Eve began, "In the last two classes, Alice has suggested that I (Eve) might be eager to eavesdrop on her conversations with my ♡close♡ friend ♡Bob♡. I assure you that that simply is in no way true."

"But such innuendo really doesn't bother me."

9.1. Opaque eavesdropping. "What really irks me is that Alice suggests that, if I were to eavesdrop (which never would happen), then I (Eve) would use **opaque eavesdropping**. By **opaque eavesdropping**, I mean that I (Eve) would intercept and observe (measure) Alice's photons, and then masquerade as Alice by sending photons in the states I had measured on to Bob."

"I assure you that, if I ever wanted to eavesdrop (which will never be the case), I would not use such a simplistic form of intrusion."

Eve really wanted to use the adjective 'stupid' instead of 'simplistic,' but restrained herself.

Eve then said indignantly, "If I ever were to eavesdrop (which would never happen), I would use more sophisticated, more intelligent, and yes ... , more deliciously devious schemes!"

9.2. Translucent eavesdropping without entanglement.
"I (Eve) could for example make my probe interact unitarily with the information carrier from Alice, and then let it proceed on to Bob in a slightly modified state. For the B92 protocol, the interaction is given by:

$$\begin{cases} |\theta_+\rangle |\psi\rangle \longmapsto U |\theta_+\rangle |\psi\rangle = |\theta'_+\rangle |\psi_+\rangle \\ |\theta_-\rangle |\psi\rangle \longmapsto U |\theta_-\rangle |\psi\rangle = |\theta'_-\rangle |\psi_-\rangle \end{cases},$$

where $|\psi\rangle$ and $|\psi_\pm\rangle$ denote respectively the state of my (Eve's) probe before and after the interaction and where $|\theta_\pm\rangle$ and $|\theta'_\pm\rangle$ denote respectively the state of Alice's photon before and after the interaction."

9.3. Translucent eavesdropping with entanglement.
"Another approach, one of the most sophisticated, would be for me (EVE) to entangle my probe with the information carrier from Alice, and then let it proceed on to Bob. For the B92 protocol, the interaction is given by:

$$\begin{cases} |\theta_+\rangle |\psi\rangle \longmapsto U |\theta_+\rangle |\psi\rangle = a |\theta'_+\rangle |\psi_+\rangle + b |\theta'_-\rangle |\psi_+\rangle \\ |\theta_-\rangle |\psi\rangle \longmapsto U |\theta_-\rangle |\psi\rangle = b |\theta'_+\rangle |\psi_-\rangle + a |\theta'_-\rangle |\psi_-\rangle \end{cases},$$

where $|\psi\rangle$ and $|\psi_\pm\rangle$ denote respectively the state of my (Eve's) probe before and after the entanglement and where $|\theta_\pm\rangle$ and $|\theta'_\pm\rangle$ denote respectively the state of Alice's photon before and after the entanglement."

9.4. Eavesdropping based on implementation weaknesses.

"On the other hand, I could also take advantage of implementation weaknesses."

"One of the great difficulties with quantum cryptography is that technology has not quite caught up with it. Many devices, such as lasers, do not emit a single quantum, but many quanta at each emission time. The implementation of quantum protocols really requires single-quantum emitters. Such single-quantum emitters are now under development. Until such emitters become available, the quantum protocols can only be approximately implemented."

"For example, for many optical implementations of quantum protocols, the laser intensity is turned down so that on the average only one photon is produced every 10 pulses. Thus, if anything is emitted at all (one chance out of 10), then the probability that it is a single photon is extremely high. However, when there is an emission, then there is a probability of $\frac{1}{200}$ that more than one photon is emitted. So it is conceivable that I (Eve) could build an eavesdropping device that would detect multiple photon transmissions, and, when so detected, would divert one of the photons for measurement. In this way, I (Eve) could conceivably read $\frac{1}{200}$ of Alice's transmission without being detected. One way of countering this type of threat is to allow for it during privacy amplification. Another is to develop devices which actually truly emit one quanta at a time."

"Finally, depending on Alice's implementation, it might also be possible for me (Eve) to gain information simply by observing Alice's transmitter without measuring its output. This may or may not be far fetched."

Eve then returned to her seat. Her face was lit up with a sinister grin of satisfaction.

10. Implementations

Before continuing our story, we should mention that quantum cryptographic protocols have been implemented over more than 30 kilometers of fiber optic cable, [25],[28],[30], [31], and most amazingly, over more than a kilometer of free space [5], [6], [13],[10], [12] in the presence of ambient sunlight. There have been a number of ambitious proposals to demonstrate the feasibility of quantum cryptography in earth to satellite communications. And as mentioned earlier, there is a clear need for the development of single-quantum emitting devices.

11. Conclusion

Much remains to be done. There has been some work on the development of multiple-user quantum cryptographic protocols for communication networks [32], [29]. The Trojan Horse attack is discussed in [15]. There also have been at least two independent claims of the proof of ultimate security, i.e., a proof that quantum cryptographic protocols are impervious to all possible eavesdropping strategies [15], [19], [20], [21], [2], [26]. For some recent experimental developments, please refer to [14], [23], [27].

Our story continues:

As Alice sat in her seat, she happened to spy in the corner of her eye an abrupt change in Eve's demeanor. Eve suddenly became agitated, lit up with excitement, and started to frantically write on her notepad. The bell rang. Eve immediately jumped up, and raced out of the lecture hall, being pushed along by the usual frantic mass of students, equally eager to get out of the classroom.

As Eve whisked past, Alice caught just a fleeting glimpse of Eve's notepad. All Alice was able to discern in that brief moment was an illegible jumble of equations and ... yes, ... the acronym "POVM."

Alice thought to herself, "Oh, well! ... Forget it! I think I'll just visit Bob this weekend."

THE END[16]

---

[16]Any resemblance of the characters in this manuscript to individuals living or dead is purely coincidental.

12. Acknowledgement

I would like to thank Howard Brandt, Lov Grover, and Hoi-Kwong Lo for their helpful suggestions. I would also like to thank the individuals who attended my talk. Their many comments and insights were of invaluable help in writing this paper. Thanks are also due to the National Institute for Standards and Technology (NIST) for providing an encouraging environment in which this paper could be completed.

References

[1] Bennett, Charles H., and Gilles Brassard, **Quantum cryptography: Public key distribution and coin tossing**, International Conference on Computers, Systems & Signal Processing, Bagalore, India, December 10-12, 1984, pp 175 - 179.
[2] Bennett, Charles H., and Peter Shor, Science, 284, (1999), 747.
[3] Biham, Eli, and Adi Shamir, "**Differential Crytanalysis of the Data Encryption Standard**," Springer-Verlag (1993).
[4] Brandt, Howard E., John M. Meyers, And Samuel J. Lomonaco,Jr., **Aspects of entangled translucent eavesdropping in quantum cryptography**, Phys. Rev. A, Vol. 56, No. 6, December 1997, pp. 4456 - 4465.
[5] Buttler, W.T., R.J. Hughes, S.K. Lamoreaux, G.L. Morgan, J.E. Nordholt, C.G. Peterson, **Daylight quantum key distribution over 1.6 km**, http://xxx.lanl.gov/abs/quant-ph/0001088.
[6] Buttler, W.T., R.J. Hughes, P.G. Kwiat, G.G. Luther, G.L. Morgan, J.E. Nordholt, C.G. Peterson, and C.M. Simmons, **Free-space quantum ket distribution**, Phys. Rev. A ,(1998). (http://xxx.lanl.gov/abs/quant-ph/9801006).
[7] Dieks, D., Phys. Lett., **92**, (1982), p 271.
[8] Diffie, W., **The first ten years in public-key cryptography**, in "Contemporary Cryptology: The Science of Information Integrity," pp 135 - 175, IEEE Press (1992).
[9] Diffie, W., and M.E. Hellman, **New directions in cryptography**, IEEE Tranactions on Information Theory, **22** (1976), pp 644 - 654.
[10] Franson, J.D., and H. Ilves, **Quantum cryptography using polarization feedback**, Journal of Modern Optics, Vol. 41, No. 12, 1994, pp 2391 - 2396.
[11] Gottesman, Daniel, and Hoi-Kwong Lo, **From quantum cheating to quantum security**, Physics Today, Vol. 53, Iss. 11, (2000), p22.
[12] Hughes, Richard J., William T. Buttler, Paul G. Kwiat, Steve K. Lamoreaux, George L. Morgan, Jane E. Nordholt, C. G. Peterson, **Practical quantum cryptography for secure free-space communications**, PRL (2000). (http://xxx.lanl.gov/abs/quant-ph/9905009).
[13] Jacobs, B.C. and J.D. Franson, **Quantum cryptography in free space**, Optics Letters, Vol. 21, November 15, 1996, p1854 - 1856.
[14] Jennewein, Thomas, Christoph Simon, Gregor Weihs, Harald Weinfurter, and Anton Zeilinger, **Quantum Cryptography with Entangled Photons**, Phys. Rev. Lett.,Vol. 84, 20, May 15, 2000, pp. 4729-4732.
[15] Lo, H.-K, and H.F. Chau, **Quantum computers render quantum key distribution unconditionally secure over arbitrarily long distance**, Science, 283, (1999), 2050. (http://xxx.lanl.gov/abs/quant-ph/9803006)
[16] Lomonaco, Samuel J., **A quick glance at quantum cryptography**, Cryptologia, Vol. 23, No. 1, January, 1999, pp1-41. (http://xxx.lanl.gov/abs/quant-ph/9811056)
[17] Lomonaco, Samuel J., Jr., **A talk on quantum cryptography, or How Alice outwits Eve**, in "Coding Theory, and Cryptography: From Geheimscheimschreiber and Enigma to Quantum Theory," (edited by David Joyner), Lecture Notes in Computer Science and Engineering, Springer-Verlag, (1999), pp. 144-174.
[18] Matsui, Mitsuru, **Linear crytanalysis method for DES cipher**, Lecture Notes in Computer Science, vol. 765, edited by T. Helleseth, Springer-Verlag (1994), pp386-397.
[19] Mayers, Dominic, Crypto'96, p343.

[20] Mayers, Dominic, and Andrew Yao, **Quantum cryptography with imperfect apparatus**, http://xxx.lanl.gov/abs/quant-ph/9809039.
[21] Mayers, Dominic, **Unconditional security in quantum cryptography**, http://xxx.lanl.gov/abs/quant-ph/9802025.
[22] Menezes, Alfred J., Paul C. van Oorschot, and Scott A. Vanstone, "Handbook of Applied Cryptography," CRC Press (1977).
[23] Naik, D.S., C. G. Peterson, A. G. White, A. J. Berglund, and P. G. Kwiat, **Entangled State Quantum Cryptography: Eavesdropping on the Ekert Protocol**, Phys. Rev. Lett.,Vol. 84, 20, May 15, 2000, pp. 4733-4736.
[24] O'Reilly, Tim, and the Electronic Frontier Foundation, "Cracking DES: Secrets of Encryption Research, Wiretap Politics & Chip Design," (1st Edition), July 1998 (US) ISBN 1-56592-520-3 (272 pages) http://www.ora.com/catalog/crackdes/
[25] Phoenix, Simon J., and Paul D. Townsend, **Quantum cryptography: how to beat the code breakers using quantum mechanics**, Comtemporay Physics, vol. 36, No. 3 (1995), pp 165 - 195.
[26] Shor, Peter W., and John Preskill, Phys. Rev. Lett., 85, (2000), 441.
[27] Tittel, W., J. Brendel, H. Zbinden, and N. Gisin, **Quantum Cryptography Using Entangled Photons in Energy-Time Bell States**, Phys. Rev. Lett.,Vol. 84, 20, May 15, 2000, pp. 4737-4740
[28] Townsend, P.D., **Secure key distribution system based on quantum cryptography**, Electronic Letters, 12 May 1994, Vol. 30, No. 10, pp 809 - 811.
[29] Townsend, Paul, Optical Fibers Technology 4, 345 (1998).
[30] Townsend, Paul D., and I Thompson, Journal of Modern Optics, **A quantum key distribution channel based on optical fibre**, Vol. 41, No. 12, (1994), pp 2425 - 2433.
[31] Townsend, P.D., J.G. Rarity, and P.R. Tapster, **Single photon interference in 10km long optical fibre interferometer,** Electronic Letters, **29** (1993), pp 634 - 635.
[32] Townsend, P.D., Nature 385, (1997), p 47.
[33] Wootters, W.K., and W.H. Zurek, **A single quantum cannot be cloned**, Nature, 299 (1982), pp982-983.

UNIVERSITY OF MARYLAND BALTIMORE COUNTY, BALTIMORE, MD 21250
E-mail address: Lomonaco@UMBC.EDU
URL: http://www.csee.umbc.edu/~lomonaco

Chapter IV

More Mathematical Connections

Topological Quantum Codes and Anyons

Alexei Kitaev

CONTENTS

1. Introduction
2. Terminology and notations
3. Surface codes based on the group $\mathbb{Z}_2$
4. Nonabelian anyons
References

1. Introduction

The topic I am going to discuss is about 15 years old; but it has an amazing number of connections in mathematics and physics.

The concept of anyons appeared in mid 1980s. Anyons are special particles (more exactly, quasi-particles, or excitations) in two-dimensional quantum systems. So far anyons have been found in two-dimensional electronic liquids exhibiting the fractional quantum Hall effect (FQHE); but they may also exist in other experimental systems. Intuitively, although very inaccurately, anyons can be understood as "quantum topological defects." Ordinary (or classical) topological defects occur in systems with an order parameter. For example, a superconducting film has an order parameter $f(x)$ – a complex-valued function whose absolute value is constant everywhere on the plane, except for the defects. Mathematically, this situation can be described by a smooth map $X \longrightarrow S^1$, where X is a plane without a discrete set of points. A single defect cannot be removed while annihilation or fusion of two defects must preserve the winding number (or some other topological number in a more general case). Anyons also obey certain conservation laws, but there is no order parameter in the surrounding space. Still one may think of this situation as if

---

2000 *Mathematics Subject Classification.* Primary 81P68, 81-06,81-01; Secondary 94B60, 57M99.

Key words and phrases. Anyons, quantum codes, homology, surfaces.

The author is on leave from L.D. Landau Institute for Theoretical Physics (Moscow).

there were an order parameter which fluctuated so wildly that only the topological numbers survived.

The most peculiar property of anyons is Aharonov-Bohm interaction. If one anyon moves around another, the quantum state *of the entire system* undergoes some unitary transformation: $|\Psi\rangle \longmapsto U|\Psi\rangle$. The operator U can be simply a number (a phase factor), in which case the anyons are called *Abelian*. For more complicated *nonabelian anyons*, the unitary operator U acts on a finite-dimensional Hilbert space $\mathcal{L}(x_1, a_1; \ldots; x_k, a_k)$, where $x_1, \ldots, x_k$ are the positions of the particles on the plane (or other surface) and $a_1, \ldots, a_k$ are their types. There is a finite set N of particle types (also known as sectors) in each anyonic system. Thus, unitary representations of all N-colored braid groups are defined. These representations, together with the fusion rules, form an interesting mathematical structure called a *unitary ribbon category*[1].

The general mathematical framework for anyons has been established after a series of remarkable results which are interesting in their own right. Abelian anyons were described by Abelian Chern-Simons theories. The study of nonabelian Chern-Simons theories led Witten to the discovery of his famous invariant of 3-manifolds, and to the construction of a nontrivial topological quantum field theory (TQFT) [2], [3]. TQFTs also appeared in conformal field theory[4]. Reshetikhin and Turaev[5] gave a mathematically rigorous definition of Witten's invariant. Other invariants, based on triangulations of 3-manifolds, were found by Kuperberg[6], Turaev and Viro[7].

Recently, it has been realized that anyons (and thus TQFTs and knot invariants) are closely related to quantum computation. In these notes, I am going to describe briefly the anyonic model[10]. From the mathematical point of view, it is a state model for surfaces which is closely related to the 3-dimensional state model[6]. From the physical point of view, it gives rise to the same types of anyons as the ones described in [8].

There are several new points though. Firstly, the model is defined as a quantum code. This sets some definite rules which make the game more interesting. Secondly, the model can be also described by a Hamiltonian, the essential properties of which are robust with respect to small local perturbations. Thirdly, (regardless of the particular model) braiding and fusion of anyons can be considered as elementary computational operations, thus posing a question of computational universality.

2. Terminology and notations

All Hilbert spaces in this paper are finite-dimensional. $L(\mathcal{H})$ stands for the algebra of linear operators on a Hilbert space $\mathcal{H}$.

A *qubit* is an object characterized by a two-dimensional Hilbert space $\mathcal{B}$ endowed with an orthonormal basis $\{|0\rangle, |1\rangle\}$. Thus, a system on n qubits is described by the Hilbert space $\mathcal{B}^{\otimes n}$ with the standard basis vectors $|h_1, \ldots, h_n\rangle$, $(h_j = 0, 1)$.

The Pauli operators σ^x, σ^y, $\sigma^z \in L(\mathcal{B})$ are given by

$$\sigma^x = \begin{pmatrix} 0 & 1 \\ 1 & 0 \end{pmatrix}, \quad \sigma^y = \begin{pmatrix} 0 & -i \\ i & 0 \end{pmatrix}, \quad \sigma^z = \begin{pmatrix} 1 & 0 \\ 0 & -1 \end{pmatrix}$$

If we consider a system of n qubits, the action of an operator σ^α ($\alpha = x, y, z$) on the j-th qubit is defined as follows

$$\sigma_j^\alpha = (I_\mathcal{B})^{\otimes(j-1)} \otimes \sigma^\alpha \otimes (I_\mathcal{B})^{\otimes(n-j)} \in L\left(\mathcal{B}^{\otimes n}\right).$$

Similarly, one can define the action of any operator $X \in L\left(\mathcal{B}^{\otimes k}\right)$ on any k qubits of n.

A *quantum code* is a linear subspace $\mathcal{L} \subseteq \mathcal{H}$. An error is an arbitrary linear operator $E \in L(\mathcal{H})$. We say that a code $\mathcal{L}$ *detects an error* $E \in L(\mathcal{H})$ if there is a complex number $c = c(E)$ such that

$$\langle \xi \mid E \mid \eta \rangle = c \langle \xi \mid \eta \rangle \quad \text{for any } |\xi\rangle, |\eta\rangle \in \mathcal{L}.$$

In physical terms, this condition means that the quantum state $|\eta\rangle$ "either" remains the same "or" moves beyond the subspace $\mathcal{L}$ as the error E is applied. To make this statement precise, we need to replace the logical alternatives by orthogonal vectors: $E|\eta\rangle = c|\eta\rangle + |\psi\rangle$, $|\psi\rangle \perp \mathcal{L}$, which is equivalent to the original definition.

The *distance* of a code $\mathcal{L} \in \mathcal{B}^{\otimes n}$ is the minimal number of qubits on which there is an error that the code does not detect.

3. Surface codes based on the group $\mathbb{Z}_2$

Let us consider a graph on a closed oriented surface M of genus g. The graph can be regarded as a cell complex, with the vertices, the edges and the faces being 0-cells, 1-cells and 2-cells, respectively. The qubits are associated with the edges of the graph. Thus each basis vector $|h_1, \ldots h_n\rangle$ corresponds to an assignment of labels $h_j \in \mathbb{Z}_2 = \{0, 1\}$ to the edges. Let us also define *stabilizer operators*

$$A_s = \prod_{j \in star(s)} \sigma_j^x, \qquad B_p = \prod_{j \in boundary(p)} \sigma_j^z$$

associated to each vertex s and to each face p of the graph. These operators are Hermitian and commute with each other. Hence the Hilbert space of the system $\mathcal{H}$ splits into common eigenspaces of the stabilizer operators. We define a quantum code $\mathcal{L} \subseteq \mathcal{H}$ to be one of these subspaces. Specifically,

$$|\xi\rangle \in \mathcal{L} \iff A_s |\xi\rangle = |\xi\rangle, \ B_p |\xi\rangle = |\xi\rangle \quad \text{for all } s, p$$

(see [**9**]).

The code $\mathcal{L}$ can be described directly in terms of the coordinate representation of code vectors. Let

$$|\xi\rangle = \sum_{h_1, \ldots, h_n} c_{h_1, \ldots, h_n} |h_1, \ldots, h_n\rangle.$$

Edge labellings $(h_1, \ldots, h_n)$ can be regarded as $\mathbb{Z}_2$-connections. The condition $B_p |\xi\rangle = |\xi\rangle$ says that $c_{h_1, \ldots, h_n} = 0$ unless $\sum_{j \in boundary(p)} h_j = 0$ for every face p, which means that the connection $(h_1, \ldots, h_n)$ is flat. The condition $A_s |\xi\rangle = |\xi\rangle$ says that the number $c_{h_1,\ldots,h_n}$ does not change if one changes h_j for all edges in a vertex star. Hence the function c is constant on the equivalence classes of flat $\mathbb{Z}_2$-connections. The equivalence classes are associated with elements of the first cohomology group $H^1(M, \mathbb{Z}_2)$; so the $dim \mathcal{L} = |H^1(M, \mathbb{Z}_2)| = 2^{2g}$. (Here $|\cdot|$ denotes the cardinality of a set.)

It is interesting to look at the action of operators (errors) on the subspace $\mathcal{L}$. Any operator $\mathcal{E} \in L(\mathcal{B}^{\otimes n})$ can be represented as a linear combination of basis operators of the form XZ, where $Z = \prod_{j \in c} \sigma_j^z$ and $X = \prod_{j \in c'} \sigma_j^x$. The set of qubits c defines (in an obvious way) a 1-chain with $\mathbb{Z}_2$ coefficients. It is also convenient to interpret c' as a 1-chain on the dual graph (or, equivalently, 1-cochain on the original graph).

Let $E = XZ$ be a basis operator, $|\eta\rangle \in \mathcal{L}$. Then $E |\eta\rangle \perp \mathcal{L}$ unless E commutes with all the stabilizer operators. E commutes with every A_s and B_p if and only if c and c' are 1-cycles (on the graph and its dual, respectively). In this case $E\mathcal{L} \subseteq \mathcal{L}$. The action of the operator E on the subspace $\mathcal{L}$ depends only on the homology classes of c and c'. In particular, this action is multiplication by a constant if and only if c and c' are homologically trivial. It follows that the code distance is the minimal length of a homologically nontrivial loop on the graph or its dual.

One can choose $2g$ basis cycles of each kind, $c_1, \ldots, c_{2g}$ and $c'_1, \ldots, c'_{2g}$, so that the intersection number between c_j and c'_k is equal to δ_{jk} (the Kronecker symbol). The corresponding operators X_j, Z_j are Hermitian and have the same commutation relations as the the Pauli operators σ_j^x, σ_j^z of $2g$ qubits. This allows one to identify the subspace $\mathcal{L}$ with the Hilbert space $\mathcal{B}^{\otimes(2g)}$ (up to a phase factor).

Changing one of the stabilizer condition to $A_s |\xi\rangle = - |\xi\rangle$ or $B_p |\xi\rangle = - |\xi\rangle$ can be interpreted as an "excitation." (This term refers to the vertex or the face where the stabilizer condition is changed. The vector $|\xi\rangle$, which satisfies the modified conditions, is called an *excited state*). Actually, it is impossible to create a single excitation; both the number of vertex excitations and the number of face excitations must be even.

The excitations reveal nontrivial properties even on the sphere: if one moves a vertex excitation around a face excitation, the quantum state is multiplied by -1. (Moving a vertex excitation along a path c is described by the operator $\prod_{j \in c} \sigma_j^z$). Such multiplication by a phase factor is a characteristic feature of *Abelian anyons*.

Suppose that one creates a pair of excitations on the a surface of genus g and then moves one of the excitations along a topologically nontrivial loop until it annihilates with the other one. This is equivalent to applying one of the basis operators X_j, Z_j to the code vector .

The excitations exhibit certain conservation laws (also called *superselection rules*): the parity of vertex excitations and the parity of face excitations are preserved by local operators.

4. Nonabelian anyons

Let G be a finite group. A *G-qubit* is an object characterized by a Hilbert space $\mathcal{G}$ with a standard orthonormal basis $\{|h\rangle : h \in G\}$. We associate a G-qubit to each edge of a oriented graph on a surface M. For simplicity, let M be a sphere. Each basis vector $|h_1,\ldots,h_n\rangle$ of the space $\mathcal{H} = \mathcal{G}^{\otimes n}$ corresponds to an assignment of group elements to the edges.

For each vertex s and group element $g \in G$, we define an operator $A^g(s)$ which acts as follows:

$$A^g(s)|h_1,\ldots,h_n\rangle = |h'_1,\ldots,h'_n\rangle, \quad h'_j = \begin{cases} gh_j & \text{if } j \text{ comes into } s \\ h_j g^{-1} & \text{if } j \text{ goes out of } s \\ h_j & \text{if } j \text{ is not incident to } s \end{cases}$$

Then we define a projector $A(s) = \frac{1}{|G|}\sum_{g \in G} A^g(s)$.

Another set of projectors is associated with the faces. Let $j_1,\ldots,j_m$ be the boundary edges of a face p listed in the clockwise order (starting from any point). Consider all basis vectors $|h_1,\ldots,h_n\rangle$ which satisfy the condition

$$h_{j_1}^{\pm 1} \cdots h_{j_m}^{\pm 1} = 1,$$

where the "+" sign corresponds to the edges directed counterclockwise. Then $B(p)$ is the projector onto the subspace spanned by these vectors.

All the operators $A(s)$, $B(p)$ commute with each other. One can define a quantum code $\mathcal{L}$ the same way as in the Abelian case. On a sphere, it is trivial: $dim \mathcal{L} = 1$. It is more interesting to study excitations which can be described as violated stabilizer conditions. We will consider excitations supported by a face and by a vertex on its boundary. Let $x_1,\ldots,x_k$ be disjoint face-vertex pairs. By definition, $|\xi\rangle \in \mathcal{L}(x_1,\ldots,x_k)$ if and only if

$$A(s)|\xi\rangle = |\xi\rangle, \quad B(p)|\xi\rangle = |\xi\rangle \quad \text{for all } s, p, \text{ accept for those in } x_1,\ldots,x_k.$$

The structure of the space $\mathcal{L}(x_1,\ldots,x_k)$ can be described as follows. Firstly, $\mathcal{L}(x_1,\ldots,x_k)$ can be embedded into a larger space $\widetilde{\mathcal{L}}(x_1,\ldots,x_k)$ which is *not* contained in $\mathcal{G}^{\otimes n}$. (The definition of this space is artificial, and depends on a set of cuts which are used to integrate G-connections that are flat everywhere except for $x_1,\ldots,x_k$). Due to the specific construction of the embedding, $\mathcal{L}(x_1,\ldots,x_k)$ is the invariant subspace of a certain action of the quantum double $D[G]$ on the space $\widetilde{\mathcal{L}}(x_1,\ldots,x_k)$. Actually, this action is very simple: $\widetilde{\mathcal{L}}(x_1,\ldots,x_k) = \mathcal{N}^{\otimes k}$, where $\mathcal{N}$ is the regular representation of the Hopf algebra $D[G]$. The regular representation splits into irreducible ones: $\mathcal{N} = \sum_a \mathcal{M}_a \otimes \mathcal{K}_a$. Here a runs through all irreducible representations; $\mathcal{M}_a$ is the corresponding representation space ($D[G]$ does not act on $\mathcal{K}_a$). Hence,

$$\mathcal{L}(x_1,\ldots,x_k) \cong \bigoplus_{a_1,\ldots,a_k} \mathcal{M}_{a_1,\ldots,a_k} \otimes (\mathcal{K}_{a_1} \otimes \cdots \otimes \mathcal{K}_{a_k}),$$

where $\mathcal{M}_{a_1,\ldots,a_k} \subseteq \mathcal{M}_{a_1} \otimes \cdots \otimes \mathcal{M}_{a_k}$ is the space of invariants. Now we can interpret the indices $a_1,\ldots,a_k$ as particle types.

Note that $\mathcal{L}(x_1,\ldots,x_k)$ is not a good quantum code. Local operators (those acting only on few G-qubits) effectively act on $\mathcal{K}_{a_1},\cdots,\mathcal{K}_{a_k}$. However, each of the spaces $\mathcal{M}_{a_1,\ldots,a_k}$ is *protected against errors*. To change a vector $|\psi\rangle \in \mathcal{M}_{a_1,\ldots,a_k}$, an error must stretch from one excitation to another.

One can act on the protected Hilbert space by moving excitations around each other. Each braid group element (i. e., a topologically different way of moving the excitations) is represented by a certain unitary operator. If one fuses two excitations into one, the Hilbert space shrinks. Actually, it splits into several Hilbert spaces corresponding to different types of the new excitation. Thus fusing two excItations is a measurement. Finally, if one creates a new pair of excitations, it always appears in a certain quantum state. All three operations, braiding, fusion, and creation of a new pair, are intrinsically fault-tolerant due to their topological nature.

An important question about anyons is whether the topological operations form a universal computational basis. This turns to be the case for $G = S_3$, despite the fact that the image of the braid group in the group of unitary operators is finite (for any given number of strands). Universality is achieved in an adaptive manner, i. e. by doing measurements during computation and by choosing the next braid group generator depending on the previous measurement outcomes.

It should be noted that anyons are not necessarily related to groups. The most general mathematical framework for anyons is a unitary ribbon category. It is an interesting question as to what unitary ribbon categories can be realized by quantum codes (under some reasonable assumptions).

References

[1] Turaev, V.G., **"Quantum invariants of knots and 3-manifolds,"** de Gruyter Studies in Mathematics, 18, Walter de Gruyter & Go., Berlin, 1994.
[2] Witten, E., **Topological quantum field theory**, Comm. Math. Phys. 117, 353-386 (1988).
[3] Witten, E., **Quantum field theory and Jones polynomial**, Comm. Math. Phys. 121,351-399 (1989).
[4] Moore, G., and N. Seiberg, **Classical and quantum conformal field theory**, Comm. Math. Phys. 123, 177-254 (1989).
[5] Reshetikhin, N., and V. Turaev, **Invariants of 3-manifolds via link polynomials and quantun groups**, Invent. Math. 103, 547-598 (1991).
[6] Kuperberg, G., **Involutary Hopf algebras and 3-manifold invariants**, Int. J. Math. 2, 41-66 (1991).
[7] Turaev, V., and O. Y. Viro, **State sum invariants of 3-manifolds and quantum 6j-symbols**, Topology 31, 865-902 (1992).
[8] Bais,F.A., P. van Driel, and M. de Wild Propitius, **Quantum symmetries in discrete gauge theories**, Phys. Lett. B280, 63 (1992).
[9] Kitaev, A. Yu., **Quantum computation: algorithms and error correction**, Russian Math. Surveys 52:6, 1191-1249 (1997).
[10] Kitaev, A. Yu., **Fault-tolerant quantum computation by anyons**, http://xxx.lanl.gov/quant-ph/9707021, (1997).

ONE MICROSOFT WAY, REDMOND, WA 98052
E-mail address: kitaev@iqi.caltech.edu

Quantum Topology and Quantum Computing

Louis H. Kauffman

ABSTRACT. This paper is a self-contained introduction to quantum topology and its relationships with quantum computing. Quantum invariants of knots and links are described in the context of quantum mechanics, and it is shown how certain unitary representations of the braid group arise naturally in the context of the Jones polynomial. Quantum computing is discussed from the point of view of quantum gates generated by unitary representations of the braid group.

CONTENTS

1. Introduction
2. A Quick Review of Quantum Mechanics
3. Dirac Brackets
4. Knot Amplitudes
5. Topological Quantum Field Theory - First Steps
6. Categorical Physics
7. Speculations on Quantum Computing
8. Summary
References

1. Introduction

This paper is a quick introduction to key relationships between the theories of knots,links, three-manifold invariants and the structure of quantum mechanics. In section 2 we review the basic ideas and principles of quantum mechanics. Section 3 shows how the idea of a quantum amplitude is applied to the construction of invariants of knots and links and how these constructions lead to relations with statistical mechanics, the Temperley Lieb algebra and representations of the Artin braid group. We give an example of unitary representations of the three strand braid

---

2000 *Mathematics Subject Classification.* Primary 81P68; Secondary 81-01.
Key words and phrases. Quantum computing, quantum topology.
This work was partially supported by the Defense Advanced Research Projects Agency (DARPA) and Air Force Materiel Command USAF under agreement number F30602-01-0522.

group that are related to the algebra of two projectors. Section 4 explains how the generalization of the Feynman integral to quantum fields leads to invariants of knots, links and three-manifolds. Section 5 is a discussion of a general categorical approach to these issues. Section 6 is a brief discussion of the relationships of quantum topology to quantum computing. This paper is intended as an introduction that can serve as a springboard for working on the interface between quantum topology and quantum computing. Section 7 summarizes the paper.

2. A Quick Review of Quantum Mechanics

To recall principles of quantum mechanics it is useful to have a quick historical recapitulation. Quantum mechanics really got started when DeBroglie introduced the notion that matter (such as an electron) is accompanied by a wave that guides its motion and produces interference phenomena just like the waves on the surface of the ocean or the diffraction effects of light going through a small aperture.

DeBroglie's idea was successful in explaining the properties of atomic spectra. In this domain, his wave hypothesis led to the correct orbits and spectra of atoms, formally solving a puzzle that had been only described in ad hoc terms by the preceding theory of Niels Bohr. In Bohr's theory of the atom, the electrons are restricted to move only in certain elliptical orbits. These restrictions are placed in the theory to get agreement with the known atomic spectra, and to avoid a paradox! The paradox arises if one thinks of the electron as a classical particle orbiting the nucleus of the atom. Such a particle is undergoing acceleration in order to move in its orbit. Accelerated charged particles emit radiation. Therefore the electron should radiate away its energy and spiral into the nucleus! Bohr commanded the electron to only occupy certain orbits and thereby avoided the spiral death of the atom - at the expense of logical consistency.

DeBroglie hypothesized a wave associated with the electron and he said that an integral multiple of the length of this wave must match the circumference of the electron orbit. Thus, not all orbits are possible, only those where the wave pattern can bite its own tail. The mathematics works out, providing an alternative to Bohr's picture.

DeBroglie had waves, but he did not have an equation describing the spatial distribution and temporal evolution of these waves. Such an equation was discovered by Erwin Schrodinger. Schrodinger relied on inspired guesswork based on DeBroglie's hypothesis, and produced a wave equation, known ever since as the Schrodinger equation. Schrodinger's equation was enormously successful, predicting fine structure of the spectrum of hydrogen and many other aspects of physics. Suddenly a new physics, quantum mechanics, was born from this musical hypothesis of DeBroglie.

Along with the successes of quantum mechanics came a host of extraordinary problems of interpretation. What is the status of this wave function of Schrodinger and DeBroglie. Does it connote a new element of physical reality? Is matter nothing but the patterning of waves in a continuum? How can the electron be a wave and still have the capacity to instantiate a very specific event at one place and one time (such as causing a bit of phosphor to glow there on your television screen)? It came to pass that Max Born developed a statistical interpretation of the wave-function wherein the wave determines a probability for the appearance of the localized particulate phenomenon that one wanted to call an electron. In this story the wave function ψ takes values in the complex numbers and the associated probability is $\psi^*\psi$, where ψ^* denotes the complex conjugate of ψ. Mathematically, this is a satisfactory recipe for dealing with the theory, but it leads to further questions about the exact character of the statistics. If quantum theory is inherently statistical, then it can give no complete information about the motion of the electron. In fact, there may be no such complete information available even in principle. Electrons manifest as particles when they are observed in a certain manner and as waves when they are observed in another complementary manner. This is a capsule summary of the view taken by Bohr, Heisenberg and Born. Others, including DeBroglie, Einstein and Schrodinger, hoped for a more direct and deterministic theory of nature.

As we shall see, in the course of this essay, the statistical nature of quantum theory has a formal side that can be exploited to understand the topological properties of such mundane objects as knotted ropes in space and spaces constructed by identifying the sides of polyhedra. These topological applications of quantum mechanical ideas are exciting in their own right. They may shed light on the nature of quantum theory itself.

In this section we review a bit of the mathematics of quantum theory. Recall the equation for a wave: $f(x,t) = sin((2\pi/l)(x - ct))$. With x interpreted as the position and t as the time, this function describes a sinusoidal wave travelling with velocity c. We define the wave number $k = 2\pi/\lambda$ and the frequency $\omega = (2\pi c/\lambda)$ where λ is the wavelength. Thus we can write $f(x,t) = sin(kx - \omega t)$. Note that the velocity, c, of the wave is given by the ratio of frequency to wave number, $c = \omega/k$. DeBroglie hypothesized two fundamental relationships, one between energy and frequency, the other between momentum and wave number. These relationships are summarized in the equations

$E = \hbar\omega$, $p = \hbar k$,

where E denotes the energy associated with a wave and p denotes the momentum associated with the wave. Here $\hbar = h/2\pi$, where h is Planck's constant. (The relation $E = \hbar\omega$ originates with Max Planck in the context of black-body radiation.)

For DeBroglie the discrete energy levels of the orbits of electrons in an atom of hydrogen could be explained by restrictions on the vibrational modes of waves associated with the motion of the electron. His choices for the energy and the momentum in relation to a wave are not arbitrary. They are designed to be consistent with the notion that the wave or wave packet moves along with the electron. That is, the velocity of the wave-packet is designed to be the velocity of the corresponding material particle.

It is worth illustrating how DeBroglie's idea works. Consider two waves whose frequencies are very nearly the same. If we superimpose them (as a piano tuner superimposes his tuning fork with the vibration of the piano string), then there will be a new wave produced by the interference of the original waves. This new wave pattern will move at its own velocity, different (and generally smaller) than the velocity of the original waves. To be specific, let $f(x,t) = sin(kx - \omega t)$ and $g(x,t) = sin(k'x - \omega t)$. Let $h(x,t) = sin(kx - \omega t) + sin(k'x - \omega t) = f(x,t) + g(x,t)$.

A little trigonometry shows that

$$h(x,t) = cos\left(\frac{(k-k')x}{2} - \frac{(\omega-\omega')t}{2}\right) sin\left(\frac{(k+k')x}{2} - \frac{(\omega+\omega')t}{2}\right).$$

If we assume that k and k' are very close and that w and w' are very close, then $(k+k')/2$ is approximately k, and $(w+w')/2$ is approximately w. Thus $h(x,t)$ can be represented by

$$H(x,t) = cos\left(\left(\frac{dk}{2}\right)x - \left(\frac{d\omega}{2}\right)t\right) f(x,t)$$

where $dk = (k-k')/2$ and $d\omega = (\omega-\omega')/2$. This means that the superposition, $H(x,t)$, behaves as the waveform $f(x,t)$ carrying a slower-moving wave-packet $G(x,t) = cos((dk/2)x - (d\omega/2)t)$. (See Figure 1.)

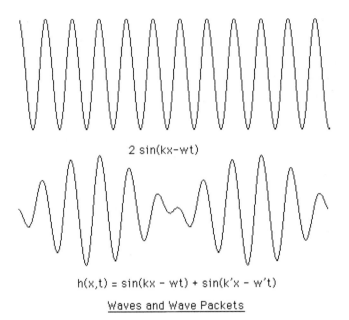

Figure 1 - Wave Packet

Since the wave packet (seen as the clumped oscillations in Figure1) has the equation $G(x,t) = cos((dk/2)x - (d\omega/2)t)$, we see that that the velocity of this wave packet is $v_g = dw/dk$. Recall that wave velocity is the ratio of frequency to wave number. Now according to DeBroglie, $E = h\omega$ and $p = hk$, where E and p are the energy and momentum associated with this wave packet. Thus we get the formula $v_g = dE/dp$. In other words, the velocity of the wave-packet is the rate of change of its energy with respect to its momentum. Now this is exactly in accord with the well-known classical laws for a material particle! For such a particle, $E = mv^2/2$ and $p = mv$. Thus $E = p^2/2m$ and $dE/dp = p/m = v$. It is this astonishing concordance between the simple wave model and the classical notions of energy and momentum that initiated the beginnings of quantum theory.

2.1. The Schrodinger Equation. Schrodinger answered the question: Where is the wave equation for DeBroglie's waves? Writing an elementary wave in complex form

$$\psi = \psi(x,t) = exp(i(kx - \omega t)),$$

we see that we can extract DeBroglie's energy and momentum by differentiating:

$$i\hbar \partial \psi / \partial t = E\psi$$

and

$$-i\hbar \partial \psi / \partial x = p\psi.$$

This led Schrodinger to postulate the identification of dynamical variables with operators so that the first equation ,

$$i\hbar \partial \psi / \partial t = E\psi$$

is promoted to the status of an equation of motion while the second equation becomes the definition of momentum as an operator:

$$p = -i\hbar \partial / \partial x.$$

Once p is identified as an operator, the numerical value of momentum is associated with an eigenvalue of this operator, just as in the example above. In our example $p\psi = \hbar k \psi$.

In this formulation, the position operator is just multiplication by x itself. Once we have fixed specific operators for position and momentum, the operators for other physical quantities can be expressed in terms of them. We obtain the energy operator by substitution of the momentum operator in the classical formula for the energy:

$$E = (1/2)mv^2 + V$$

$$E = p^2/2m + V$$

$$E = -(\hbar^2/2m)\partial^2/\partial x^2 + V.$$

Here V is the potential energy, and its corresponding operator depends upon the details of the application.

With this operator identification for E, Schrodinger's equation

$$i\hbar \partial \psi / \partial t = -(\hbar^2/2m)\partial^2 \psi / \partial x^2 + V\psi$$

is an equation in the first derivatives of time and in second derivatives of space. In this form of the theory one considers general solutions to the differential equation and this in turn leads to excellent results in a myriad of applications.

In quantum theory, observation is modelled by the concept of eigenvalues for corresponding operators. The quantum model of an observation is a projection of the wave function into an eigenstate. An energy spectrum $\{E_k\}$ corresponds to wave functions ψ satisfying the Schrodinger equation, such that there are constants E_k with $E\psi = E_k\psi$. An observable (such as energy) E is a Hermitian operator on a Hilbert space of wave functions. Since Hermitian operators have real eigenvalues, this provides the link with measurement for the quantum theory.

It is important to notice that there is no mechanism postulated in this theory for how a wave function is sent into an eigenstate by an observable. Just as mathematical logic need not demand causality behind an implication between propositions, the logic of quantum mechanics does not demand a specified cause behind an observation. The absence of an assumption of causality in logic does not obviate the possibility of causality in the world. Similarly, the absence of causality in quantum observation does not obviate causality in the physical world. Nevertheless, the debate over the interpretation of quantum theory has often led its participants into asserting that causality has been demolished in physics.

Note that the operators for position and momentum satisfy the equation

$$xp - px = \hbar i.$$

This corresponds directly to the equation obtained by Heisenberg, on other grounds, that dynamical variables can no longer necessarily commute with one another. In this way, the points of view of DeBroglie, Schrodinger and Heisenberg came together, and quantum mechanics was born. In the course of this development, interpretations varied widely. Eventually, physicists came to regard the wave function not as a generalized wave packet, but as a carrier of information about possible observations. In this way of thinking $\psi^*\psi$ (ψ^* denotes the complex conjugate of ψ.) represents the probability of finding the particle (A particle is an observable with local spatial characteristics.) at a given point in spacetime.

3. Dirac Brackets

We now discuss Dirac's notation, $<b|a>$, [7]. In this notation $<a|$ and $|b>$ are covectors and vectors respectively. $<b|a>$ is the evaluation of $|a>$ by $<b|$. Hence it is a scalar, and in ordinary quantum mechanics, it is a complex number. One can think of this as the amplitude for the state to begin in a and end in b. That is, there is a process that can mediate a transition from state a to state b. Except for the fact that amplitudes are complex valued, they obey the usual laws of probability. This means that if the process can be factored into a set of all possible intermediate states $c_1, c_2, \ldots, c_n$, then the amplitude for $a \longrightarrow b$ is the sum of the amplitudes for $a \longrightarrow c_i \longrightarrow b$. Meanwhile, the amplitude for $a \longrightarrow c_i \longrightarrow b$ is the product of the amplitudes of the two subconfigurations $a \longrightarrow c_i$ and $c_i \longrightarrow b$. Formally we have

$$< b|a > = \Sigma_i < b|c_i >< c_i|a >$$

where the summation is over all the intermediate states $i = 1, ..., n$.
In general, the amplitude for mutually disjoint processes is the sum of the amplitudes of the individual processes. The amplitude for a configuration of disjoint processes is the product of their individual amplitudes.

Dirac's division of the amplitudes into bras $< b|$ and kets $|a >$ is done mathematically by taking a vector space V (a Hilbert space, but it can be finite dimensional) for the kets: $|a >$ belongs to V. The dual space V^* is the home of the bras. Thus $< b|$ belongs to V^* so that $< b|$ is a linear mapping $< b| : V \longrightarrow C$ where C denotes the complex numbers. We restore symmetry to the definition by realizing that an element of a vector space V can be regarded as a mapping from the complex numbers to V. Given $|a >: C \longrightarrow V$, the corresponding element of V is the image of 1 (in C) under this mapping. In other words, $|a > (1)$ is a member of V. Now we have $|a >: C \longrightarrow V$ and $< b| : V \longrightarrow C$. The composition $< b| \circ |a > = < b|a >: C \longrightarrow C$ is regarded as an element of C by taking the specific value $< b|a > (1)$. The complex numbers are regarded as the VACUUM, and the entire amplitude $< b|a >$ is a vacuum to vacuum amplitude for a process that includes the creation of the state a, its transition to b, and the annihilation of b to the vacuum once more.[1]

Dirac notation has a life of its own. Let $P = |y >< x|$. Let $< x||y > = < x|y >$. Then $PP = |y >< x||y >< x| = |y >< x|y >< x| = < x|y > P$. Up to a scalar multiple, P is a projection operator. That is, if we let $Q = P/< x|y >$, then $QQ = PP/< x|y >< x|y > = < x|y > P/< x|y >< x|y > = P/< x|y > = Q$. Thus $QQ = Q$. In this language, the completeness of intermediate states becomes the statement that a certain sum of projections is equal to the identity: Suppose that $\Sigma_i |c_i >< c_i| = 1$ (summing over i) with $< c_i|c_i > = 1$ for each i. Then

$$< b|a > = < b||a > = < b|\Sigma_i|c_i >< c_i||a > = \Sigma_i < b||c_i >< c_i||a >$$

$$< b|a > = \Sigma_i < b|c_i >< c_i|a >$$

Iterating this principle of expansion over a complete set of states leads to the most primitive form of the Feynman integral [8]. Imagine that the initial and final states a and b are points on the vertical lines $x = 0$ and $x = n+1$ respectively in the $x - y$ plane, and that $(c(k)i(k), k)$ is a given point on the line $x = k$ for $0 < i(k) < m$. Suppose that the sum of projectors for each intermediate state is complete. That is, we assume that following sum is equal to one, for each k from 1 to $n - 1$:

$$|c(k)1 >< c(k)1| + ... + |c(k)m >< c(k)m| = 1.$$

Applying the completeness iteratively, we obtain the following expression for the amplitude $< b|a >$:

$$< b|a > = \Sigma\Sigma\Sigma...\Sigma < b|c(1)i(1) >< c(1)i(1)|c(2)i(2) > ... < c(n)i(n)|a >$$

---

[1]This is analogous to Julian Schwinger's algebra of measurement. See [13] and [31].

where the sum is taken over all $i(k)$ ranging between 1 and m, and k ranging between 1 and n. Each term in this sum can be construed as a combinatorial path from a to b in the two dimensional space of the $x - y$ plane. Thus the amplitude for going from a to b is seen as a summation of contributions from all the paths connecting a to b. See Figure 2.

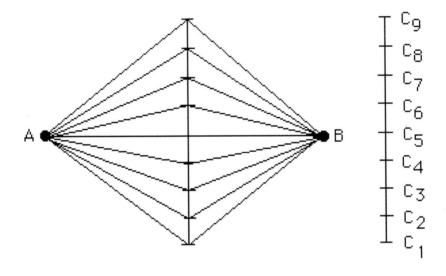

Figure 2 - Intermediates

Feynman used this description to produce his famous path integral expression for amplitudes in quantum mechanics. His path integral takes the form

$$\int dP exp(iS)$$

where i is the square root of minus one. The integral is taken over all paths from point a to point b, and S is the action for a particle to travel from a to b along a given path. For the quantum mechanics associated with a classical (Newtonian) particle the action S is given by the integral along the given path from a to b of the difference $T - V$ where T is the classical kinetic energy and V is the classical potential energy of the particle.

The beauty of Feynman's approach to quantum mechanics is that it shows the relationship between the classical and the quantum in a particularly transparent manner. Classical motion corresponds to those regions where all nearby paths contribute constructively to the summation. This classical path occurs when the variation of the action is null. To ask for those paths where the variation of the action is zero is a problem in the calculus of variations, and it leads directly to Newton's equations of motion. Thus with the appropriate choice of action, classical and quantum points of view are unified.

The drawback of this approach lies in the unavailability at the present time of an appropriate measure theory to support all cases of the Feynman integral.

On the other hand it is easy to see that a discretization of the Schrodinger equation leads to a sum over paths that is an exact solution to the discretization. To see this first write the time derivative as a difference quotient and get

$$\psi(x, t + \Delta t) = (1 - (i/\hbar)\Delta t E)\psi$$

where $E = -(\hbar^2/2m)\partial^2/\partial x^2 + V$.

Now approximate $\partial^2 \psi/\partial x^2$ by

$$(\psi(x - \delta x, t) - 2\psi(x, t) + \psi(x + \Delta x, t))/(\Delta x)^2.$$

Putting this into the equation, we get a temporal recursion of the form

$$\psi(x, t + \Delta t) = A\psi(x - \Delta x, t) + B\psi(x, t) + A\psi(x + \Delta x, t)$$

where

$$A = i\hbar \Delta t/(\Delta x)^2$$

and

$$B = 1 - iV(x)/\hbar - 2i\hbar \Delta t/(\Delta x)^2.$$

If we take $\psi(x, t)$ to be the sum over all lattice paths (in the spacetime lattice with steps Δx and Δt) where each path receives a product of weights A and B as defined above, then the recursion equation for the next time step of ψ is a tautology. In this sense it is easy to see that the discretized Schrodinger equation has a discrete path integral as its solution. Note that these lattice paths have exactly three possibilities entering $(x, t + \Delta t)$ from the past, namely $(x - \Delta x, t)$, (x, t) and $(x + \Delta x, t)$. Thus the particle travelling on the x-axis is executing a one-step random walk. See Figure 3.

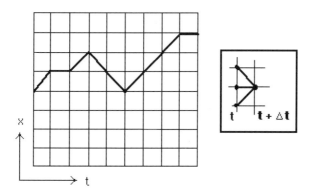

Figure 3 - Discrete Walks

It is also worth noting that the equation

$$\psi(x, t + \Delta t) = (1 - (i/\hbar)\Delta t E)\psi$$

is the infinitesimal step of the formal equation

$$\psi(x,t) = exp(-i\hbar t E \psi_0),$$

describing the wave function at later times as the result of a unitary evolution from an initial time. In devising algorithms for quantum computing the condition of unitary state evolution is the primary constraint that must be obeyed.

To summarize, Dirac notation shows at once how the probabilistic interpretation for amplitudes is tied with the vector space structure of the space of states of the quantum mechanical system. Our strategy for bringing forth relations between quantum theory and topology is to pivot on the Dirac bracket. The Dirac bracket intermediates between notation and linear algebra. In a very real sense, the connection of quantum mechanics with topology is an amplification of Dirac notation.

4. Knot Amplitudes

At the end of section 1, we said,

> The connection of quantum mechanics with topology is an amplification of Dirac notation.

Consider first a circle in a spacetime plane with time represented vertically and space horizontally. The circle represents a vacuum to vacuum process that includes the creation of two "particles", and their subsequent annihilation. See Figures 4 and 5.

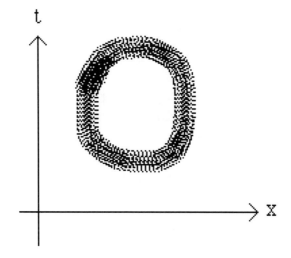

Figure 4 - Circle in Spacetime

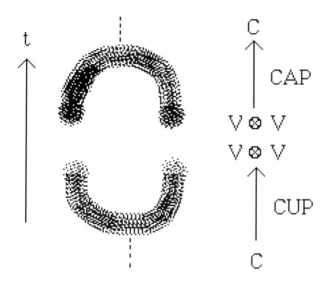

Figure 5 - Creation and Annihilation

In accord with our previous description, we could divide the circle into these two parts (creation(a) and annihilation (b)) and consider the amplitude $<b|a>$. Since the diagram for the creation of the two particles ends in two separate points, it is natural to take a vector space of the form $V \otimes V$ as the target for the bra and as the domain of the ket.

We imagine at least one particle property being catalogued by each dimension of V. For example, a basis of V could enumerate the spins of the created particles. If $\{e_a\}$ is a basis for V, then $\{e_a \otimes e_b\}$ forms a basis for $V \otimes V$. The elements of this new basis constitute all possible combinations of the particle properties. Since such combinations are multiplicative, the tensor product is the appropriate construction.

In this language, the creation ket is a map *cup*,

$$cup = |a> : C \longrightarrow V \otimes V ,$$

and the annihilation bra is a mapping *cap*,

$$cap = <b| : V \otimes V \longrightarrow C.$$

The first hint of topology comes when we realize that it is possible to draw a much more complicated simple closed curve in the plane that is nevertheless decomposed with respect to the vertical direction into many cups and caps. In fact, any non-self-intersecting differentiable curve can be rigidly rotated until it is in general position with respect to the vertical. It will then be seen to be decomposed into these minima and maxima. Our prescriptions for amplitudes suggest that we regard any such curve as an amplitude via its description as a mapping from C to C.

Each simple closed curve gives rise to an amplitude, but any simple closed curve in the plane is isotopic to a circle, by the Jordan Curve Theorem. If these are topological amplitudes, then they should all be equal to the original amplitude for the circle. Thus the question: What condition on creation and annihilation will insure topological amplitudes? The answer derives from the fact that all isotopies of the simple closed curves are generated by the cancellation of adjacent maxima and minima as illustrated below.

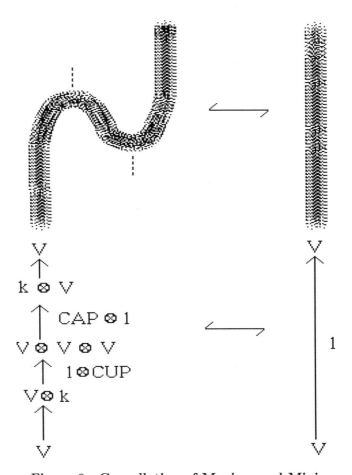

Figure 6 - Cancellation of Maxima and Minima

In composing mappings, it is necessary to use the identifications $(V \otimes V) \otimes V = V \otimes (V \otimes V)$ and $V \otimes k = k \otimes V = V$. Thus in the illustration above, the composition on the left is given by

$$V = V \otimes k \xrightarrow{1 \otimes cup} V \otimes (V \otimes V)$$

$$= (V \otimes V) \otimes V \xrightarrow{cap \otimes 1} k \otimes V = V.$$

This composition must equal the identity map on V (denoted 1 here) for the amplitudes to have a proper image of the topological cancellation. This condition is said very simply by taking a matrix representation for the corresponding operators.

Specifically, let $\{e_1, e_2, ..., e_n\}$ be a basis for V. Let $e_{ab} = e_a \otimes e_b$ denote the elements of the tensor basis for $V \otimes V$. Then there are matrices M_{ab} and M^{ab} such that

$$cup(1) = \Sigma M^{ab} e_{ab}$$

with the summation taken over all values of a and b from 1 to n. Similarly, cap is described by

$$cap(e_{ab}) = M_{ab}.$$

Thus the amplitude for the circle is

$$cap[cup(1)] = cap \Sigma M^{ab} e_{ab} = \Sigma M^{ab} M_{ab}.$$

In general, the value of the amplitude on a simple closed curve is obtained by translating it into an "abstract tensor expression" in the M_{ab} and M^{ab}, and then summing over these products for all cases of repeated indices.

Returning to the topological conditions, we see that they are just that the matrices (M_{ab}) and (M^{ab}) are inverses in the sense that $\Sigma M_{ai} M^{ib} = \delta_a^b$ and $\Sigma M^{ai} M_{ib} = \delta_b^a$ where δ_a^b denotes the (identity matrix) Kronecker delta that is equal to one when its two indices are equal to one another and zero otherwise.

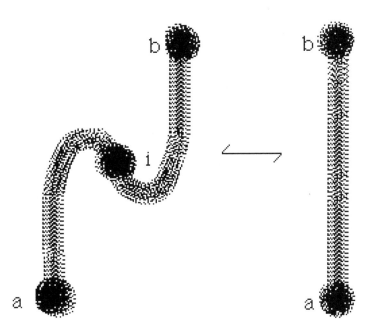

Figure 7 - Algebraic Cancellation of Maxima and Minima

In Figure 7, we show the diagrammatic representative of the equation $\Sigma M_{ai} M^{ib} = \delta_a^b$.

In the simplest case *cup* and *cap* are represented by 2×2 matrices. The topological condition implies that these matrices are inverses of each other. Thus the problem of the existence of topological amplitudes is very easily solved for simple closed curves in the plane.

Now we go to knots and links. Any knot or link can be represented by a picture that is configured with respect to a vertical direction in the plane. The picture will decompose into minima (creations) maxima (annihilations) and crossings of the two types shown below. (Here I consider knots and links that are unoriented. They do not have an intrinsic preferred direction of travel.) See Figure 2. In Figure 2, next to each of the crossings we have indicated mappings of $V \otimes V$ to itself, called R and R^{-1} respectively. These mappings represent the transitions corresponding to these elementary configurations.

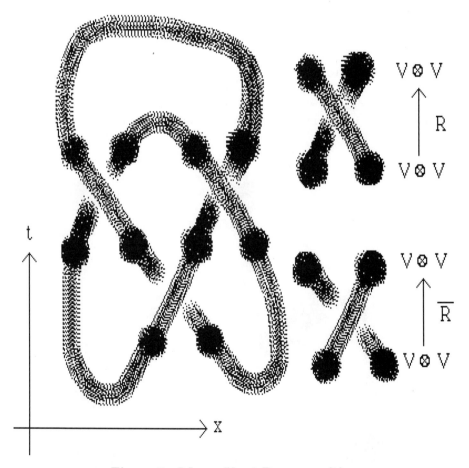

Figure 8 - Morse Knot Decomposition

That R and R^{-1} really must be inverses follows from the isotopy shown in Figure 9 (This is the second Reidemeister move.)

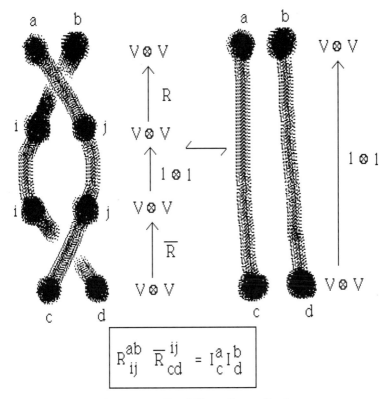

Figure 9 - Braiding Cancellation

We now have the vocabulary of *cup*, *cap*, R and R^{-1}. Any knot or link can be written as a composition of these fragments, and consequently a choice of such mappings determines an amplitude for knots and links. In order for such an amplitude to be topological, we want it to be invariant under the list of local moves on the diagrams shown in Figure 11. These moves are an augmented list of the Reidemeister moves, adjusted to take care of the fact that the diagrams are arranged with respect to a given direction in the plane. The equivalence relation generated by these moves is called regular isotopy. It is one move short of the relation known as ambient isotopy. The missing move is the first Reidemeister move shown in Figure 10.

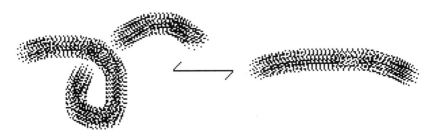

Figure 10 - Framing Cancellation

In the first Reidemeister move, a curl in the diagram is created or destroyed. Ambient isotopy (generated by all the Reidemeister moves) corresponds to the full topology of knots and links embedded in three dimensional space. Two link diagrams are ambient isotopic via the Reidemeister moves if and only if there is a continuous family of embeddings in three dimensions leading from one link to the other. The moves give us a combinatorial reformulation of the spatial topology of knots and links.

By ignoring the first Reidemeister move, we allow the possibility that these diagrams can model framed links, that is links with a normal vector field or, equivalently, embeddings of curves that are thickened into bands. It turns out to be fruitful to study invariants of regular isotopy. In fact, one can usually normalize an invariant of regular isotopy to obtain an invariant of ambient isotopy. We shall see an example of this phenomenon with the bracket polynomial in a few paragraphs.

As the reader can see, we have already discussed the algebraic meaning of moves 0. and 2. The other moves translate into very interesting algebra. Move 3., when translated into algebra, is the famous Yang-Baxter equation. The Yang-Baxter equation occurred for the first time in problems related to exactly solved models in statistical mechanics. (See [24].) All the moves taken together are directly related to the axioms for a quasi-triangular Hopf algebra, also known as a quantum group. We shall not go into this connection here.

There is an intimate connection between knot invariants and the structure of generalized amplitudes, as we have described them in terms of vector space mappings associated with link diagrams. This strategy for the construction of invariants is directly motivated by the concept of an amplitude in quantum mechanics. It turns out that the invariants that can actually be produced by this means (that is by assigning finite dimensional matrices to the caps, cups and crossings) are incredibly rich. They encompass, at present, all of the known invariants of polynomial type, i.e., the Alexander polynomial, Jones polynomial, and their generalizations.

It is now possible to indicate the construction of the Jones polynomial via the bracket polynomial as an amplitude, by specifying its matrices. The cups and the caps are defined by $(M_{ab}) = (M^{ab}) = M$, where M is the 2×2 matrix

$$M = \begin{bmatrix} 0 & iA \\ -iA^{-1} & 0 \end{bmatrix},$$

where with $ii = -1$.
Note that $MM = I$, where I is the identity matrix. Note also that the amplitude for the circle is

$$\Sigma M_{ab} M^{ab} = \Sigma M_{ab} M_{ab} = \Sigma M_{ab}^2$$

$$= (iA)^2 + (-iA^{-1})^2 = -A^2 - A^{-2}.$$

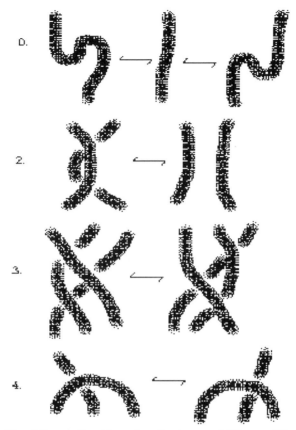

Figure 11 - Moves for Regular Isotopy of Morse Diagrams

The matrix R is then defined by the equation

$$R^{ab}_{cd} = AM^{ab}M_{cd} + A^{-1}\delta^a_c\delta^b_d,$$

Since, diagrammatically, we identify R with a (right handed) crossing, this equation can be written diagrammatically as the generating identity for the bracket polynomial:

Taken together with the loop value of $-A^2-A^{-2}$ that is a consequence of this matrix choice, these equations can be regarded as a recursive algorithm for computing the amplitude. This algorithm is the bracket state model for the (unnormalized) Jones polynomial [15]. This model can be studied on its own grounds.

We end this section with some comments about this algorithm and its properties.

4.1. The Bracket Model.
If we were to start with just the calculational formulas as indicated above but with arbitrary coefficients A and B for the two smoothings, and an arbitrary loop value d, then it is easy to see that the resulting method of calculating a three variable polynomial (in the commuting variables A, B and d) from a link diagram is well-defined, although not necessarily invariant under the Reidemeister moves. It is then an interesting exercise to see that asking for invariance under just the second Reidemeister move essentially forces $B = A^{-1}$ and $d = -A^2 - A^{-2}$. Thus the parameters arising from the algebra that we have sketched actually come directly from the topology. It is equally easy to see the the resulting Laurent polynomial is a well defined invariant of regular isotopy. Lets denote that invariant by $<K>$, the (unnormalized) bracket polynomial of K. In this version of the bracket we have $<O> = -A^2 - A^{-2}$ where O denotes a circle in the plane. If we define $f_K(A) = (-A^3)^{-w(K)} <K> / <O>$ where $w(K)$ denotes the sum of the signs of the crossings in an oriented link K ([15], [19]), then $f_K(A)$ is an invariant of ambient isotopy, and the original Jones polynomial [14] $V_K(t)$ is given by the formula $V_K(t) = f_K(t^{-1/4})$. The bracket model for the Jones polynomial is quite useful both theoretically and in terms of practical computations. One of the neatest applications is to simply compute $f_K(A)$ for the trefoil knot T and determine that $f_K(A)$ is not equal to $f_K(A^{-1})$. This shows that the trefoil is not ambient isotopic to its mirror image (See Figure 12), a fact that is quite tricky to prove by classical methods.

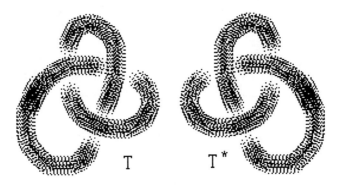

Figure 12 - Trefoil and Mirror Image

4.2. The Temperley Lieb Algebra and Representations of the Artin Braid Group.
The very close relationship between elementary quantum mechanics and topology is very well illustrated by the structure and representations of the Temperley Lieb algebra, an algebra generated by projectors that figures in the original construction of the Jones polynomial and in related structures for the bracket polynomial model of the Jones polynomial.

The Temperley Lieb algebra TL_n [15] is an algebra over a commutative ring k with generators $\{1, U_1, U_2, ..., U_{n-1}\}$ and relations

$$U_i^2 = \delta U_i,$$

$$U_i U_{i\pm 1} U_i = U_i,$$

$$U_i U_j = U_j U_i, |i - j| > 1,$$

where δ is a chosen element of the ring k. These equations give the multiplicative structure of the algebra. The algebra is a free module over the ring k with basis the equivalence classes of these products modulo the given relations.

Figure 13 illustrates a diagrammatic interpretation of this algebra that is intimately linked with the bracket polynomial. In this interpretation, the multiplicative generators of the module are collections of strands connecting n top points and n bottom points. Top points can be connected either to top or to bottom points. Bottom points can be connected to either bottom or to top points. All connections are made in the plane with no overlapping lines and no lines going above the top row of points or below the bottom row of points. Multiplication is accomplished by connecting the bottom row of one configuration with the top row of another. In Figure 13 we have illustrated the types of special configurations that correspond to the U_i, and we have shown that δ is interpreted as a closed loop.

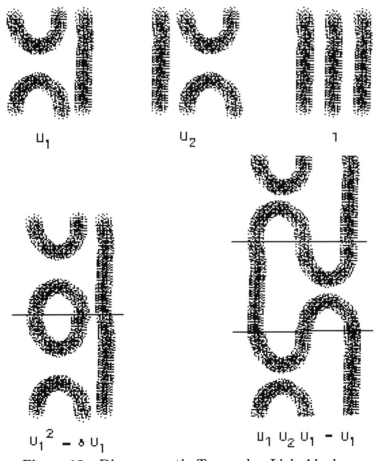

Figure 13 - Diagrammatic Temperley Lieb Algebra

The relationship with the bracket polynomial comes through the basic bracket identity. This identity, interpreted in the context of the diagrammatic Temperley Lieb algebra, becomes a representation ρ of the Artin braid group B_n on n strands to the Temperley Lieb algebra TL_n defined by the formulas

$$\rho(\sigma_i) = AU_i + A^{-1}1$$

$$\rho(\sigma_i^{-1}) = A^{-1}U_i + A1.$$

Here σ_i denotes the braid generator that twists strands i and $i+1$ as shown in Figure 13.1. For this representation of the Temperley Lieb algebra, the loop value δ is $-A^2 - A^{-2}$ and the ring k is $Z[A, A^{-1}]$, the ring of Laurent polynomials in A with integer coefficients.

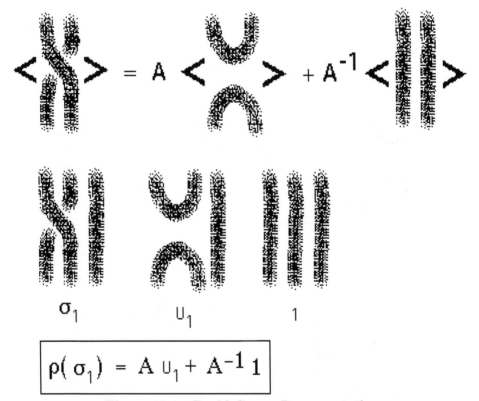

Figure 13.1 - Braid Group Representation

One way to make a matrix representation of the Temperley Lieb algebra (and a corresponding representation of the braid group) is to use the matrix M that we discussed for the matrix model of the bracket polynomial. Since M represents the structure of a cap or a cup and since the basic Temperley Lieb element is a cup-cap, we can combine two copies of M to form a matrix

$$U^{ab}_{cd} = M^{ab}M_{cd}$$

with the property that $U^2 = \delta U$ where $\delta = -A^2 - A^{-2}$. Then U_i is a tensor product of identity matrices corresponding to the vertical lines in the diagram for this element and one factor of U for the placement of the cup-cap at the locations i and $i+1$. This representation of the Temperley Leib algebra is very useful for knot theory, and it is conjectured to be a faithful representation. One may also conjecture that the corresponding braid group representation is faithful.

The representation of the braid group that we have described is however not a unitary representation except when $A^2 = -1$, a value that is not of interest in the knot theory. In order to find elementary unitary representations of the braid group, one has to go deeper. (See [**10**].)

It is useful to think of the Temperley Lieb algebra as generated by projections $e_i = U_i/\delta$ so that $e_i^2 = e_i$ and $e_i e_{i\pm 1} e_i = \tau e_i$, where $\tau = \delta^{-2}$ and e_i and e_j commute for $|i - j| > 1$.

With this in mind, consider elementary projectors $e = |A><A|$ and $f = |B><B|$. We assume that $<A|A> = <B|B> = 1$ so that $e^2 = e$ and $f^2 = f$. Now note that

$$efe = |A><A|B><B|A><A| = <A|B><B|A>e = \tau e$$

Thus
$$efe = \tau e$$
where $\tau = <A|B><B|A>$.

This algebra of two projectors is the simplest instance of a representation of the Temperley Lieb algebra. In particular, this means that a representation of the three-strand braid group is naturally associated with the algebra of two projectors, a simple toy model of quantum physics!

Quite specifically, if we let $|A> = (a, b)$ and $<A| = (a, b)^t$ the transpose of this row vector, then

$$e = |A><A| = \begin{bmatrix} a^2 & ab \\ ab & b^2 \end{bmatrix}$$

is a standard projector matrix when $a^2 + b^2 = 1$. To obtain a specific representation, let

$$e_1 = \begin{bmatrix} 1 & 0 \\ 0 & 0 \end{bmatrix}$$

and

$$e_2 = \begin{bmatrix} a^2 & ab \\ ab & b^2 \end{bmatrix}.$$

It is easy to check that

$$e_1 e_2 e_1 = a^2 e_1$$

and that

$$e_2 e_1 e_2 = a^2 e_2.$$

We then have $U_i = \delta e_i$ for $i = 1, 2$ so that $a^2 = \delta^{-2}$. Since $a^2 + b^2 = 1$ this means that $\delta^{-2} + b^2 = 1$, whence

$$b^2 = 1 - \delta^{-2}.$$

Therefore b is real when δ^2 is greater than or equal to 1.

We are interested in the case where $\delta = -A^2 - A^{-2}$ and *A is a unit complex number*. Under these circumstances the braid group representation

$$\rho(\sigma_i) = AU_i + A^{-1}1$$

will be unitary whenever U_i is a real symmetric matrix. Thus we will obtain a unitary representation of the three-strand braid group B_3 when $\delta^2 \geq 1$. Specifically, let $A = e^{i\theta}$. Then $\delta = -2cos(2\theta)$, so the condition $\delta^2 \geq 1$ is equivalent to $cos^2(2\theta) \geq 1/4$. Thus we get the specific range of angles $|\theta| \leq \pi/6$ and $|\theta - \pi| \leq \pi/6$ that give unitary representations of the three-strand braid group. We will discuss these representations in relation to quantum computing and the Jones polynomial elsewhere [26]. The point here is that while from the point of view of topology it is helpful to release the restriction of unitarity, it is nevertheless quite interesting to search for such representations.

5. Topological Quantum Field Theory - First Steps

In order to further justify this idea of the amplification of Dirac notation, consider the following scenario. Let M be a 3-dimensional manifold. Suppose that F is a closed orientable surface inside M dividing M into two pieces M_1 and M_2. These pieces are 3-manifolds with boundary. They meet along the surface F. Now consider an amplitude $< M_2|M_1 > = Z(M)$. The form of this amplitude generalizes our previous considerations, with the surface F constituting the distinction between the preparation M_1 and the detection M_2. This generalization of the Dirac amplitude $< b|a >$ amplifies the notational distinction consisting in the vertical line of the bracket to a topological distinction in a space M. The amplitude $Z(M)$ will be said to be a topological amplitude for M if it is a topological invariant of the 3-manifold M. Note that a topological amplitude does not depend upon the choice of surface F that divides M.

From a physical point of view the independence of the topological amplitude of the particular surface that divides the 3-manifold is the most important property. An amplitude arises in the condition of one part of the distinction carved in the 3-manifold acting as the observed and the other part of the distinction acting as the observer. If the amplitude is to reflect physical (read topological) information about the underlying manifold, then it should not depend upon this particular decomposition into observer and observed. The same remarks apply to 4-manifolds and interface with ideas in relativity. We mention 3-manifolds because it is possible to describe many examples of topological amplitudes in three dimensions. The matter of 4-dimensional amplitudes is a topic of current research. The notion that an amplitude be independent of the distinction producing it is prior to topology. Topological invariance of the amplitude is a convenient and fundamental way to produce such independence.

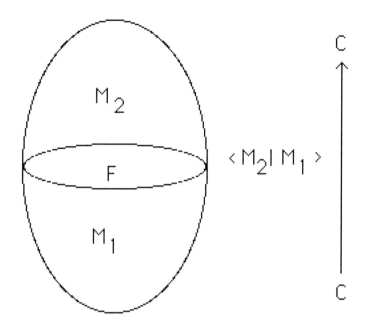

Figure 14 - Three Manifold as Vacuum-Vacuum Expectation

This sudden jump to topological amplitudes has its counterpart in mathematical physics. In [**32**] Edward Witten proposed a formulation of a class of 3-manifold invariants as generalized Feynman integrals taking the form $Z(M)$, where

$$Z(M) = \int dA exp[(ik/4\pi)S(M,A)].$$

Here M denotes a 3-manifold without boundary, and A is a gauge field (also called a gauge potential or gauge connection) defined on M. The gauge field is a one-form on M with values in a representation of a Lie algebra. The group corresponding to this Lie algebra is said to be the gauge group for this particular field. In this integral the action $S(M, A)$ is taken to be the integral over M of the trace of the Chern-Simons three-form $CS = AdA + (2/3)AAA$. (The product is the wedge product of differential forms.)

Instead of integrating over paths, the integral $Z(M)$ integrates over all gauge fields modulo gauge equivalence. This generalization from paths to fields is characteristic of quantum field theory. Quantum field theory was designed in order to accomplish the quantization of electromagnetism. In quantum electrodynamics the classical entity is the electromagnetic field. The question posed in this domain is to find the value of an amplitude for starting with one field configuration and ending with another. The analogue of all paths from point a to point b is all fields from field A to field B.

Witten's integral $Z(M)$ is, in its form, a typical integral in quantum field theory. In its content, $Z(M)$ is highly unusual. The formalism of the integral, and its internal logic supports the existence of a large class of topological invariants of 3-manifolds and associated invariants of knots and links in these manifolds.

Invariants of three-manifolds were initiated by Witten as functional integrals in [32] and at the same time defined in a combinatorial way by Reshetikhin and Turaev in [30]. The Reshetikhin-Turaev definition proceeds in a way that is quite similar to the definition that we gave for the bracket model for the Jones polynomial in section 2. (See also [20].) It is an amazing fact that Witten's definition seems to give the very same invariants. We are not in a position to go into the details of this correspondence here. However, one theme is worth mentioning: For k large, the Witten integral is approximated by those gauge connections A for which $S(M, A)$ has zero variation with respect to change in A. These are the so-called flat connections. It is possible in many examples to calculate this contribution via both the functional integral and by the combinatorial definition of Reshetikhin and Turaev. In all cases, the two methods agree. (See e.g. [12].) This is one of the pieces of evidence in a puzzle that everyone expects will eventually justify the formalism of the functional integral. Note how this case corresponds exactly to the relation of classical and quantum physics as it was discussed in Section 1.

In order to obtain invariants of knots and links from Witten's integral, one adds an extra bit of machinery to the brew. The new machinery is the Wilson loop. The Wilson loop is an exponentiated version of integrating the gauge field along a loop K. We take this loop K in three space to be an embedding (a knot) or a curve with transversal self-intersections. It is usually indicated by the symbolism $tr(Pexp(\int_K A))$. Here the P denotes path ordered integration - that is we are integrating and exponentiating matrix valued functions, and one must keep track of the order of the operations. The symbol tr denotes the trace of the resulting matrix.

With the help of the Wilson loop function on knots and links, Witten [32] writes down a functional integral for link invariants in a 3-manifold M:

$$Z(M, K) = \int dA exp[(ik/4\pi)S(M, A)] tr(Pexp(\int_K A)).$$

Here $S(M, A)$ is the Chern-Simons Lagrangian, as in the previous discussion.

If one takes the standard representation of the Lie algebra of $SU(2)$ as 2×2 complex matrices then it is a fascinating exercise to see that the formalism of $Z(S, K)$ (S^3 denotes the three-dimensional sphere.) produces the original Jones polynomial with the basic properties as discussed in section 1. See Witten's paper [32] or [19],[21] for discussions of this part of the heuristics.

This approach to link invariants crosses boundaries between different methods. There are close relations between $Z(S^3, K)$ and the invariants defined by Vassiliev [3], to name one facet of this complex crystal.

This deep relationship between topological invariants in low dimensional topology and quantum field theory in the sense of Witten's functional integral is really still in its infancy. There will be many surprises in the future as we discover that what has so far been uncovered is only the tip of an iceberg.

6. Categorical Physics

We have seen that in quantum topology and topological quantum field theory, the Dirac notational viewpoint on quantum mechanics has become amplified into a framework that embraces amplitudes associated with topological spaces and with embeddings of one space within another (e.g. knots and links in three dimensional space). The brackets, kets and bras are generalized to become maps of vector spaces associated with these topological spaces in a category that allows tensor products. (Thus we associated many tensor products of a single vector space V with itself in analyzing knots.) The correct formal notion is that of a tensor category. But I will omit the precise definition in this informal discussion. On the other hand, the notion of category is worth examining in this context.

A category is a set with two types of elements called objects and morphisms. A morphism f is associated with two objects A and B, and is written $f : A \longrightarrow B$, where we say that f is a morphism from A to B. In a category, if there is a morphism $f : A \longrightarrow B$ and a morphism $g : B \longrightarrow C$, then there is a morphism $g \circ f : A \longrightarrow C$ called the *composition* of f and g. Composition of morphisms is associative, and every object A has a morphism $I(A) : A \longrightarrow A$ such that if $f : A \longrightarrow B$ is any other morphism then $I(B)f = fI(A) = f$. These properties comprise the definition of a category.

Given a category C and another category $C\prime$, we say that $F : C \longrightarrow C\prime$ is a *functor* from C to $C\prime$ if F takes objects to objects, morphisms to morphisms, F applied to an identity morphism in C is an identity morphism in $C\prime$, and F applied to a composition of morphisms in C is equal to the composition of the corresponding morphisms in $C\prime$. In other words, $F(I(A)) = I(F(A))$ for any object A in C, and $F(ab) = F(a)F(b)$ for any composable morphisms a and b in C. A functor is a structural mapping from one category to another.

The morphisms in a category are not necessarily functions from some set to another set. Rather they are directed structural relations that are of significance in a particular domain. A case in point is our discussion of knots where we associated linear mappings to cups, caps, crossings and compositions of these forming all sorts of knots and tangles. The cups, caps and crossings can be regarded by themselves as the generating morphisms for a tensor category whose objects are just ordered collections of points (including the empty collection!) corresponding to the endpoints of arcs. Composition of morphisms corresponds to attaching endpoints together in the fashion that we described in that section. We call this category the (unoriented) tangle category. The association of linear mappings to elements in the tangle category that we so carefully described in our section on knots and links comprises a functor from the tangle category to the category of vector spaces and linear mappings.

Quantum amplitudes are calculated in the vector space category. The functor that we described from tangles to vector spaces tells us how to do quantum mechanics on the tangle category. But this quantum mechanics is a generalization of the usual quantum mechanics. The underlying topological spaces (here the knots and links) have quantum states, but they themselves are classical (at least in the sense that our abstraction of a knot from the physical rope embodies properties from classical physics). These same issues necessarily come up when trying to marry quantum mechanics and relativity theory since one wants to bring an underlying topological manifold (with changing topology and metric) into the discussion.

Furthermore, the issue of measurement is directly related to cutting the spaces apart or making distinctions in the underlying space. Thus in our example with three manifolds in the last section, we divided the three manifold into parts M1 and M2 and then looked at the amplitude $< M_2 | M_1 >$. In this view, either half of the manifold can be regarded as an observer of the other half.

This description of states of affairs is very similar to the time-honored discussion of the relationship of ordinary language and the classical description of measuring apparatus in relation to quantum mechanical calculations. Thus we could begin to formalize quantum mechanics as a special sort of functor whose domain category is a classical category analogous to knots, links and manifolds, and whose range category is an appropriate tensor category where amplitudes and observables can be computed. The classical category then gets structured in a non-classical way by this functor.

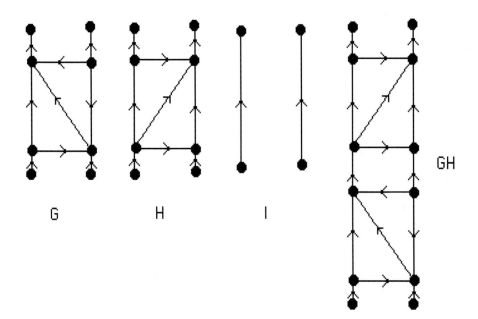

Figure 15 - Composition in the Graph Category

Here is another example of a structure of the sort that I just described. Consider the set of finite directed multi-graphs. Call a node in a graph G an input node if it has exactly one directed edge emanating from it and no edges entering it. Call a node an output node if it has exactly one edge entering it and no edges leaving it. Let $DG(n)$ denote the set of digraphs of this kind that have n inputs and n outputs. Further assume that each such graph is equipped with an ordering of its inputs and an ordering of its outputs. Thus G in $DG(n)$ will have inputs labelled 1,2,3,..., n and outputs labelled 1,2,3,...,n. Given G and H in $DG(n)$, we define their composition GH by attaching the k-th output of G to the k-th input of H (by removing corresponding nodes and amalgamating the two directed edges at those nodes to a single edge). As shown below, the graph I_n consisting of n parallel edges is an identity for this composition.

The upshot is that $G(n)$ is a category where the digraphs are the morphisms and the one object is the ordered set $\{1, 2, ..., n\}$. Juxtaposition of graphs gives a tensor structure and a mapping $DG(n) \otimes DG(m) \longrightarrow DG(n+m)$. With a little work, all the $DG(n)$'s can be put together in one category DG. When $n = 0$, we have digraphs without inputs or outputs, analogous to knots and links. Clearly the categories $DG(n)$ are analogous to n-strand tangles as we have discussed them in the section on knot amplitudes.

A functor on $DG(n)$ that takes the category to vector spaces and linear maps can be constructed by associating a linear mapping or matrix to each different species of directed node in the graphs under consideration. Then composition of graphs will correspond to matrix multiplication in much the same manner as our discussion for knots and links. To give a simple example, lets work in $DG(1)$. Then each graph has one input and one output. Then we regard the input line of the graph as corresponding to the left index of a matrix and the output line as corresponding to the right index of a matrix. If G is the graph and $F(G)$ the corresponding matrix then $G \circ H$ corresponds to $F(G)F(H)$ if we regard the tying of the output line of G to the input line of H as connoting summation over all possibilities for the common index and take the sum of the products of the matrix entries for $F(G)$ and $F(H)$. This defines a functor from $DG(1)$ to the category of matrices where the morphisms are the matrices and composition is the matrix product.

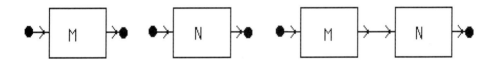

In multiplying matrices M and N, we have $(MN)_{ij} = \Sigma_k M_{ik} N_{kj}$. In the graph category, the internal edge corresponds to the index susceptible to summation. With this interpretation, many elementary formulas and patterns of quantum mechanics become simple matters of diagrammatics. For example, if we are computing $tr(MP)$ (tr denotes trace) where P is a projection operator $P = |A><A|$, then it is easy to see in the graph category that $tr(MP) = <A|M|A>$. (See Figure 16.)

This example also indicates how to conceptualize measurement in the graph category. An elementary measurement consists in inserting a projector P into a link in the graph. The effect of such an insertion is non-local, since the amplitudes are computed via the functor to the matrix category and consequently involve summations over all states of the graph (where a state consists in assignments of indices to all the internal lines of the graph and the amplitude is computed by summing over all the products of the resulting matrix elements).

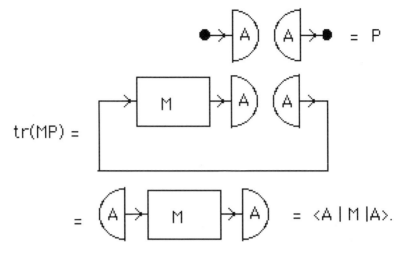

Figure 16 - $Trace(MP) = <A|M|A>$

The graph is a classical but abstract description of a set of relationships. The functor that computes amplitudes from the graph does a non-local computation involving the graph as a whole. If we imagine that the universe is a large network analogous to such a graph, then it will be necessary to understand how one part of the network becomes an observable for the rest, and how this classical level of description intertwines with the quantum amplitude functor.

7. Speculations on Quantum Computing

In this paper I have concentrated on giving a picture of the general framework of quantum topology and how it is related to a very general, in fact categorical, view of quantum mechanics. Many algorithms in quantum topology are configured without regard to unitary evolution of the amplitude, since the constraint has been topological invariance rather than conformation to physical reality. This gives rise to a host of problems (that we shall discuss elsewhere) of attempting to reformulate topological amplitudes as quantum computations. A particular case in point is the bracket model for the Jones polynomial. It would be of great interest to see a reformulation of this algorithm that would make it a quantum computation in the strict sense of quantum computing.

To see how this can be formulated, consider the vacuum-vacuum computation of a link amplitude as we have described it in section 4. See Figures 8 and 17. Particularly in Figure 17 we have indicated an amplitude where the temporal decomposition consists first in a composition of cups (creations), then braiding and then caps (annihilations). Thus we can write the amplitude in the form

$$Z_K = <CUP|M|CAP>$$

where $<CUP|$ denotes the composition of cups, M is the composition of elementary braiding matrices and $|CAP>$ is the composition of caps. We then regard $<CUP|$ as the preparation of this state and $|CAP>$ as the detection of this state. In order to view Z_K as a quantum computation, we need that M is a unitary operator. This will be the case if the R-matrices (the solutions to the Yang-Baxter equation used in the model for this amplitude) are unitary. In this case, each R-matrix can be viewed as a quantum gate (or possibly a composition of quantum gates) and the vacuum-vacuum diagram for the knot is interpreted as a quantum computer. This quantum computer will probabilistically compute the values of the states in the state sum for Z_K.

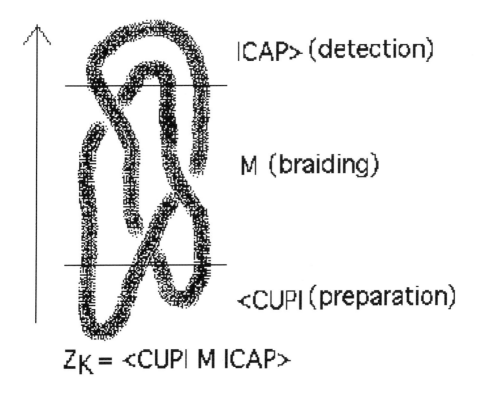

Figure 17 - $Z_K = <CUP|M|CAP>$

Many questions are raised by this formulation of a quantum computer associated with a given Morse link diagram. First of all, unitary solutions to the Yang-Baxter equation (or unitary representation of the Artin braid group) that also give link invariants are not so easy to come by. Secondly, it is not clear what the practical value of such a computation will be for understanding a given link invariant. Nevertheless, it is to be expected that a close relationship between quantum link invariants and quantum computing will be fruitful for both fields.

There are other ideas in the topology that deserve comparison with the quantum states. For example, topological entanglement in the sense of linking and braiding is intuitively related to the entanglement of quantum states. (See [**29**] and [**29**].) This is actually the case for the quantum topological states associated with the bracket polynomial, and undoubtedly would figure strongly in a quantum computing model of this algorithm. For this and many other reasons it is worthwhile to make the comparison between quantum topology and quantum computation.

8. Summary

We have, in this short paper, given an almost unbroken line of argument from the beginnings of quantum mechanics to the construction of topological quantum field theories and link invariants associated with quantum amplitudes.

One of the prospects for these new invariants is the possibility of their application in quantum gravity. See [5] for an account of these developments. Many other applications are possible, and the subject is just beginning. For a survey of past and present applications of knots and links, we refer the reader to [23] [22],[25].

In relating quantum computing with quantum topology, the key themes are unitarity and measurement. The commonality of quantum topology and quantum computing will be in a part of quantum topology that has direct bearing on physical reality. Much is now surely unforeseen. For a good survey of quantum computing, we recommend [1], and for another view of topological issues see [9] and [10].

References

[1] D. Aharonov, **Quantum computation**, (1998), http://xxx.lanl.gov/abs/quant-phys/9812037 15, Dec 1998.
[2] M.F. Atiyah, "**The Geometry and Physics of Knots**," Cambridge University Press, (1990).
[3] D. Bar-Natan, **On the Vassiliev Knot invariants**, Topology, Vol.34, No.2, (1995), pp. 423-472.
[4] R.J. Baxter, "**Exactly Solved Models in Statistical Mechanics**," Acad. Press, (1982).
[5] J. Baez and J.P. Muniain, "**Gauge Fields, Knots and Gravity**," World Sci. Press, (1994).
[6] L.C. Biedenharn and J.D. Louck, **Angular Momentum in Quantum Physics- Theory and Application**, in "Encyclopedia of Mathematics and its Applications," Cambridge University Press, (1979).
[7] P.A.M. Dirac, "**Principles of Quantum Mechanics**," Oxford University Press, (1958).
[8] R. Feynman and A.R. Hibbs, "**Quantum Mechanics and Path Integrals**," McGraw Hill, (1965).
[9] M. Freedman, **Topological Views on Computational Complexity**, Documenta Mathematica - Extra Volume ICM, (1998), pp. 453-464.
[10] M. Freedman, M. Larsen and Z. Wang, **A modular functor which is universal for quantum computation**, (2000), http://xxx.lanl.gov/abs/quant-ph/0001108.
[11] V.F.R. Jones, **A polynomial invariant for links via von Neumann algebras**, Bull.Amer.Math.Soc., 129, (1985), pp. 103-112.
[12] D.S. Freed and R.E. Gompf, **Computer Calculation of Witten's 3-Manifold Invariant**, Commun. Math. Phys., 141, (1991), pp. 79-117.
[13] Gottfried, "**Quantum Mechanics: Volume I. Fundamentals**," Addison-Wesley, (1989).
[14] V.F.R. Jones, **A new knot polynomial and von Neumann algebras**, Notices of AMS, 33, (1986), pp 219-225.
[15] L.H. Kauffman, **State Models and the Jones Polynomial**, Topology, 26, (1987), pp 395-407.
[16] L.H. Kauffman, "**On Knots**," Annals of Mathematics Studies Number 115, Princeton University Press, (1987).
[17] L.H. Kauffman, **New invariants in the theory of knots**, Amer. Math. Monthly, Vol.95, No.3, March 1988, pp 195-242.
[18] L.H. Kauffman, "**Statistical mechanics and the Jones polynomial**," AMS Contemp. Math. Series (1989), Vol. 78, pp. 263-297.
[19] L.H. Kauffman, "**Knots and Physics**," World Scientific Pub., (1991, 1993, and 2001).
[20] L.H. Kauffman and D.E.Radford, **Invariants of 3-manifolds derived from finite dimensional Hopf algebras**, Journal of Knot Theory and its Ramifications, Vol.4, No.1, (1995), pp. 131-162.

[21] L. H. Kauffman, **Functional integration and the theory of knots**, J. Math. Phys., Vol. 36, No.5, (1995), pp. 2402-2429.
[22] L. H. Kauffman, **Witten's Integral and the Kontsevich Integral**, in "Particles Fields and Gravitation," edited by Jakub Rembielinski, AIP Proceedings No. 453, American Inst. of Physics Pub., (1998), pp. 368 - 381.
[23] L.H. Kauffman (Editor), **"Knots and Applications,"** World Scientific Pub. Co., (1995).
[24] L.H. Kauffman, **Knots and Statistical Mechanics,** in "The Interface of Knots and Physics," AMS Proceedings of Symposia in Applied Mathematics, edited by L. Kauffman, Vol.51, (1996), pp. 1-87.
[25] L.H. Kauffman, **Spin Networks and Topology**, in "The Geometric Universe," edited by Huggett et al., Oxford University Press, (1998), pp. 277-290.
[26] L.H. Kauffman, **Quantum Computing and the Jones Polynomial,** Contemporary Math Series, American Mathematical Society, Providence, Rhode Island, (to appear 2001).
[27] D. Lidar and O. Biham, **Simulating ising spin glasses on a quantum computer**, quant-ph/9611038v6, 23 Sept. 1997.
[28] S.J. Lomonaco, Jr., **A Rosetta Stone for quantum mechanics with an introduction to quantum computation,** in this AMS Proceedings of Applied Mathematics (PSAPM).
[29] S.J. Lomonaco, Jr., **An entangled tale of quantum entanglement,** in "Quantum Computation," in this AMS Proceedings of Applied Mathematics (PSAPM).
[30] N.Yu. Reshetikhin and V.G. Turaev, **Invariants of 3-manifolds via link polynomials and quantum groups**, Invent. Math., Vol. 103, (1991), pp. 547-597.
[31] J. Schwinger, **"Quantum Mechanics: Symbolism of Quantum Measurement,"** Springer-Verlag, (2001).
[32] E. Witten, **Quantum field theory and the Jones polynomial**, Commun.Math.Phys., 121, (1989), pp 351-399.

DEPARTMENT OF MATHEMATICS, STATISTICS AND COMPUTER SCIENCE, UNIVERSITY OF ILLINOIS AT CHICAGO, 851 SOUTH MORGAN STREET, CHICAGO, IL 60607-7045
E-mail address: [kauffman@uic.edu
URL: http://math.uic.edu/~kauffman

An Entangled Tale of Quantum Entanglement

Samuel J. Lomonaco, Jr.

ABSTRACT. These lecture notes give an overview from the perspective of Lie group theory of some of the recent advances in the rapidly expanding research area of quantum entanglement.

This paper is a written version of the last of eight one hour lectures given in the American Mathematical Society (AMS) Short Course on Quantum Computation held in conjunction with the Annual Meeting of the AMS in Washington, DC, USA in January 2000.

Contents

1. Introduction
2. A Story of Two Qubits, or How Alice & Bob Learn to Live with Quantum Entanglement and Love It.
3. Lest we forget, quantum entanglement is ...
4. Back to Alice and Bob: Local Moves and the Fundamental Problem of Quantum Entanglement (FPQE)
5. A momentary digression: Two different perspectives
6. The Group of Local Unitary Transformations and the Restricted FPQE
7. Summary and List of Objectives
8. If you are unfamiliar with ... , then make a quantum jump to Appendices A and B
9. Definition of Quantum Entanglement Invariants
10. The Lie Algebra $\ell(2^n)$ of $\mathbb{L}(2^n)$

2000 *Mathematics Subject Classification.* [2000]Primary 81P68, 81-01,81-02, 81R05, 81R12; Secondary 22E70,17B81,16W25.

Key words and phrases. Quantum mechanics, quantum computation, quantum entanglement, quantum information, quantum algorithms.

This work was partially supported by Army Research Office (ARO) Grant #P-38804-PH-QC, by the National Institute of Standards and Technology (NIST), by the Defense Advanced Research Projects Agency (DARPA) and Air Force Materiel Command USAF under agreement number F30602-01-0522, and by the L-O-O-P Fund. The author gratefully acknowledges the hospitality of the University of Cambridge Isaac Newton Institute for Mathematical Sciences, Cambridge, England, where some of this work was completed. I would also like to thank the other AMS Short Course lecturers, Howard Brandt, Dan Gottesman, Lou Kauffman, Alexei Kitaev, Peter Shor, Umesh Vazirani and the many Short Course participants for their support.

11. Definition of the Infinitesimal Action
12. What is the meaning of the infinitesimal action Ω?
13. The significance of the infinitesimal action $\Omega = ad_*$
14. Achieving two of our objectives, ... finally
15. Example 1. The entanglement classes of $n = 1$ qubits
16. Example 2. The entanglement classes of $n = 2$ qubits
17. Example n. The entanglement classes of n qubits, $n > 2$
18. Conclusion
19. Appendix A. Some Fundamental Concepts from the Theory of Differential Manifolds
20. Appendix B. Some Fundamental Concepts from the Theory of Lie Groups

References

1. Introduction

These lecture notes were written for the American Mathematical Society (AMS) Short Course on Quantum Computation held 17-18 January 2000 in conjunction with the Annual Meeting of the AMS in Washington, DC in January 2000.

The objective of this lecture is to discuss quantum entanglement from the perspective of the theory of Lie groups. More specifically, the ultimate objective of this paper is to quantify quantum entanglement in terms of Lie group invariants, and to make this material accessible to a larger audience than is currently the case. These notes depend extensively on the material presented in AMS Short Course Lecture I [42]. It is assumed that the reader is familiar with the material on density operators and quantum entanglement found in sections 5 and 7 of [42].

Of necessity, the scope of this paper is eventually restricted to the study of qubit quantum systems, and to a specific problem called the *Restricted Fundamental Problem in Quantum Entanglement* (*RFPQE*). References to the broader scope of quantum entanglement are given toward the end of the paper.

1.1. Preamble.

At first sight, a physics research lab dedicated to the pursuit of quantum entanglement might look something like the drawing found in Figure 1, i.e., like an indecipherable, incoherent jumble of wires, fiber optic cable, lasers, bean splitters, lenses. Perhaps some large magnets for NMR equipment, or some supercooling equipment for rf SQUIDs are tossed in for good measure. Whatever ... It is indeed a most impressive collection of adult "toys."

However, to a mathematician, such a lab appears very much like a well orchestrated collection of intriguing mathematical "toys," just beckoning with new tantalizing mathematical challenges.

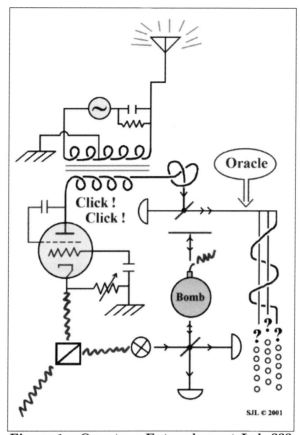

Figure 1. Quantum Entanglement Lab ???

1.2. A Sneak Preview.

In the hope of piquing your curiosity to read on, we give the following brief preview of what is to come:

The *Restricted Fundamental Problem of Quantum Entanglement (RFPQE)* reduces to the mathematical problem of determining the orbits of the big adjoint action of the group of local unitary transformations $\mathbb{L}(2^n)$ on the Lie algebra $u(2^n)$ of the unitary group $\mathbb{U}(2^n)$, as expressed by the following formula:

$$\boxed{\mathbb{L}(2^n) \times u(2^n) \xrightarrow{Ad} u(2^n)}$$

where "Ad" denotes the big adjoint operator, and where the remaining symbols are defined in the table below.

$\mathbb{L}(2^n) = \bigotimes_{1}^{n} SU(2)$	Local Unitary Group
$\ell(2^n) = \boxplus_{1}^{n} su(2)$	Lie Algebra of $\mathbb{L}(2^n)$
$\mathbb{U}(2^n)$	Unitary Group
$u(2^n)$	Lie Algebra of $\mathbb{U}(2^n)$

We attack this problem by lifting the above big adjoint action to the **induced infinitesimal action**

$$\ell(2^n) \xrightarrow{ad_*} \mathbf{Vec}((\mathbf{u}(2^n)))$$

which, for a 3 qubit density operator ρ, is explicitly given by[1]

$$ad_v(i\rho) = \sum_{q_1,q_2=0}^{3} \left(a^{(1)} \cdot x_{*q_1q_2} \times \frac{\partial}{\partial x_{*q_1q_2}} + a^{(2)} \cdot x_{q_1*q_2} \times \frac{\partial}{\partial x_{q_1*q_2}} + a^{(3)} \cdot x_{q_1q_2*} \times \frac{\partial}{\partial x_{q_1q_2*}} \right)$$

where $v \in \ell(2^3)$ and $i\rho \in \mathbf{u}(2^3)$ are given by

$$\begin{cases} v &= a^{(1)} \cdot \xi_{*00} + a^{(2)} \cdot \xi_{0*0} + a^{(3)} \cdot \xi_{00*} \\ \\ i\rho &= \sum_{r_1,r_2,r_3=0}^{3} x_{r_1r_2r_3}\xi_{r_1r_2r_3} \end{cases}$$

and where $\mathbf{Vec}((\mathbf{u}(2^n)))$ denotes the Lie algebra of vector fields on $\mathbf{u}(2^n)$.

The induced infinitesimal action can then be used to quantify and to classify quantum entanglement through the construction of a complete set of quantum entanglement invariants.

In the pages to follow, we make every effort to make the above sneak preview more transparent and understandable. Our goal is to present the underlying intuitions without getting lost in an obscure haze of technicalities. However, presenting this topic is much like tiptoeing through a mine field. One false move, and everything explodes into a dense jungle and clutter of technicalities. We leave it to the reader to determine how successful this endeavor is.

1.3. How our view of quantum entanglement has dramatically changed over this past century.

Finally, we close this introduction with a brief historical perspective.

Over the past twentieth century, the scientific community's view of quantum entanglement has dramatically changed. It continues to do so even today.

Initially, quantum entanglement was viewed as an unnecessary and unwanted wart on quantum mechanics. Einstein, Podolsky, and Rosen[18] tried to surgically remove it. Bell[2],[3] showed that such surgery can not be performed without destroying the very life of physical reality.

Today, quantum entanglement is viewed as a useful resource within quantum mechanics. It is now viewed as a commodity to be utilized and traded, much as would be a commodity on the stock exchange.

---

[1]This expression will be explained later in the paper. I hope that this will make you curious enough to read on?

Quantum entanglement appears to be one of the physical phenomena at the central core of quantum computation. Many believe that it is quantum entanglement that somehow enables us to harness the vast parallelism of quantum superposition.

But what is quantum entanglement?

How do we measure, quantify, classify quantum entanglement? When is the quantum entanglement of two quantum systems the same? different? When is the quantum entanglement of one quantum system greater than that of another?

It is anticipated that answers to the above questions will have a profound impact on the development of quantum computation. Finding answers to these questions is challenging, intriguing, and indeed very habit forming.

2. A Story of Two Qubits, or How Alice & Bob Learn to Live with Quantum Entanglement and Love It.

Our entangled tale of quantum entanglement begins with Alice and Bob's first encounter with quantum entanglement.

Alice and Bob, who happen to be good friends (as attested, time and time again, by the open literature on quantum computation), meet one day. A discussion ensues. The topic, of course, is quantum entanglement. Fortunately or unfortunately, depending on how one looks at it, their discussion explodes into a heated argument. After a lengthy debate, they agree that the only way to resolve their conflict is to purchase the real McCoy, i.e., a pair of entangled qubits. So they rush to the nearest TOYS FOR AGING CHILDREN STORE to see what they can find.

Almost immediately upon entering the store, they happen to spy on one of the store shelves, an elaborately decorated box labelled:

$$\begin{array}{|c|}\hline \mathbb{Q}.\mathbb{E}., \text{ Inc.} \\ \hline \textbf{Two Entangled Qubits} \\ \mathcal{Q}_{AB} \\ \text{Consisting of qubits} \\ \mathcal{Q}_A \text{ and } \mathcal{Q}_B \\ \hline \end{array}$$

On the back of the box is the content label, required by federal law, which reads:

		U.S. Certified **Contents** [EPR Pair] (*)		
Q Sys	**Hilb. Sp.**	**State**	**Unitary Transf.**	**State Space**
$\mathcal{Q}_{AB}$	$\mathcal{H}_{AB}$	$\rho_{AB} = \begin{pmatrix} \frac{1}{2} & 0 & 0 & -\frac{1}{2} \\ 0 & 0 & 0 & 0 \\ 0 & 0 & 0 & 0 \\ -\frac{1}{2} & 0 & 0 & \frac{1}{2} \end{pmatrix}$	$\mathbb{U}(2^2)_{AB}$	$u(2^2)_{AB}$
$\mathcal{Q}_A$	$\mathcal{H}_A$	$\rho_A = \begin{pmatrix} \frac{1}{2} & 0 \\ 0 & \frac{1}{2} \end{pmatrix}$	$\mathbb{U}(2)_A$	$u(2)_A$
$\mathcal{Q}_B$	$\mathcal{H}_B$	$\rho_B = \begin{pmatrix} \frac{1}{2} & 0 \\ 0 & \frac{1}{2} \end{pmatrix}$	$\mathbb{U}(2)_B$	$u(2)_B$
		(*) Caveat Emptor: Not legally responsible for the effects of decoherence.		

Alice and Bob hurriedly purchase the two qubit quantum system $\mathcal{Q}_{AB}$. Outside the store, they rip open the box. Alice grabs the qubit labelled $\mathcal{Q}_A$. Bob then takes the remaining qubit $\mathcal{Q}_B$.

Alice and Bob then immediately[2] depart for their separate destinations. Alice flies to Queensland, Australia to continue with her Ph.D. studies at the University of Queensland. She arrives just in time to attend the first class lecture on quantum mechanics. Bob, on the other hand, flies to Vancouver, British Columbia to continue with his Ph.D. studies at the University of British Columbia. He just barely arrives in time to hear the first lecture in a course on differential geometry and Lie groups.

Soon after her quantum mechanics lecture, Alice begins to have second thoughts about their joint purchase of two entangled qubits. She quickly reaches for her cellphone, calls Bob, and nervously fires off in rapid succession three questions:

"Did we get our money's worth of quantum entanglement?"
"How much quantum entanglement did we actually purchase?"
"Are we the victims of a modern day quantum entanglement scam?"

After the phone conversation, Bob is indeed deeply concerned. In desperation, he calls the U.S. Quantum Entanglement Protection Agency, which refers him to the U.S. National Institute of Quantum Entanglement Standards and Technology (NI$_{\mathbb{QE}}$ST) in Gaithersburg, Maryland.

---

[2] For some unknown reason, everyone involved with the quantum world is always in a hurry. Perhaps such haste is caused by concerns in regard to decoherence?

After a long conversation, a representative of NI$_{\mathrm{QE}}$ST agrees to send Alice and Bob, free of charge, the NI$_{\mathrm{QE}}$ST Quantum Entanglement Standards Kit. On hanging up, the NI$_{\mathrm{QE}}$ST representative takes the NI$_{\mathrm{QE}}$ST standard entangled two qubit quantum system $\mathcal{Q}'_{A'B'}$ off the shelf, places $\mathcal{Q}'_{A'}$ together with a User's Manual into a box marked "Alice." He/She also places the remaining qubit $\mathcal{Q}'_{B'}$ together with a User's Manual into a second box labeled "Bob," and then sends the two boxes by overnight mail to Alice and Bob respectively.

The very next day (in different time zones, of course) Alice and Bob each receive their respective packages, take out their respective qubits, and read the enclosed user's manuals.

The NI$_{\mathrm{QE}}$ST User's Manual reads as follows:

Q.E. YARDSTICK 1. An EPR pair $\mathcal{Q}_{AB}$ possess the same quantum entanglement as the NI$_{\mathrm{QE}}$ST standard EPR pair $\mathcal{Q}'_{A'B'}$ if it is possible for you, Alice and Bob, to use your own local reversible operations (either individually or collectively) to transform $\mathcal{Q}_{AB}$ and $\mathcal{Q}'_{A'B'}$ into one another. If this is possible, then $\mathcal{Q}_{AB}$ and $\mathcal{Q}'_{A'B'}$ are of the **same entanglement type**, written

$$\mathcal{Q}_{AB} \underset{loc}{\sim} \mathcal{Q}'_{A'B'}$$

Q.E. YARDSTICK 2. An EPR pair $\mathcal{Q}_{AB}$ **possesses more quantum entanglement than** the NI$_{\mathrm{QE}}$ST standard EPR pair $\mathcal{Q}'_{A'B'}$ if it is possible for you, Alice and Bob, (either individually or collectively) to apply your own local reversible and irreversible operations to your respective qubits to transform $\mathcal{Q}_{AB}$ into $\mathcal{Q}'_{A'B'}$. In this case, we write

$$\mathcal{Q}_{AB} \underset{loc}{\geq} \mathcal{Q}'_{A'B'}$$

CAVEAT. Quantum entanglement may be irrevocably lost if Quantum Entanglement Yardstick 2 is applied.

In summary, the above story about Alice and Bob has raised the following questions:
- **Question:** What type of entanglement do Alice and Bob collectively possess?
- **Question:** Is the quantum entanglement of $\mathcal{Q}_{AB}$ the same as the quantum entanglement of $\mathcal{Q}'_{A'B'}$?
- **Question:** Is the quantum entanglement of $\mathcal{Q}_{AB}$ greater than the quantum entanglement of $\mathcal{Q}'_{A'B'}$?

3. Lest we forget, quantum entanglement is ...

Before we continue with our story of Alice and Bob, now is a good opportunity to restate the definition of quantum entanglement found in [**42**]. Readers not familiar with this definition or related concepts should refer to sections 5 and 7 of [**42**].

DEFINITION 1. *Let* Q_1, Q_2, ... , Q_n *be quantum systems with underlying Hilbert spaces* $\mathcal{H}_1$, $\mathcal{H}_2$, ... , $\mathcal{H}_n$, *respectively. And let* Q *denote the global quantum system consisting of all the quantum systems* Q_1, Q_2, ... , Q_n, *where* $\mathcal{H} = \bigotimes_{j=1}^{n} \mathcal{H}_j$ *denotes the underlying Hilbert space of* Q. *Finally let the density operator* ρ *on the Hilbert space* $\mathcal{H}$ *denote the state of the global quantum system* Q. *Then* Q *is said to be* **entangled** *with respect to the Hilbert space decomposition*

$$\mathcal{H} = \bigotimes_{j=1}^{n} \mathcal{H}_j$$

if it can not be written in the form

$$\rho = \sum_{k=1}^{K} \lambda_k \left(\bigotimes_{j=1}^{n} \rho_{(j,k)} \right) ,$$

for some positive integer K, *where the* λ_k's *are positive real numbers such that*

$$\sum_{k=1}^{K} \lambda_k = 1 ,$$

and where each $\rho_{(j,k)}$ *is a density operator on the Hilbert space* $\mathcal{H}_j$. *If* ρ *is a pure state, then* Q *is* **entangled** *if* ρ *can not be written in the form*

$$\rho = \bigotimes_{j=1}^{n} \rho_j ,$$

where ρ_j *is a density operator on the Hilbert space* $\mathcal{H}_j$.

4. Back to Alice and Bob: Local Moves and the Fundamental Problem of Quantum Entanglement (FPQE)

Although the story of Alice and Bob was told with two qubits, the same story could have been told instead with three people, Alice, Bob, Cathy, and three qubits. Or for that matter, it could have equally been told for n people with n qubits. From now on, we will consider the more general story of n people and n qubits.

What Alice, Bob, Cathy, et al were trying to understand can be stated most succinctly as the Fundamental Problem of quantum entanglement, namely:

Fundamental Problem of Quantum Entanglement (FPQE). Let ρ and ρ' be density operators representing two different states of a quantum system Q. Is it possible to move Q from state ρ to state ρ' by applying only **local moves**?

> But what is meant by the phrase "**local move**" ?

We define the **standard local moves** as:

DEFINITION 2. *The **standard local moves** are:*
- Local unitary transformations of the form
$$\bigotimes_{k=1}^{n} U_k \in \bigotimes_{k=1}^{n} \mathbb{U}(\mathcal{H}_k)$$

  For example, for bipartite quantum systems, unitary transformations of the form $U_A \otimes I$, $I \otimes U_B$, $U_A \otimes U_B$
- Measurement of local observables of the form
$$\bigotimes_{k=1}^{n} \mathcal{O}_k \in \bigotimes_{k=1}^{n} Observables(\mathcal{H}_k)$$

  For example, for bipartite quantum systems[3], measurement of local observables of the form $\mathcal{O}_A \otimes I$, $I \otimes \mathcal{O}_B$, $\mathcal{O}_A \otimes \mathcal{O}_B$

We also define the **extended local moves** as

DEFINITION 3. *The **extended local moves** are:*
- Extended local unitary transformations of the form
$$\bigotimes_{k=1}^{n} \mathbb{U}\left(\mathcal{H}_k \otimes \widetilde{\mathcal{H}}_k\right),$$

  where $\mathcal{H}_1, \widetilde{\mathcal{H}}_1, \ldots, \mathcal{H}_n, \widetilde{\mathcal{H}}_n$ are distinct non-overhapping Hilbert spaces
- Measurement of extended local observables of the form
$$\bigotimes_{k=1}^{n} Observables\left(\mathcal{H}_k \otimes \widetilde{\mathcal{H}}_k\right),$$

  where $\mathcal{H}_1, \widetilde{\mathcal{H}}_1, \ldots, \mathcal{H}_n, \widetilde{\mathcal{H}}_n$ are distinct non-overlapping Hilbert spaces

DEFINITION 4. *Moves based on unitary transformation are called **reversible**. Those based on measurement are called **irreversible**.*

The Horodecki's[26], [27], [28], Jonathan[34], [35], Linden[38], [39], Nielsen[45], [46], [47], [48], Plenio[34], [35], Popescu[38], [39], [65], [41] have made some progress in understanding the FPQE in terms of all four of the above local moves. For the rest of the talk, we restrict our discussion to reversible standard local moves.

---

[3]A bipartite quantum system is a global quantum system consisting of two quantum systems.

5. A momentary digression: Two different perspectives

Before continuing, it should be mentioned that physics and mathematics approach quantum mechanics from two slightly different but equivalent viewpoints. To avoid possible confusion, we describe below the minor terminology differences that arise from these two slightly different perspectives.

Physics describes the state of a quantum system in terms of a traceless Hermitian operator ρ, called the density operator. Observables are Hermitian operators $\mathcal{O}$. Quantum states change via unitary transformations U according to the rubric

$$\rho \longmapsto U\rho U^\dagger \ .$$

On the other hand, mathematics describes the state of a quantum system in terms of a skew Hermitian operator $i\rho$, also called the density operator. Observables are skew Hermitian operators $i\mathcal{O}$. Quantum dynamics are defined via the rule

$$i\rho \longmapsto Ad_U(i\rho) \ ,$$

where U is a unitary operator lying in the Lie group of unitary transformations $\mathbb{U}(N)$, and where Ad denotes the big adjoint operator. Please note that both density operators $i\rho$ and observables $i\mathcal{O}$ lie in the Lie algebra $u(N)$ of the unitary group $\mathbb{U}(N)$.

These minor, but nonetheless annoying differences are summarized in the table below.

Physics	Math
Hilbert Space $\mathcal{H}$, $Dim(\mathcal{H}) = N$, Unitary Group, Lie Group $\mathbb{U}(N)$	
Observables: $\mathcal{O}$ Density Ops: ρ $N \times N$ Hermitian Ops $A^\dagger = \overline{A}^T = A$	Observables: $i\mathcal{O}$ Density Ops: $i\rho$ $N \times N$ skew Hermitian Ops $\in u(N)$ $(iA)^\dagger = \overline{(iA)}^T = -iA$ where $u(N)$ = Lie algebra of $\mathbb{U}(N)$
Dynamics via $U \in \mathbb{U}(N)$ $\|\psi\rangle \longmapsto U\|\psi\rangle$ $\rho \longmapsto U\rho U^\dagger$	Dynamics via $U \in \mathbb{U}(N)$ $\|\psi\rangle \longmapsto U\|\psi\rangle$ $i\rho \longmapsto Ad_U(i\rho)$ where $Ad_U(i\rho) = U(i\rho)U^{-1}$ is the Big adjoint rep.

We will use the two different terminologies and conventions interchangeably. Which terminology we are using should be clear from context.

REMARK 1. *From [42] we know that an element $i\rho$ of the Lie algebra $\mathbf{u}(N)$ is a physical density operator if and only if ρ is positive semi-definite and of trace 1. Thus, the set*

$$\mathbf{density}\,(N) = \{i\rho \in \mathbf{u}(N) \mid \rho \text{ is positive semi-definite of trace 1}\}$$

of physical density operators is a convex subset of the Lie algebra $\mathbf{u}(N)$.

6. The Group of Local Unitary Transformations and the Restricted FPQE

For the sake of clarity of exposition and for the purpose of avoiding minor technicalities, from this point on we consider only qubit quantum systems, i.e., quantum systems consisting of qubits. The reader, if he/she so wishes, should be able to easily rephrase the results of this paper in terms of more general quantum systems.

Moreover, from this point on, we limit the scope of this talk to the study of quantum entanglement from the perspective of the standard local unitary transformations, i.e., from the perspective of standard reversible local moves as defined in section 5 of this paper. To emphasize this point, we define the group of local unitary transformations $\mathbb{L}(2^n)$ as follows:

DEFINITION 5. *The **group of local unitary transformations** $\mathbb{L}(2^n)$ is the subgroup of $\mathbb{U}(2^n)$ defined by*

$$\mathbb{L}(2^n) = \bigotimes_{1}^{n} \mathbb{SU}(2),$$

where $\mathbb{SU}(2)$ denotes the special unitary group.

Henceforth, the phrase **"local move"** will mean an element of the group $\mathbb{L}(2^n)$ of **local unitary transformations**. From this point on,

$$\boxed{\textbf{Local Moves } = \mathbb{L}(2^n)}$$

Thus, for the rest of this paper we consider only the **Restricted Fundamental Problem of Quantum Entanglement (RFPQE)**, which is defined as follows:

Restricted Fundamental Problem of Quantum Entanglement (RFPQE). *Let $i\rho$ and $i\rho'$ be density operators lying in the Lie algebra $\mathbf{u}(2^n)$. Does there exist a local move U, i.e., a $U \in \mathbb{L}(2^n)$ such that*

$$i\rho' = U(i\rho)U^\dagger = Ad_U(i\rho) \ ?$$

We will need the following definition:

DEFINITION 6. *Two elements $i\rho$ and $i\rho'$ in $u(2^n)$ are said to be **locally equivalent** (or, **of the same entanglement type**), written*

$$i\rho \underset{loc}{\sim} i\rho'$$

provided there exists a $U \in \mathbb{L}(2^n)$ such that

$$i\rho' = Ad_U(i\rho) = U(i\rho)U^{-1}$$

The equivalence class

$$[i\rho]_E = \left\{ i\rho' \mid i\rho \underset{loc}{\sim} i\rho' \right\}$$

*is called an **entanglement class** (or, an **orbit** of the big adjoint action of $\mathbb{L}(2^n)$ on the Lie algebra $\mathbf{u}(2^n)$). Finally, let*

$$\mathbf{u}(2^n)/\mathbb{L}(2^n)$$

*denote the **set of entanglement classes**.*

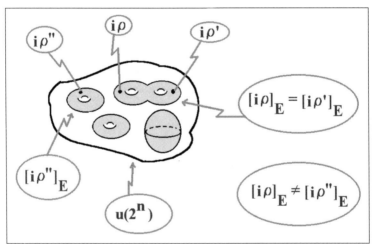

Figure 2. Quantum entanglement classes.

The entanglement classes of the Lie algebra $\mathbf{u}(2^n)$ are just the **orbits** of the big adjoint action of $\mathbb{L}(2^n)$ on $\mathbf{u}(2^n)$. Two states are entangled in the same way if and only if they lie in the same entanglement class, i.e., in the same orbit.

REMARK 2. *Local unitary transformations can not entangle quantum systems with respect to the above tensor product decomposition. However, **global unitary transformations** (i.e., unitary transformations lying in $\mathbb{U}(2^n) - \mathbb{L}(2^n)$) are those unitary transformations which can and often do produce interactions which entangle quantum systems.*

But what is quantum entanglement?

7. Summary and List of Objectives

We are now in a position to state clearly the main objectives of this paper. Namely, in regard to the **Restricted Fundamental Problem of Quantum Entanglement (RFPQE)**, our objectives are threefold:

Objective 1. Given a density operator $i\rho$, devise a means of determining the dimension of its entanglement class $[i\rho]_E$. This will be accomplished for $n = 1$ and 2 qubits by determining the dimension of the tangent plane $T_{i\rho}[i\rho]_E$ to the manifold $[i\rho]_E$ at the point $i\rho$.

Objective 2. Given two states $i\rho$ and $i\rho'$, devise a means of determining whether they belong to the same entanglement class or to different entanglement classes. This will be accomplish for $n = 1$ and 2 qubits by constructing a **complete set of quantum entanglement invariants**, i.e., invariants that completely specify all the orbits (i.e., all the entanglement classes). In this sense, quantum entanglement will be completely quantified for $n = 1$ and 2 qubits. In particular, a finite set $\{f_1, f_2, \ldots, f_K\}$ of real valued functions on $\mathbf{u}(2^n)$ will be constructed which distinguishes all entanglement classes, i.e.,

$$i\rho \underset{loc}{\sim} i\rho' \iff f_k(i\rho) = f_k(i\rho') \text{ for every } k.$$

Objective 3. Briefly outline the intriguing mathematical research opportunities for quantifying quantum entanglement for $n = 3$ qubits, and beyond. Surprisingly, very little is known about quantum entanglement for $n > 3$ qubits[4].

8. If you are unfamiliar with ... , then make a quantum jump to Appendices A and B

This section is meant to play the role of a litmus test for the reader. If the reader feels reasonably comfortable with the concepts listed below, then it is suggested that the reader proceed to the next section of this paper. If not, it is strongly suggested that the reader read Appendices A and B of this paper before proceeding to the next section.

Let $\mathbb{G}$ be a Lie group, and let $\mathfrak{g}$ denote its Lie algebra.

8.1. Litmus Test 1. The exponential map.

The reader should be familiar with the exponential map

$$\exp : \mathfrak{g} \longrightarrow \mathbb{G} ,$$

which for matrix Lie Groups is given by the power series

$$\exp(M) = \sum_{k=0}^{\infty} \frac{1}{k!} M^k$$

---

[4]**Warning:** This research area is very addictive. :-)

8.2. Litmus Test 2. The Lie bracket.

The reader should be familiar with the Lie bracket
$$[-,-] : \mathfrak{g} \times \mathfrak{g} \longrightarrow \mathfrak{g} ,$$
which for matrix Lie groups is given by the commutator
$$[A, B] = AB - BA$$

8.3. Litmus Test 3. The Lie algebra under three different guises.

The Lie algebra $\mathfrak{g}$ of the Lie group $\mathbb{G}$ can be viewed in each of the following mathematically equivalent ways:

- As $T_I \mathbb{G}$, i.e., as the tangent space to the Lie group $\mathbb{G}$ at the identity I.
- As $\mathbf{Vec}_R(\mathbb{G})$, i.e., as the Lie algebra of right invariant smooth vector fields on the Lie group $\mathbb{G}$.
- As $\mathbf{Der}_R(\mathbb{G})$, i.e., as the Lie algebra of all right invariant derivations (i.e., right invariant directional derivatives) on the algebra $C^\infty(\mathbb{G})$ of all smooth real valued functions on $\mathbb{G}$.

In summary,
$$\boxed{\mathfrak{g} = T_I \mathbb{G} = \mathbf{Vec}_R(\mathbb{G}) = \mathbf{Der}_R(\mathbb{G})}$$

If you feel comfortable with the above three litmus tests, then please proceed to the next section.

9. Definition of Quantum Entanglement Invariants

Let $C^\infty(\mathbf{u}(2^n))$ denote the algebra of smooth (C^∞) real valued functions on the Lie algebra $\mathbf{u}(2^n)$, i.e.,
$$C^\infty(\mathbf{u}(2^n)) = \{f : \mathbf{u}(2^n) \longrightarrow \mathbb{R} \mid f \text{ is smooth}\}$$

DEFINITION 7. *A function $f \in C^\infty(\mathbf{u}(2^n))$ is called a **(quantum) entanglement invariant** if f is invariant under the big adjoint action of $\mathbb{L}(2^n)$, i.e., if*
$$f(Ad_U(i\rho)) = f(i\rho)$$
for all $U \in \mathbb{L}(2^n)$, and for all $i\rho$ in $\mathbf{u}(2^n)$. The collection of all (quantum) entanglement invariants forms an algebra, which we denote by
$$C^\infty(\mathbf{u}(2^n))^{\mathbb{L}(2^n)} .$$

DEFINITION 8. *A subset $\{f_1, f_2, \ldots, f_m\}$ of $C^\infty(\mathbf{u}(2^n))^{\mathbb{L}(2^n)}$ is called a **complete set of entanglement invariants** if*
$$i\rho \underset{loc}{\sim} i\rho' \text{ iff } f_k(i\rho) = f_k(i\rho') \text{ for all } f_k \text{ in } \{f_1, f_2, \ldots, f_m\} .$$

DEFINITION 9. *Let $\mathcal{P}(\mathbf{u}(2^n))$ be the subalgebra of $C^\infty(\mathbf{u}(2^n))$ of all functions $f \in C^\infty(\mathbf{u}(2^n))$ which are polynomial functions, i.e., of all functions f for which $f(v)$ is a polynomial function of the entries in v. We define the **algebra of polynomial entanglement invariants** as*

$$\mathcal{P}(\mathbf{u}(2^n))^{\mathbb{L}(2^n)} = \mathcal{P}(\mathbf{u}(2^n)) \cap C^\infty(\mathbf{u}(2^n))^{\mathbb{L}(2^n)}$$

The following theorem can be found in standard reference for Lie groups, such as for example [9], [22], [30], [53], [69].

THEOREM 1. *$\mathcal{P}(\mathbf{u}(2^n))^{\mathbb{L}(2^n)}$ is a finitely generated algebra.*

DEFINITION 10. *A minimal set of generators of $\mathcal{P}(\mathbf{u}(2^n))^{\mathbb{L}(2^n)}$ is called a **basic set of entanglement invariants**.*

10. The Lie Algebra $\ell(2^n)$ of $\mathbb{L}(2^n)$

To understand and work with the big adjoint action

$$\mathbb{L}(2^n) \times \mathbf{u}(2^n) \xrightarrow{Ad} \mathbf{u}(2^n)$$

we will need to lift this action to the corresponding infinitesimal action of the Lie algebra $\ell(2^n)$ of $\mathbb{L}(n)$. The Lie algebra $\ell(2^n)$ will play a crucial role in our achieving objectives stated in the previous section.

DEFINITION 11. *The Lie algebra $\ell(2^n)$ is the (real) Lie algebra given by the following Kronecker sum*

$$\ell(2^n) = \underbrace{\mathbf{su}(2) \boxplus \mathbf{su}(2) \boxplus \cdots \boxplus \mathbf{su}(2)}_{n \text{ terms}},$$

*where $\mathbf{su}(2)$ denotes the Lie algebra of the special unitary group $\mathbb{SU}(2)$, and where the **Kronecker sum** '$A \boxplus B$' of two matrices (or operators) A and B is defined by*

$$A \boxplus B = A \otimes \mathbf{1} + \mathbf{1} \otimes B,$$

with '$\mathbf{1}$' denoting the identity matrix (or operator).

A basis[5] of the (real) Lie algebra $\mathbf{u}(2^n)$ is given by

$$\{\xi_{k_1 k_2 \ldots k_n} \mid k_1, k_2, \ldots, k_n = 0, 1, 2, 3\},$$

where

$$\xi_{k_1 k_2 \ldots k_n} = -\frac{i}{2} \sigma_{k_1} \otimes \sigma_{k_2} \otimes \cdots \otimes \sigma_{k_n},$$

and where

$$\sigma_1 = \begin{pmatrix} 0 & 1 \\ 1 & 0 \end{pmatrix}, \sigma_2 = \begin{pmatrix} 0 & -i \\ i & 0 \end{pmatrix}, \sigma_3 = \begin{pmatrix} 1 & 0 \\ 0 & -1 \end{pmatrix},$$

denote the Pauli spin matrices, and where

$$\sigma_0 = \begin{pmatrix} 1 & 0 \\ 0 & 1 \end{pmatrix}$$

---

[5]For more information, please refer to Appendix B.

denotes the 2×2 identity matrix.

It follows that a basis of the Lie algebra $\ell(2^n)$ as a subalgebra of $\mathbf{u}(2^n)$ is:

$$\{\xi_{k_1 k_2 \ldots k_n} \mid k_1, k_2, \ldots, k_n = 0, 1, 2, 3, \text{ where exactly one } k_j \neq 0\}$$

For example,
- $\{\xi_1, \xi_2, \xi_3\}$ is a basis of $\ell(1)$
- $\{\xi_{01}, \xi_{02}, \xi_{03}, \xi_{10}, \xi_{20}, \xi_{30}\}$ is a basis of $\ell(2)$
- $\{\xi_{001}, \xi_{002}, \xi_{003}, \xi_{010}, \xi_{020}, \xi_{030}, \xi_{100}, \xi_{200}, \xi_{300}\}$ is a basis of $\ell(3)$

Thus, we have the following proposition:

PROPOSITION 1. *The Lie algebra $\ell(2^n)$ of the Lie group $\mathbb{L}(2^n)$ of local unitary transformations is of dimension $3n$, i.e.,*

$$Dim(\ell(2^n)) = 3n$$

11. Definition of the Infinitesimal Action

We now show how quantum entanglement invariants can be found by lifting the big adjoint action to the Lie algebra $\ell(2^n)$, where the problem becomes a linear one.

The big adjoint action

$$\mathbb{L}(2^n) \times \mathbf{u}(2^n) \xrightarrow{Ad} \mathbf{u}(2^n)$$

induces an **infinitesimal action**

$$\ell(2^n) \xrightarrow{\Omega} \mathbf{Vec}(\mathbf{u}(2^n))$$

as follows.

Let $v \in \ell(2^n)$. We define the vector field $\Omega(v)$ on $\mathbf{u}(2^n)$ by constructing a tangent vector $\Omega(v)|_{i\rho}$ for each $i\rho \in \mathbf{u}(2^n)$.

Let $\gamma_v(t)$ be the smooth curve in $\mathbf{u}(2^n)$ defined by

$$\gamma_v(t) = Ad_{\exp(tv)}(i\rho) \ .$$

Then $\gamma_v(t)$ is a curve which passes through $i\rho$ at time $t = 0$. We define $\Omega(v)|_{i\rho}$ as the tangent vector to $\gamma_v(t)$ at $t = 0$.

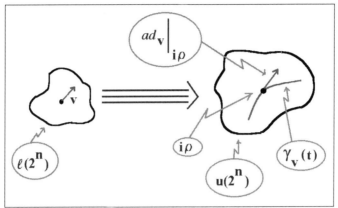

Figure 3. Induced infinitesimal action

12. What is the meaning of the infinitesimal action Ω?

But what is the meaning of the above defined infinitesimal action

$$\ell(2^n) \xrightarrow{\Omega} \mathbf{Vec}(\mathbf{u}(2^n)) \ ?$$

Each $\Omega(i\rho)|_{i\rho}$ is a direction in $\mathbf{u}(2^n)$ from which we can move away from $i\rho$ without leaving the quantum entanglement class $[i\rho]_E$. Movement in all directions not in $\text{Im}(\Omega)|_{i\rho}$ will force us to immediately leave $[i\rho]_E$.

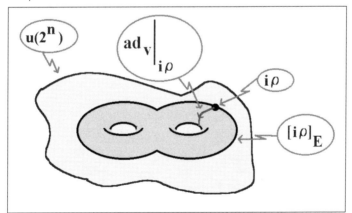

Figure 4. Moving in a direction that stays within the entanglement class.

As the reader might expect, the infinitesimal action Ω can naturally be expressed in terms of the small adjoint operator ad. In particular, we have:

PROPOSITION 2. $\Omega(v) = ad_v$ for all v in the Lie algebra $\ell(2^n)$

PROOF. Let v be an arbitrary element of the Lie algebra $\ell(2^n)$, and let $i\rho$ be an arbitrary element of the Lie algebra $\mathbf{u}(2^n)$.

By definition, $\Omega(v)(i\rho)$ is the tangent vector at $t=0$ to the curve $\gamma_v(t)$ in $\mathbf{u}(2^n)$ given by

$$\gamma_v(t) = Ad_{\exp(tv)}(i\rho) \ .$$

Hence,

$$\begin{aligned}
\Omega(v)(i\rho) &= \tfrac{d}{dt} Ad_{\exp(tv)}(i\rho)\big|_{t=0} \\
&= \tfrac{d}{dt} \exp(ad_{tv})(i\rho)\big|_{t=0} \\
&= \tfrac{d}{dt} \exp(t \cdot ad_v)(i\rho)\big|_{t=0} \\
&= \tfrac{d}{dt} \left(1 + t \cdot ad_v + o(t^2)\right)(i\rho)\big|_{t=0} \\
&= ad_v(i\rho)
\end{aligned}$$

$\square$

The above formula will prove to be useful when we actually calculate the entanglement invariants of examples given in later sections.

Notation Convention. *As a consequence of Proposition 2, we will henceforth denote the infinitesimal action Ω by the more mathematically transparent notation ad_*.*

13. The significance of the infinitesimal action $\Omega = ad_*$

As stated in the appendices, the Lie algebra $\mathbf{Vec}(\mathbf{u}(2^n))$ of all smooth vector fields on $\mathbf{u}(2^n)$ can be identified with the Lie algebra of derivations $\mathbf{Der}(C^\infty \mathbf{u}(2^n))$.

The significance of the infinitesimal action

$$ad_* : \ell(2^n) \longrightarrow \mathbf{Vec}(\mathbf{u}(2^n))$$

is best expressed in terms of the following theorem:

THEOREM 2. *Let*[6]

$$\{v_1, v_2, \ldots, v_{3n}\}$$

be a basis for the Lie algebra $\ell(2^n)$. Then a smooth real valued function

$$f : \mathbf{u}(2^n) \longrightarrow \mathbb{R}$$

---

[6]Proposition 1 states that the dimension of the Lie algebra $\ell(2^n)$ is $3n$.

is an entanglement invariant if and only if it satisfies the following system of partial differential equations

$$\begin{cases} ad_{v_1} f &= 0 \\ ad_{v_2} f &= 0 \\ \phantom{ad_{v_1} f} \vdots & \vdots \\ ad_{v_{3n}} f &= 0 \end{cases}$$

The intuition underlying the above theorem is that $ad_{v_1}, ad_{v_2}, \ldots, ad_{v_{3n}}$ are the linearly independent directions in which we can move without leaving the entanglement class we are currently occupying. Hence, if f is an entanglement invariant, then its rate of change (i.e., its directional derivative) in each of the directions $ad_{v_1}, ad_{v_2}, \ldots, ad_{v_{3n}}$ must be zero, and vice versa.

This theorem provides us with a means of determining a complete set of entanglement invariants. All that we need to do is to solve the above system of partial differential equations. However, finding a complete set of solutions to this system of partial differential equations is a most daunting task. We will solve this system only for $n = 1$ and 2 qubits.

14. Achieving two of our objectives, ... finally

We now show how the infinitesimal action

$$\ell(2^n) \xrightarrow{ad_*} \textbf{Vec}\,(\textbf{u}(2^n))$$

can be used to achieve the first two objectives listed in section 8 of this paper.

Objective 1. Given an arbitrary density operator $i\rho$, devise a means of determining the dimension of its entanglement class $[i\rho]_E$.

Objective 1 is achieved as follows:

We begin by noting that $\textbf{Vec}\,(\textbf{u}(2^n))|_{i\rho}$ is the same as the tangent space $T_{i\rho}(\textbf{u}(2^n))$ to $\textbf{u}(2^n)$ at the point $i\rho$, and that $\text{Im}(ad_*)|_{i\rho}$ is the same as the tangent space $T_{i\rho}([i\rho]_E)$ to the entanglement class $[i\rho]_E$ at $i\rho$. Hence, the dimension of $[i\rho]_E$ is the same as the dimension as its tangent space at $i\rho$, i.e.,

$$Dim\,([i\rho]_E) = Dim\,(T_{i\rho}([i\rho]_E)) = Dim\left(\text{Im}\,(ad_*)|_{i\rho}\right)$$

The task of finding the dimension of the entanglement class $[i\rho]_E$ reduces to that of computing the dimension of the vector space $\text{Im}\,(ad_*)|_{i\rho}$. We will give examples of this dimension calculation in the next two sections.

We next use the infinitesimal action to achieve:

Objective 2. Given two states $i\rho$ and $i\rho'$, devise a means of determining whether they belong to the same entanglement class or to different entanglement classes.

as follows:

We begin by noting that $\mathbf{Vec}\left(\mathbf{u}\left(2^n\right)\right)$ can be identified with the Lie algebra $Der\left(C^\infty \mathbf{u}(2^n)\right)$ of derivations on $\mathbf{u}\left(2^n\right)$. Next we recall that $\text{Im}\left(ad_*\right)$ consists of all directions in $\mathbf{u}\left(2^n\right)$ that we can move without leaving an entanglement class that we are in. If

$$f \in (C^\infty\left(\mathbf{u}(2^n)\right))^{\mathbb{L}(2^n)}$$

is an entanglement invariant, then f will not change if we move in any direction within $\text{Im}\left(ad_*\right)$. As a result we have the following theorem:

THEOREM 3. *Let $v_1, v_2, \ldots, v_{3n}$ be a vector space basis of the (real) Lie algebra $\ell\left(2^n\right)$. Then*

$$f \in (C^\infty\left(\mathbf{u}(2^n)\right))^{\mathbb{L}(2^n)} \Longleftrightarrow ad_{v_j} f = 0$$

for all j, where ad_{v_j} is interpreted as a differential operator in $Der\left(C^\infty \mathbf{u}(2^n)\right)$.

In other words, the task of finding entanglement invariants reduces to that of solving a system of linear partial differential equations. We will give examples of this calculation in the examples found in the next two sections of this paper.

15. Example 1. The entanglement classes of $n = 1$ qubits

We now make use of the methods developed in the previous section to study the entanglement classes associated with $n = 1$ qubits. This is a trivial but nonetheless instructive case. As we shall see, there is no entanglement in this case. But there are many entanglement classes!

For this example, the local unitary group $\mathbb{L}\left(2^1\right)$ is the same as the special unitary group $\mathbb{SU}\left(2^1\right)$. The corresponding Lie algebra $\ell\left(2^1\right)$ is the same as the Lie algebra $\mathbf{su}(2)$. Each density operator $i\rho$ lies in the Lie algebra $\mathbf{u}\left(2^1\right)$.

As stated in Proposition 2 of Section 11, the infinitesimal action is nothing more than the little adjoint action

$$\begin{array}{ccc} \ell\left(2^1\right) & \xrightarrow{ad_*} & \mathbf{Vec}\left(\mathbf{u}\left(2\right)\right) \\ v & \longmapsto & ad_v \end{array}$$

We can now use the bases[7]

$$\left\{\xi_1 = -\tfrac{1}{2}\sigma_1,\ \xi_2 = -\tfrac{1}{2}\sigma_2,\ \xi_3 = -\tfrac{1}{2}\sigma_3\right\}$$
and
$$\left\{\xi_0 = -\tfrac{1}{2}\sigma_0,\ \xi_1 = -\tfrac{1}{2}\sigma_1,\ \xi_2 = -\tfrac{1}{2}\sigma_2,\ \xi_3 = -\tfrac{1}{2}\sigma_3\right\}$$

of the respective Lie algebras $\ell\left(2^1\right)$ and $\mathbf{u}(2)$ to find a more useful expression for ad_v.

---

[7] See Section 10.

Each element $v \in \ell(2^1)$ can be uniquely expressed in the form
$$v = a \cdot \xi,$$
where $a = (a_1, a_2, a_3) \in \mathbb{R}^3$ and $\xi = (\xi_1, \xi_2, \xi_3)$. Thus,
$$ad_v = ad_{a \cdot \xi} = a \cdot ad_\xi,$$
where
$$ad_\xi = (ad_{\xi_1}, ad_{\xi_2}, ad_{\xi_3}).$$

Moreover, each element $i\rho \in \mathbf{u}(2)$ can be uniquely written in terms of the basis of $\mathbf{u}(2)$ as
$$i\rho = x_0 \xi_0 + x \cdot \xi,$$
where $x = (x_1, x_2, x_3)$ and $\xi = (\xi_1, \xi_2, \xi_3)$.

In terms of the basis of $\mathbf{u}(2)$,
$$ad_{\xi_j} = \begin{cases} \begin{pmatrix} 0 & 0 \\ 0 & L_j \end{pmatrix} = 0 \oplus L_j & \text{if } j = 1, 2, 3 \\ \begin{pmatrix} 0 & 0 \\ 0 & 0 \end{pmatrix} = 0 & \text{if } j = 0 \end{cases},$$
where
$$L_1 = \begin{pmatrix} 0 & 0 & 0 \\ 0 & 0 & -1 \\ 0 & 1 & 0 \end{pmatrix}, \quad L_2 = \begin{pmatrix} 0 & 0 & 1 \\ 0 & 0 & 0 \\ -1 & 0 & 0 \end{pmatrix}, \quad L_3 = \begin{pmatrix} 0 & -1 & 0 \\ 1 & 0 & 0 \\ 0 & 0 & 0 \end{pmatrix}$$
is the basis[8] of the Lie algebra $\mathbf{so}(3)$ of the special orthogonal group $\mathbb{SO}(3)$ given in Appendix B.

Let
$$\left\{ \frac{\partial}{\partial x_0}, \frac{\partial}{\partial x_1}, \frac{\partial}{\partial x_2}, \frac{\partial}{\partial x_3} \right\}$$
denote the basis[9] of $\mathbf{Vec}(\mathbf{u}(2))$ induced by the chart
$$\mathbf{u}(2) \xrightarrow{\pi} \mathbb{R}^4$$
$$i\rho = \sum_{j=0}^{3} x_j \xi_j \longmapsto (x_0, x_1, x_2, x_3) = (x_0, x)$$
In other words, for each j, $\partial/\partial x_j$ denotes the vector field on $\mathbf{u}(2)$ defined at each point $i\rho$ as the tangent vector to the curve $\pi^{-1}(x_0, \ldots, x_j + t, \ldots, x_3) = i\rho + t\xi_j$ at $t = 0$.

---

[8]This follows from the following calculation:
$$\begin{aligned} ad_{\xi_j}(\xi_k) &= ad_{-i\sigma_j/2}(-i\sigma_k/2) = [-i\sigma_j/2, -i\sigma_k/2] = -\tfrac{1}{4}[\sigma_j, \sigma_k] \\ &= -\tfrac{1}{2} i \epsilon_{jkp} \sigma_p = \epsilon_{jkp} \xi_p \end{aligned}$$
where $L_j = (\epsilon_{jkp})$, and where
$$\epsilon_{jkp} = \begin{cases} 1 & \text{if } jkp \text{ is an even permutation of } 123 \\ -1 & \text{if } jkp \text{ is an odd permutation of } 123 \\ 0 & \text{otherwise} \end{cases}$$

[9]For those unfamiliar with this basis, please refer to Appendix A page 337.

Then,
$$ad_v(i\rho) = (x_0, x) \cdot (0 \oplus a \cdot L) \cdot \begin{pmatrix} \partial/\partial x_0 \\ \partial/\partial x_1 \\ \partial/\partial x_2 \\ \partial/\partial x_3 \end{pmatrix},$$
$$= x \cdot (a \cdot L) \cdot \nabla$$
$$= a \cdot x \times \nabla,$$

where '×' denotes the vector cross product, and where

$$L = (L_1, L_2, L_3) \text{ and } \nabla = \begin{pmatrix} \partial/\partial x_1 \\ \partial/\partial x_2 \\ \partial/\partial x_3 \end{pmatrix}.$$

We can now achieve objective 1.

Objective 1. *Given an arbitrary density operator $i\rho$ in $\mathbf{u}(2)$, find the dimension of an arbitrary entanglement class $[i\rho]_E$.*

From the above discussion, it follows that the image $\text{Im}(ad_*)$ of the infinitesimal action ad_* is spanned by the three vector fields

$$\begin{cases} ad_{\xi_1} = x_3 \frac{\partial}{\partial x_2} - x_2 \frac{\partial}{\partial x_3} \\ ad_{\xi_2} = x_1 \frac{\partial}{\partial x_3} - x_3 \frac{\partial}{\partial x_1} \\ ad_{\xi_3} = x_2 \frac{\partial}{\partial x_1} - x_1 \frac{\partial}{\partial x_2} \end{cases},$$

defined on $\mathbf{u}(2^n) = \left\{ i\rho = \sum_{j=0}^{3} x_j \xi \mid x_0, x_1, x_2, x_3 \in \mathbb{R} \right\}$. In particular, the tangent space $T_{i\rho}([i\rho]_E) = (\text{Im}\, ad_*)|_{i\rho}$ of the entanglement class $[i\rho]_E$ at the point $i\rho$ is spanned by

$$ad_{\xi_1}|_{i\rho}, \; ad_{\xi_2}|_{i\rho}, \; ad_{\xi_3}|_{i\rho}$$

As can be easily verified by the reader, the above three vectors span a two dimensional space if $|x| \neq 0$ and a zero dimensional vector space if $|x| = 0$.

Since $(\text{Im}\, ad_*)|_{i\rho}$ is the tangent space $T_{i\rho}([i\rho]_E)$ of $[i\rho]_E$ at the point $i\rho$, and since the dimension of $[i\rho]_E$ is the same as the dimension of its tangent space $T_{i\rho}([i\rho]_E)$ at $i\rho$, it follows that the dimension of the entanglement class $[i\rho]_E$ is given by:

$$Dim\,[i\rho]_E = \begin{cases} 2 & \text{if } |x| \neq 0 \\ 0 & \text{if } |x| = 0 \end{cases}$$

We are now ready to achieve objective 2:

Objective 2. *Given two states $i\rho$ and $i\rho'$, devise a means of determining whether they belong to the same entanglement class or to different entanglement classes.*

We achieve this objective by determining a complete set of entanglement invariants[10] for one qubit quantum systems, i.e., by determining a set of entanglement invariants $\{f_1, f_2, \ldots, f_k\}$ such that

$$i\rho \underset{loc}{\sim} i\rho' \text{ if and only if } f_j(i\rho) = f_j(i\rho') \text{ for all } j \ .$$

We begin by recalling that the Lie algebra $\mathbf{Vec}\,(\mathbf{u}\,(2))$ of vector fields on $\mathbf{u}\,(2)$ can be identified with the Lie algebra $\mathbf{Der}\,(C^\infty \mathbf{u}(2))$ of all derivations on the smooth real valued functions on $\mathbf{u}(2)$. Thus, the elements of $\mathrm{Im}\,(ad_*)$ can be viewed as directional derivatives, directional derivatives in those directions in which we can move and still remain in the same entanglement class.

From theorem 2, it immediately follows that a real valued function $f : \mathbf{u}(2) \longrightarrow \mathbb{R}$ is an entanglement invariant if and only it is a solution of the system of partial differential equations (PDEs):

$$\begin{cases} ad_{\xi_1} f = 0 \\ ad_{\xi_2} f = 0 \\ ad_{\xi_3} f = 0 \end{cases}$$

Since from above we know that $ad_{\xi_j}(i\rho) = x \cdot L_j \cdot \nabla$, we can write the above system of PDEs more explicitly as:

$$\begin{cases} x_3 \frac{\partial f}{\partial x_2} - x_2 \frac{\partial f}{\partial x_3} = 0 \\ x_1 \frac{\partial f}{\partial x_3} - x_3 \frac{\partial f}{\partial x_1} = 0 \\ x_2 \frac{\partial f}{\partial x_1} - x_1 \frac{\partial f}{\partial x_2} = 0 \end{cases},$$

where, as before, $i\rho = x_0 \xi_0 + x \cdot \xi$.

From theorem 2, we know that a complete set of quantum entanglement invariants for one qubit systems is the same as a complete functionally independent set of solutions of the above system of PDEs. Thus, solving the above system of PDEs by standard methods found in the theory of differential equations, we find that

$$\left\{ f(x) = \sqrt{x_1^2 + x_2^2 + x_3^2} \right\}$$

is a complete set of entanglement invariants.

A functionally equivalent complete set of entanglement invariants is

$$\left\{ f' = x_1^2 + x_2^2 + x_3^2 \right\} ,$$

which is also a basic set of entanglement invariants.

---

[10]As we shall see, in this particular case of $n = 1$ qubits, the complete set of entanglement invariants consists of only one invariant.

15.1. The Bloch "sphere".

As a result of the previous calculation, we have a complete set of entanglement invariants, namely

$$f(x) = \sqrt{x_1^2 + x_2^2 + x_3^2} = |x|$$

We have completely classified all the entanglement classes for 1 qubit quantum systems. For in this case,

$$[i\rho]_E = [i\rho'] \iff f(i\rho) = f(i\rho') \ .$$

As a consequence of this result, the induced foliation of the space **density** (2^1) of all physical density operators lying in in the Lie algebra $\mathbf{u}(2^1)$ can be visualized in terms of the 3-ball of radius 1 in $\mathbb{R}^3$, called the **Bloch "sphere."**

Recall from remark on page 11 that the space **density** (2^1) of density operators representing the states of all one qubit quantum systems is

density $(2^1) = \{i\rho \in \mathbf{u}(2^1) \mid \rho \text{ is positive semi-definite and of trace one}\}$,

which is a convex subset of the of the Lie algebra $\mathbf{u}(2^1)$. In this special case of $n = 1$ qubit, it is a straight forward exercise to show that

density $(2^1) = \{i\rho = x_0\xi_0 + x \cdot \xi \mid |x| \leq 1 \text{ and } x_0 = -1\}$.

Thus, the convex subset **density** (2^1) of $\mathbf{u}(2^1)$ of all physical density operators $i\rho$ in $\mathbf{u}(2^1)$ can naturally be identified with the 3-ball of radius one via the one-to-one correspondence

$$i\rho = x_0\xi_0 + x_1\xi_1 + x_2\xi_2 + x_3\xi_3 \longleftrightarrow (x_1, x_2, x_3)$$

as illustrated in Figure 5.

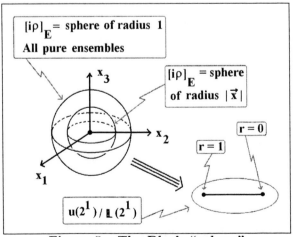

Figure 5. The Bloch "sphere"

It follows that each entanglement class $[i\rho]_E$ is simply a sphere of radius $f(i\rho) = |x|$. The sphere of radius one is the entanglement class of all pure ensembles. All other spheres represent entanglement classes of mixed ensembles. The "sphere" of radius 0 (i.e., the origin) represents the entanglement class of the maximally mixed ensemble.

Thus the space of entanglement classes lying in the space formed by identifying the elements of the convex set **density** (2^1) via the action of the local transformation group $\mathbb{L}(2^1)$, namely

$$\mathbf{density}\,(2^1)/\mathbb{L}(2^1),$$

is simply a closed[11] line segment.

In terms of this picture, it is easy to visualize the tangent space $T_{i\rho}([i\rho]_E)$ to $[i\rho]_E$ at $i\rho$. Moreover, it is easy to visualize the normal bundle of $[i\rho]_E$. For the normal vector field is simply

$$\left. x_1 \frac{\partial}{\partial x_1} + x_2 \frac{\partial}{\partial x_2} + x_3 \frac{\partial}{\partial x_3} \right|_{[i\rho]_E}$$

Unfortunately, for quantum systems of more than one qubit, such a visualization is by no means as easy. The dimension of **density** (2^n) grows exponentially as a function of the number n of qubits. Moreover, the number of entanglement classes also grows exponentially as a function of n.

16. Example 2. The entanglement classes of $n = 2$ qubits

As previously stated, the entanglement of two qubit quantum systems is much more complex than that of one qubit quantum systems. With each additional qubit, the entanglement becomes exponentially more complex than before. Perhaps this is a strong hint as to where the power of quantum computation is coming from?

For this example, the local unitary group $\mathbb{L}(2^2)$ is the Lie group $\mathbb{SU}(2^1) \otimes \mathbb{SU}(2^1)$. The corresponding Lie algebra $\ell(2^2)$ is Kronecker sum[12] $\mathbf{su}(2) \boxplus \mathbf{su}(2)$. Each density operator $i\rho$ lies in the Lie algebra $\mathbf{u}(2^2)$.

We can now use the bases[13]

$$\{\xi_{10}, \xi_{20}, \xi_{30}, \xi_{01}, \xi_{02}, \xi_{03},\}$$
and
$$\{\xi_{ij} \mid i, j = 0, 1, 2, 3\}$$

---

[11]The adjective "closed" means that the line segment contains both its endpoints.

[12]We remind the reader that the Kronecker sum $A \boxplus B$ of two matrices (operators) A and B is defined as

$$A \boxplus B = A \otimes \mathbf{1} + \mathbf{1} \otimes B$$

where $\mathbf{1}$ denotes the identity matrix (operator).

[13]See Section 10.

of the respective Lie algebras $\ell(2^2)$ and $\mathbf{u}(2^2)$ to find a more useful expression for ad_v, where

$$\xi_{ij} = -\frac{i}{2}\sigma_i \otimes \sigma_j \ .$$

Each element $v \in \ell(2^2)$ can be uniquely expressed in the form

$$v = (a \cdot \xi) \otimes I_4 + I_4 \otimes (b \cdot \xi) = a \cdot \xi \boxplus b \cdot \xi \ ,$$

where $a = (a_1, a_2, a_3)$ and $b = (b_1, b_2, b_3,)$ lie in $\mathbb{R}^3$, where $\xi = (\xi_1, \xi_2, \xi_3)$, and where I_4 is the 4×4 identity matrix. Thus,

$$\begin{aligned} ad_v &= ad_{\left(\sum_{j=1}^3 (a_j \xi_{j0} + b_j \xi_{0j})\right)} \\ &= ad_{(a \cdot \xi \boxplus b \cdot \xi)} \\ &= ad_{(a \cdot \xi) \otimes I_4} + ad_{I_4 \otimes (b \cdot \xi)} \\ &= I_4 \otimes (a \cdot ad_\xi) + (b \cdot ad_\xi) \otimes I_4 \end{aligned}$$

where

$$ad_\xi = (ad_{\xi_1}, ad_{\xi_2}, ad_{\xi_3}) \ .$$

But as in example 1,

$$ad_{\xi_j} = \begin{cases} \begin{pmatrix} 0 & 0 \\ 0 & L_j \end{pmatrix} = 0 \oplus L_j & \text{if } j = 1, 2, 3 \\ \begin{pmatrix} 0 & 0 \\ 0 & 0 \end{pmatrix} = 0 & \text{if } j = 0 \end{cases},$$

where

$$L_1 = \begin{pmatrix} 0 & 0 & 0 \\ 0 & 0 & -1 \\ 0 & 1 & 0 \end{pmatrix}, L_2 = \begin{pmatrix} 0 & 0 & 1 \\ 0 & 0 & 0 \\ -1 & 0 & 0 \end{pmatrix}, L_3 = \begin{pmatrix} 0 & -1 & 0 \\ 1 & 0 & 0 \\ 0 & 0 & 0 \end{pmatrix}$$

is the basis of the Lie algebra $\mathbf{so}(3)$ of the special orthogonal group $\mathbb{SO}(3)$ given in Appendix B on page 344.

Let

$$\{\partial/\partial x_{jk} \mid j, k = 0, 1, 2, 3\}$$

denote the basis of $\mathbf{Vec}\left(\mathbf{u}\left(2^2\right)\right)$ induced by the chart

$$\mathbf{u}\left(2^2\right) \xrightarrow{\pi} \mathbb{R}^{16}$$

$$i\rho = \sum_{i,j=0}^3 x_{ij}\xi_{ij} \longmapsto (x_{00}, x_{0*}, x_{10}, x_{1*}, x_{20}, x_{2*}, x_{30}, x_{3*})$$

where

$(x_{00}, x_{0*}, x_{10}, x_{1*}, x_{20}, x_{2*}, x_{30}, x_{3*})$

$= (x_{00}, \ x_{01}, x_{02}, x_{03}, \ x_{10}, \ x_{11}, x_{12}, x_{13}, \ x_{20}, \ x_{21}, x_{22}, x_{23}, \ x_{30}, \ x_{31}, x_{32}, x_{33})$

In other words, for each pair (j,k), $\partial/\partial x_{jk}$ denotes the vector field on $\mathbf{u}\left(2^2\right)$ defined at each point $i\rho$ as the tangent vector to the curve

$$\pi^{-1}\left(x_{00},\ldots,x_{jk}+t,\ldots,x_{33}\right) = i\rho + t\xi_{jk}$$

at $t=0$.

In terms of the above chart, $ad_v\left(i\rho\right)$ can be written as

$$\left(x_{00},x_{0*},x_{10},x_{1*},x_{20},x_{2*},x_{30},x_{3*}\right)\cdot\left[I_4\otimes\left(0\oplus a\cdot L\right)+\left(0\oplus b\cdot L\right)\otimes I_4\right]\cdot\begin{pmatrix}\partial/\partial x_{00}\\ \partial/\partial x_{0*}\\ \partial/\partial x_{10}\\ \partial/\partial x_{1*}\\ \partial/\partial x_{20}\\ \partial/\partial x_{2*}\\ \partial/\partial x_{30}\\ \partial/\partial x_{3*}\end{pmatrix},$$

which simplifies to

$$ad_v\left(i\rho\right) = \sum_{q=0}^{3}\left(a\cdot x_{q*}\times\frac{\partial}{\partial x_{q*}}+b\cdot x_{*q}\times\frac{\partial}{\partial x_{*q}}\right),$$

where '$\times$' denotes the vector cross product[14].

We can now achieve objective 1.

Objective 1. *Given an arbitrary density operator $i\rho$ in $\mathbf{u}\left(2\right)$, find the dimension of an arbitrary entanglement class $[i\rho]_E$.*

From the above discussion, it follows that the image $\operatorname{Im}\left(ad_*\right)$ of the infinitesimal action ad_* is spanned by the six vector fields

$$\begin{cases}ad_{\xi_{01}} = \sum_{q=0}^{3}\left(x_{q2}\frac{\partial}{\partial x_{q3}}-x_{q3}\frac{\partial}{\partial x_{q2}}\right)\\[6pt]ad_{\xi_{02}} = \sum_{q=0}^{3}\left(x_{q3}\frac{\partial}{\partial x_{q1}}-x_{q1}\frac{\partial}{\partial x_{q3}}\right)\\[6pt]ad_{\xi_{03}} = \sum_{q=0}^{3}\left(x_{q1}\frac{\partial}{\partial x_{q2}}-x_{q2}\frac{\partial}{\partial x_{q1}}\right)\end{cases}$$

$$\begin{cases}ad_{\xi_{10}} = \sum_{q=0}^{3}\left(x_{2q}\frac{\partial}{\partial x_{3q}}-x_{3q}\frac{\partial}{\partial x_{2q}}\right)\\[6pt]ad_{\xi_{20}} = \sum_{q=0}^{3}\left(x_{3q}\frac{\partial}{\partial x_{1q}}-x_{1q}\frac{\partial}{\partial x_{3q}}\right)\\[6pt]ad_{\xi_{30}} = \sum_{q=0}^{3}\left(x_{1q}\frac{\partial}{\partial x_{2q}}-x_{2q}\frac{\partial}{\partial x_{1q}}\right)\end{cases}$$

---

[14]The vector cross product is computed according to the right-hand rule.

In particular, the tangent space $T_{i\rho}([i\rho]_E) = (\operatorname{Im} ad_*)|_{i\rho}$ to the entanglement class $[i\rho]_E$ at the point $i\rho$ is spanned by

$$ad_{\xi_{01}}|_{i\rho},\ ad_{\xi_{02}}|_{i\rho},\ ad_{\xi_{03}}|_{i\rho},\ ad_{\xi_{10}}|_{i\rho},\ ad_{\xi_{20}}|_{i\rho},\ ad_{\xi_{30}}|_{i\rho}$$

We leave it as an exercise for the reader to verify that the above six vector fields are linearly independent almost every where. Thus, it follows that almost all entanglement classes are of dimension six.

However, there are notable exceptions. Consider the Bell basis state[15] $|\psi\rangle = \frac{1}{\sqrt{2}}(|00\rangle - |11\rangle)$. The corresponding density operator $i\rho$ is

$$i\rho = \frac{i}{2}\begin{pmatrix} 1 & 0 & 0 & -1 \\ 0 & 0 & 0 & 0 \\ 0 & 0 & 0 & 0 \\ -1 & 0 & 0 & 1 \end{pmatrix} = \left(-\frac{1}{2}\right)\xi_{00} + \left(\frac{1}{2}\right)\xi_{11} + \left(-\frac{1}{2}\right)\xi_{22} + \left(-\frac{1}{2}\right)\xi_{33},$$

where

$$\xi_{jk} = -\frac{i}{2}\sigma_j \otimes \sigma_k.$$

Hence,

$$x_{jk} = \begin{cases} -\frac{1}{2} & \text{if } j = k = 0, 2, 3 \\ \frac{1}{2} & \text{if } j = k = 1 \\ 0 & \text{if } j \neq k \end{cases}$$

Thus, in this case $\operatorname{Im}(ad_*)|_{i\rho}$ is spanned by

$$\begin{cases} ad_{\xi_{01}}|_{i\rho} &= \frac{1}{2}\left(\frac{\partial}{\partial x_{23}} - \frac{\partial}{\partial x_{32}}\right) \\[4pt] ad_{\xi_{02}}|_{i\rho} &= \frac{1}{2}\left(\frac{\partial}{\partial x_{31}} + \frac{\partial}{\partial x_{13}}\right) \\[4pt] ad_{\xi_{03}}|_{i\rho} &= \frac{1}{2}\left(-\frac{\partial}{\partial x_{12}} - \frac{\partial}{\partial x_{21}}\right) \end{cases}$$

$$\begin{cases} ad_{\xi_{10}}|_{i\rho} &= \frac{1}{2}\left(\frac{\partial}{\partial x_{32}} - \frac{\partial}{\partial x_{23}}\right) \\[4pt] ad_{\xi_{20}}|_{i\rho} &= \frac{1}{2}\left(\frac{\partial}{\partial x_{13}} + \frac{\partial}{\partial x_{31}}\right) \\[4pt] ad_{\xi_{30}}|_{i\rho} &= \frac{1}{2}\left(-\frac{\partial}{\partial x_{21}} - \frac{\partial}{\partial x_{12}}\right) \end{cases}$$

Hence,

$$Dim\,[i\rho_{Bell}]_E = Dim\left[(\operatorname{Im} ad_*)|_{i\rho_{Bell}}\right] = 3$$

This only confirms the conventional wisdom that the entanglement class of the Bell basis states is truly exceptional.

---

[15] It should be noted that all four 2 qubit Bell basis states lie in the same entanglement class. It is this fact that makes quantum teleportation possible.

We are now ready for objective 2:

Objective 2. *Given two states $i\rho$ and $i\rho'$, devise a means of determining whether they belong to the same entanglement class or to different entanglement classes.*

The complete functionally independent set of solutions to the above system of PDEs (hence, a complete set of entanglement invariants) was found by Linden and Popescu in[39]. These invariants are as described below[16]. For further details please refer to their work [39].

Let $i\rho$ be an arbitrary element of the Lie algebra $\mathbf{u}\left(2^2\right)$. Then in terms of the earlier described chart π,

$$i\rho = \sum_{j,k=0}^{3} x_{jk}\xi_{jk} ,$$

where $\{\xi_{jk}\}$ denotes the basis of $\mathbf{u}\left(2^2\right)$ described earlier.

We will change our notation slightly. Let x_{**} denote the 3×3 matrix

$$x_{**} = (x_{jk})_{j,k=1,2,3} ,$$

and let x_{0*} and x_{*0} denote the vectors

$$\begin{cases} x_{0*} = (x_{01}, x_{02}, x_{02}) \\ x_{*0} = (x_{10}, x_{20}, x_{30}) \end{cases}$$

Finally, let Z denote the matrix

$$Z = x_{**}x_{**}^T ,$$

where the superscript 'T' denotes the transpose. Then the nine algebraically independent polynomial functions listed in the table

$Tr(Z)$	$Tr(Z^2)$	$\det(x_{**})$
$x_{0*}x_{0*}^T$	$x_{0*}Zx_{0*}^T$	$x_{0*}Z^2x_{0*}^T$
$x_{0*}x_{**}x_{*0}^T$	$x_{0*}Zx_{**}x_{*0}^T$	$x_{0*}Z^2x_{**}x_{*0}^T$

together with a tenth algebraically dependent function

$$x_{0*} \cdot \left(Zx_{0*}^T\right) \times \left(Z^2x_{0*}^T\right)$$

form a complete set of entanglement invariants.

The tenth function, which is algebraically dependent on the other nine, is needed to determine the sign of the components of $i\rho$.

---

[16]We are using a notation different from that found in [39].

17. Example n. The entanglement classes of n qubits, $n > 2$

Finally, we come to Objective 3.

Objective 3. Briefly outline the intriguing mathematical research opportunities for quantifying quantum entanglement for $n = 3$ qubits, and beyond. Surprisingly, very little is known about quantum entanglement for $n > 3$ qubits.

For n qubits $(n > 2)$, the same methods lead to the following formula for the infinitesimal action

$$ad_v(i\rho) = \sum_{q_1,q_2\cdots q_{n-1}=0}^{3} \sum_{k=1}^{n} a^{(k)} \cdot x_{q_1 q_2 \cdots q_{k-1} * q_{k+1} \cdots q_{n-1}} \times \frac{\partial}{\partial x_{q_1 q_2 \cdots q_{k-1} * q_{k+1} \cdots q_{n-1}}}$$

where $v \in \ell(2^n)$ and $i\rho \in \mathbf{u}(2^n)$ are given by

$$\begin{cases} v = \sum_{k=1}^{n} a^{(k)} \cdot \xi_{\underbrace{00\cdots 0 * 0 \cdots 0}_{* \text{ in } k\text{-th position}}} \\ i\rho = \sum_{r_1,r_2,\cdots,r_n=0}^{3} x_{r_1 r_2 \cdots r_n} \xi_{r_1 r_2 \cdots r_n} \end{cases}$$

However, finding a complete set of solutions to the corresponding system of PDEs appears to be by no means an easy task. We leave this apparently daunting problem to future researchers with a closing caveat:

Beware! This research problem is very addictive.

18. Conclusion

There is much more that could be said about quantum entanglement. This paper presents only a small part of the big picture. But hopefully this exposition will provide the reader with some insight into this rapidly growing research field. Since this paper was written, research in quantum entanglement has literally had an explosive expansion, and even now continues to do so. We refer the reader to the references at the end of this paper, which represent only a few of the many papers in this rapidly expanding field.

19. Appendix A. Some Fundamental Concepts from the Theory of Differential Manifolds

19.1. Differential manifolds, tangent bundles, and vector fields.

DEFINITION 12. *A topological space M^m is an **m-dimensional manifold** if it is locally homeomorphic to $\mathbb{R}^m$, i.e., if there exists an open cover $\mathcal{W} = \{W_\alpha\}$ of M^m such that for each $W_\alpha \in \mathcal{W}$, there is associated a homeomorphism*

$$W_\alpha \xrightarrow{\varphi_\alpha} \mathbb{R}^m$$
$$x \longmapsto (x_1, x_2, \ldots, x_m)$$

which maps W_α onto an open subset of $\mathbb{R}^m$. We call

$$(\varphi_\alpha, W_\alpha)$$

*a **chart on** M^m, and*

$$\Phi = \{\, (\varphi_\alpha, W_\alpha) \,\}$$

*an **Atlas** on M^m.*

*An Atlas is said to be **smooth** (C^∞), if whenever*

$$\varphi_\beta \varphi_\alpha^{-1} : \varphi_\alpha(W_\alpha \cap W_\beta) \longrightarrow \varphi_\beta(W_\alpha \cap W_\beta)$$

*is defined, it is a smooth (C^∞) map of $\varphi_\alpha(W_\alpha \cap W_\beta) \subseteq \mathbb{R}^n$ into $\varphi_\beta(W_\alpha \cap W_\beta) \subseteq \mathbb{R}^n$. A **smooth** ($C^\infty$) **manifold** is a topological manifold with a smooth atlas.*

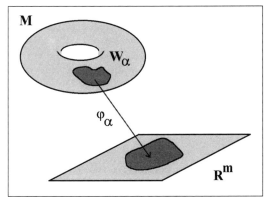

Figure 6. A chart $\varphi_\alpha \colon M \longrightarrow \mathbb{R}^4$ on a manifold M.

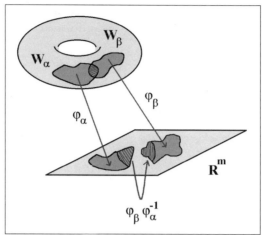

Figure 7. An atlas is smooth if every $\varphi_\beta \varphi_\alpha^{-1}$ is smooth when defined.

DEFINITION 13. *Let M and N be smooth manifolds. Then a map*

$$f : M \longrightarrow N$$

*is said to be **smooth** if for every $x \in M$ there exist charts $(\varphi_\alpha, W_\alpha)$ of M and (ψ_β, V_β) of N containing x and $f(x)$ respectively such that*

$$\psi_\beta f \varphi_\alpha^{-1} : \varphi_\alpha(W_\alpha) \longrightarrow \varphi_\beta(V_\beta)$$

is smooth.

DEFINITION 14. *Let x be an element of a smooth manifold M, and let $\gamma_1(t)$ and $\gamma_2(t)$ be smooth curves in M which pass through x, i.e., such that there exists $t_1, t_2 \in \mathbb{R}$ for which*

$$\gamma_1(t_1) = x = \gamma_2(t_2)$$

*Then γ_1 and γ_2 are said to be **tangentially equivalent** at x, written*

$$\gamma_1 \underset{x}{\sim} \gamma_2 ,$$

if they are tangent at the point x, i.e., if there is a chart $(\varphi_\alpha, W_\alpha)$ on M containing x such that

$$\frac{d}{dt}(\varphi_\alpha \circ \gamma_1)(t)|_{t=t_1} = \frac{d}{dt}(\varphi_\alpha \circ \gamma_2)(t)|_{t=t_2}$$

REMARK 3. *It can easily be shown that the relation $\underset{x}{\sim}$ is independent of the chart selected.*

DEFINITION 15. *A **tangent vector** (x, v) (also written simply as v) to M at x is a tangential equivalence class at x. The tangent space of M^n at x, denoted by $T_x M^n$, is the set of tangent vectors to M at x. $T_x M$ can be shown to be an n-dimensional vector space.*

Let

$$TM = \bigcup_{x \in M} T_x M ,$$

and let π be the map
$$TM \xrightarrow{\pi} M$$
$$(x, v) \longmapsto x$$
If $\varphi_\alpha : W_\alpha \longrightarrow \mathbb{R}^m$ is a chart on M, then
$$\varphi_\alpha \pi : \pi^{-1} W_\alpha \longrightarrow \mathbb{R}^m$$
can be shown to be a chart on TM. In this way, TM becomes a smooth manifold and π becomes a smooth map. TM together with the map π is called the **tangent bundle** of M.

DEFINITION 16. *A **vector field** v on a smooth manifold M^m is a smooth map*
$$v : M^m \longrightarrow TM^m$$

Let $\mathbf{Vec}(M^m)$ be the set of all vector fields on the smooth manifold M^m. This is easily seen to be a vector space where, for example, the sum $u + v$ of two vector fields is defined by
$$(u + v)|_x = u|_x + v|_x$$
for all $x \in M^m$.

We will now consider the charts of the tangent bundle TM in a more explicit way.

Let
$$W_\alpha \xrightarrow{\varphi_\alpha} \mathbb{R}^m$$
$$x \longmapsto (x_1, x_2, \ldots, x_m)$$
be a chart on the smooth manifold M^m, and let a be an arbitrary point in U_α. Thus,
$$\varphi_\alpha(a) = (a_1, a_2, \ldots, a_m) \ .$$

For each j ($j = 1, 2, \ldots, m$) consider the smooth curve
$$\gamma_j(t) = \varphi_\alpha^{-1}(a_1, a_2, \ldots, a_j + t, \ldots, a_m)$$
in U_α which passes through the point a at time $t = 0$. Then for each such j, let
$$\left.\frac{\partial}{\partial x_j}\right|_a \in T_a M$$
denote the tangent vector to the curve γ_j at a. It can be shown that
$$\left.\frac{\partial}{\partial x_1}\right|_a, \left.\frac{\partial}{\partial x_2}\right|_a, \ldots, \left.\frac{\partial}{\partial x_m}\right|_a$$
is a vector space basis of the tangent space $T_a M$.

Moreover, since this construction is with respect to an arbitrary point a in W_α, it can be shown that we have actually constructed for each j a smooth vector field
$$\frac{\partial}{\partial x_j} \in \mathbf{Vec}(TW_\alpha) \subseteq \mathbf{Vec}(TM)$$

In fact, it can be shown that

$$\frac{\partial}{\partial x_1}, \frac{\partial}{\partial x_2}, \ldots, \frac{\partial}{\partial x_m}$$

is a basis of $\mathbf{Vec}(TW_\alpha)$, and hence a local basis of $\mathbf{Vec}(TM)$.

We can now express each chart $\left(\varphi_\alpha\pi, \pi^{-1}W_\alpha\right)$ explicitly as:

$$\pi^{-1}W_\alpha \xrightarrow{\varphi_\alpha\pi} \mathbb{R}^{2m}$$

$$(x, \mu_1\frac{\partial}{\partial x_1} + \mu_2\frac{\partial}{\partial x_2} + \ldots + \mu_m\frac{\partial}{\partial x_m}) \longmapsto (x_1, x_2, \ldots, x_m,\ \mu_1, \mu_2, \ldots, \mu_m)$$

where μ_j's on the left denote functions of $x \in M$, and where μ_j's on the right denote functions of $(x_1, x_2, \ldots, x_m) \in \mathbb{R}^m$.

DEFINITION 17. *Let M and N be smooth manifolds, let $f : M \longrightarrow N$ be a smooth map, and let a be an arbitrary point of M. We define a vector space morphism*

$$df|_a : T_aM \longrightarrow T_{f(a)}N$$

as follows:

For each $v \in T_aM$, there is a representative smooth curve $\gamma_v(t)$ in M which passes through the point a and which has v as its tangent vector at the point a. It follows that $f \circ \gamma_v(t)$ is a smooth curve in N passing through the point $f(a)$. We define

$$df|_a(v) \in T_{f(a)}N$$

as the tangent vector to $f \circ \gamma_v(t)$ at the point $f(a)$. It is then a simple exercise to show that $df|_a$ is a vector space morphism.

Since a was an arbitrary point of M, this leads to the definition of a smooth map $df : TM \longrightarrow TN$, called the differential of f, such that the following diagram is commutative:

$$\begin{array}{ccc} TM & \xrightarrow{df} & TN \\ \downarrow & & \downarrow \\ M & \xrightarrow{f} & N \end{array}.$$

REMARK 4. *In local coordinates, df maps the tangent vector*

$$v|_x = \sum_{i=1}^{m} \mu_i \frac{\partial}{\partial x_i}$$

to the tangent vector

$$df(v|_x) = \sum_{j=1}^{n} \left(\sum_{i=1}^{m} \mu_i \frac{\partial f_j}{\partial x_i}\right) \frac{\partial}{\partial y_j}.$$

Thus, the matrix expression of the linear transformation df is just the Jacobian matrix

$$\left(\frac{\partial f_j}{\partial x_i}(x)\right)_{m \times n}.$$

19.2. Exponentiation of vector fields.

DEFINITION 18. *Let M be a smooth manifold, and let $v \in \mathbf{Vec}(M)$ be a smooth vector field on M. A curve $\gamma(t)$ in M is said to be an **integral curve** of v if $v|_{\gamma(t)}$ is the tangent vector to $\gamma(t)$ for each t for which $\gamma(t)$ is defined.*

In terms of local coordinates, an integral curve $\gamma(t)$ of a smooth vector field

$$v(x) = \sum_{i=1}^{n} \mu_i(x_1, x_2, \ldots, x_m) \frac{\partial}{\partial x_i}$$

is a solution to the system of ordinary differential equations

$$\frac{dx_i}{dt} = \mu_i(x_1, x_2, \ldots, x_m) \, , \, i = 1, 2, \ldots, m$$

Since v is smooth, its coefficients $\mu_i(x_1, x_2, \ldots, x_m)$ are smooth functions. Consequently, it follows from the standard existence and uniqueness theorems for systems of ordinary differential equations that there exists a unique solution for each set of initial conditions.

Thus, for each x in M, there exists a unique maximal integral curve $\gamma_v(t, x)$ passing through x at time $t = 0$. We call $\gamma_v(t, x)$ the **flow** generated by the vector field v. The vector field v is called the **infinitesimal generator** of the flow. It can be easily shown that

$$\gamma_v(t, \gamma_v(s, x)) = \gamma_v(t + s, x) \ .$$

Hence, we are justified in adopting the following suggestive notation:

$$e^{tv} x$$

for the flow $\gamma_v(t, x)$.

In terms of our new notation, the properties of the flow can be expressed as
1) $e^{sv} e^{tv} x = e^{(s+t)v} x$
2) $e^{0 \cdot v} x = x$
3) $\frac{d}{dt}\left(e^{tv} x\right) = v|_{e^{tv} x}$.

19.3. Vector fields viewed as directional derivatives.

We now show how vector fields can be viewed as partial differential operators.

DEFINITION 19. *A **derivation** D on an algebra $\mathcal{A}$ is a map*

$$D : \mathcal{A} \to \mathcal{A}$$

such that
1) *(Linearity)* $D(\alpha f + \beta g) = \alpha D f + \beta D g$
2) *(Leibnitz Rule)* $D(fg) = (Df) g + f (Dg)$

DEFINITION 20. *A **Lie algebra** $\mathbb{A}$ is a vector space together with a binary operation*

$$[-, -] : \mathbb{A} \times \mathbb{A} \to \mathbb{A} \, ,$$

*called a **Lie bracket** for $\mathbb{A}$, such that*

1) *(Bilinearity)*
$$[\lambda_1 a_1 + \lambda_2 a_2,\ b] = \lambda_1 [a_1, b] + \lambda_2 [a_2, b]$$
$$[a,\ \lambda_1 b_1 + \lambda_2 b_2] = \lambda_1 [a, b_1] + \lambda_2 [a, b_2]$$

2) *(Skew-Symmetry)*
$$[a, b] = -[b, a]$$

3) *(Jacobi Identity)*
$$[a, [b, c]] + [c, [a, b]] + [b, [c, a]] = 0$$

PROPOSITION 3. *The set of derivations $Der(\mathcal{A})$ on an algebra $\mathcal{A}$ is a Lie algebra with Lie bracket given by:*
$$[D_1, D_2] = D_1 \circ D_2 - D_2 \circ D_1$$

Let $\mathbf{C}^\infty(M)$ denote the **algebra of real valued functions** on the smooth manifold M. Then, it follows that $\mathbf{Der}(\mathbf{C}^\infty(M))$ is a Lie algebra.

We will now show how to identify the elements of $\mathbf{Vec}(M)$ with derivations in $\mathbf{Der}(\mathbf{C}^\infty(M))$, and thereby show that $\mathbf{Vec}(M)$ is more than a vector space. It is actually a Lie algebra.

Each smooth vector field v on M can be thought of as a directional derivative in the direction v as follows: Let $v \in \mathbf{Vec}(M)$ and let $f \in \mathbf{C}^\infty(M)$. Define $v(f)$ as:
$$v(f)|_x = \frac{d}{dt} f\left(e^{tv} x\right)\Big|_{t=0}$$

Thus, we have:

PROPOSITION 4. $\mathbf{Vec}(M)$ *is a Lie algebra of derivations on the algebra $\mathbf{C}^\infty(M)$.*

It is enlightening, to view the above in terms of local coordinates. From this perspective,
$$v = \sum_{i=1}^{m} \mu_i(x) \frac{\partial}{\partial x_i} \ .$$
Thus, if we use the chain rule and the fact that
$$\frac{d}{dt}\left(e^{tv} x\right) = v|_{e^{tv} x} \ ,$$
we have
$$\frac{d}{dt} f\left(e^{tv} x\right) = \sum_{i-1}^{m} \xi^i\left(e^{tv} x\right) \frac{\partial f}{\partial x^i}\left(e^{tv} x\right) = \left(\sum_{i=1}^{m} \xi^i \frac{\partial}{\partial x^i}\right)(f)\Big|_{e^{tv} x} \ .$$
Hence,
$$v = \sum_{i=1}^{m} \mu_i \frac{\partial}{\partial x_i} \in Vec(U_\alpha) \subseteq Vec(M)$$
acts as a first order partial differential operator, thereby justifying the notation.

So viewing v as a first order partial differential operator, we can write
$$\frac{d}{dt} f\left(e^{tv} x\right) = v(f)|_{e^{tv}x} ,$$
and, in particular,
$$\left.\frac{d}{dt} f\left(e^{tv} x\right)\right|_{t=0} = v(f)(x) ,$$
where $v(f)(x)$ now denotes (locally) $\left(\sum_{i=1}^m \mu_i \frac{\partial}{\partial x_i}\right) f$ evaluated at x.

20. Appendix B. Some Fundamental Concepts from the Theory of Lie Groups

20.1. Lie groups.

DEFINITION 21. *A **Lie group** $\mathbb{G}$ is a group which is a smooth manifold whose differential structure is compatible with the group operations, i.e., such that*
  1) *The multiplication map of $\mathbb{G}$*
$$\mathbb{G} \times \mathbb{G} \longrightarrow \mathbb{G}$$
$$(g_1, g_2) \longmapsto g_1 g_2$$
  and,
  2) *The inverse map of G*
$$\mathbb{G} \longrightarrow \mathbb{G}$$
$$g \longmapsto g^{-1}$$
  are smooth functions.

*Every closed subgroup $\mathbb{H}$ of $\mathbb{G}$ can be shown to be a Lie group, inheriting its Lie group structure from from that of $\mathbb{G}$. It is called a **Lie subgroup** of $\mathbb{G}$.*

DEFINITION 22. *A **one parameter subgroup** of a Lie group $\mathbb{G}$ is a smooth morphism from the additive Lie group of reals $\mathbb{R}, +$ to the group $\mathbb{G}$.*

20.2. Some examples of Lie groups.

Let V denote an n-dimensional vector space over the real numbers $\mathbb{R}$ with the standard vector inner product which we denote by $\langle\,,\,\rangle$.

- $\mathbb{GL}(n, \mathbb{R})$ is the **real general linear group** of all automorphisms of the vector space V. This can be identified with the group of all nonsingular $n \times n$ matrices over the reals.
- $\mathbb{O}(n)$ is the **real orthogonal group** of all automorphisms which preserve the inner product $\langle\,,\,\rangle$. This can be identified with the group of orthogonal matrices, i.e., matrices A of the form
$$A^T = A^{-1}$$
  where the superscript "T" denotes the matrix transpose.

- $SL(n, \mathbb{R})$ is the **real special linear group** of all real $n \times n$ matrices of determinant 1. $SL(n, \mathbb{R})$ is the group of all rigid motions in hyperbolic n-space.
- $SO(n) = O(n) \cap SL(n, \mathbb{R})$ is the **special orthogonal group** of all orthogonal real $n \times n$ matrices of determinant 1. This group can be identified with the group of all rotations in $\mathbb{R}^n$ about a fixed point such as the origin.

Let W denote an n-dimensional vector space over the complex numbers $\mathbb{C}$ with the standard sesquilinear inner product which we also denote by $\langle \, , \, \rangle$.

- $GL(n, \mathbb{C})$ is the **complex general linear group** of all automorphisms of the vector space W. This can be identified with the group of all nonsingular $n \times n$ matrices over the complexes.
- $SL(n, \mathbb{C})$ is the **complex special linear group** of all complex $n \times n$ matrices of determinant 1.
- $U(n)$ is the **unitary group** of all $n \times n$ unitary matrices over the complex numbers $\mathbb{C}$, i.e., all $n \times n$ complex matrices A such that

$$A^\dagger = A^{-1}$$

where $A^\dagger$ denotes the conjugate transpose.
- $SU(n) = U(n) \cap SL(n, \mathbb{C})$ is the special unitary group of all unitary matrices of determinant 1.

20.3. The Lie algebra of a Lie group.

DEFINITION 23. *Let $\mathbb{G}$ be a Lie group. For each element $h \in \mathbb{G}$, we define the* **right multiplication map**, *written R_h, as*

$$\mathbb{G} \xrightarrow{R_h} \mathbb{G}$$
$$g \longmapsto gh$$

The map R_h is an autodiffeomorphism of $\mathbb{G}$. We let

$$dR_h : T\mathbb{G} \longrightarrow T\mathbb{G}$$

denote the corresponding differential of this diffeomorphism.

Finally, a vector field $v \in \mathbf{Vec}(\mathbb{G})$ is said to be **right invariant** *if*

$$(dR_h)\left(v|_g\right) = v|_{gh}$$

DEFINITION 24. *Let $\mathbf{Vec}_R(\mathbb{G})$ denote the set of right invariant vector fields on $\mathbb{G}$. Then $\mathbf{Vec}_R(\mathbb{G})$ as a subset of the Lie algebra $\mathbf{Vec}(\mathbb{G})$ inherits the structure of a Lie algebra. We call $\mathbf{Vec}_R(\mathbb{G})$ the* **Lie algebra** *of the Lie group $\mathbb{G}$.*

Let I denote the identity element of the Lie group $\mathbb{G}$. Since a right invariant vector field $v \in \mathbf{Vec}_R(\mathbb{G})$ is completely determined by its restriction to the tangent space $T_I \mathbb{G}$ via

$$v|_g = (dR_g)(v|_I) ,$$

we can, and do, identify the Lie algebra $\mathbf{Vec}_R(\mathbb{G})$ with the tangent space $T_I \mathbb{G}$, i.e.,

$$\mathbf{Vec}_R(\mathbb{G}) = T_I \mathbb{G} .$$

The tangent bundle $T\mathbb{G}$ of a Lie group $\mathbb{G}$ is trivial. For it can be shown that $T\mathbb{G}$ is bundle isomorphic to $\mathbb{G} \times T_I \mathbb{G}$. However, the Lie group structure of $\mathbb{G}$ induces some additional and useful structure induced on the Lie algebra $Vec_R(\mathbb{G}) = T_I\mathbb{G}$, namely, the exponential map.

DEFINITION 25. *We define the **exponential map** $\exp$ from the Lie algebra $\mathbf{Vec}_R(\mathbb{G})$ to the Lie group G as*

$$\mathbf{Vec}_R(\mathbb{G}) \xrightarrow{\exp} \mathbb{G}$$

$$v \longmapsto \left(e^{vt}\right) I\big|_{t=1}$$

In other words, we simply follow the flow $\gamma_v(t,g) = e^{vt}g$ from the identity I to the point $e^v I$ in $\mathbb{G}$.

It can be shown that the exponential map

$$\exp : \mathbf{Vec}_R(\mathbb{G}) \longrightarrow \mathbb{G}$$

is a local diffeomorphism. It also follows that, for each $v \in \mathbf{Vec}_R(\mathbb{G})$, $\exp(tv)$ is a one parameter subgroup of G. In fact, all one parameter subgroups are of this form.

20.4. Some examples of Lie algebras.

20.4.1. *Example: The Lie algebra $\mathbf{u}(N)$ of the unitary group $\mathbb{U}(N)$.*

In this case, $\mathbf{u}(N)$ is the Lie algebra of all $N \times N$ skew Hermitian[17] matrices over $\mathbb{C}$. This can be seen as follows:

The Lie algebra $\mathbf{u}(N)$ is the tangent space $T^I \mathbb{U}(N)$ to $\mathbb{U}(N)$ at the $N \times N$ identity matrix I. Hence, $\mathbf{u}(N)$ consists of all tangent vectors $\dot{U}(0) = \frac{d}{dt}U(t)\big|_{t=0}$ of all curves $U(t)$ in $\mathbb{U}(N)$ which pass through I at $t = 0$, i.e., which satisfy $U(0) = I$.

Since $U(t)$ is unitary, i.e., since

$$U(t)\overline{U}(t)^T = I,$$

we find by differentiating the above formula that

$$\dot{U}(t)\overline{U}(t)^T + U(t)\overline{\dot{U}(t)}^T = 0 .$$

Setting $t = 0$, we have

$$\overline{\dot{U}(0)}^T = -\dot{U}(0) .$$

Thus all matrices in $\mathbf{u}(N)$ are skew Hermitian.

Let M be an arbitrary skew $N \times N$ Hermitian matrix. Then

$$U(t) = \exp(tM)$$

---

[17]A square matrix M is skew Hermitian if $\overline{M}^T = -M$.

is a curve in $\mathbb{U}(N)$ which passes through I at $t = 0$ for which $\dot{U}(0) = M$. Hence, $\mathbf{u}(N)$ is the Lie algebra of all $N \times N$ skew Hermitian matrices over $\mathbb{C}$.

Let
$$\sigma_1 = \begin{pmatrix} 0 & 1 \\ 1 & 0 \end{pmatrix}, \sigma_2 = \begin{pmatrix} 0 & -i \\ i & 0 \end{pmatrix}, \sigma_3 = \begin{pmatrix} 1 & 0 \\ 0 & -1 \end{pmatrix}$$
denote the Pauli spin matrices, and let
$$\sigma_0 = \begin{pmatrix} 1 & 0 \\ 0 & 1 \end{pmatrix}$$
denote the 2×2 identity matrix. Then the following is a basis of the Lie algebra $\mathbf{u}(2^n)$
$$\{\xi_{j_1 j_2 \cdots j_n} \mid j_1, j_2, \ldots, j_n = 0, 1, 2, 3\},$$
where
$$\xi_{j_1 j_2 \cdots j_n} = -\frac{i}{2} \sigma_{j_1} \otimes \sigma_{j_2} \otimes \ldots \otimes \sigma_{j_n}.$$

REMARK 5. *Please note that, although $\mathbf{u}(N)$ is a Lie algebra of complex matrices, it is nonetheless a real Lie algebra. Thus, the above basis $\{\xi_{j_1 j_2 \cdots j_n}\}$ of $\mathbf{u}(2^n)$ is a basis of $\mathbf{u}(2^n)$ over the reals $\mathbb{R}$. But the matrices in $\mathbf{u}(2^n)$ are still matrices of complex numbers!*

20.4.2. Example: The Lie algebra $\mathbf{su}(N)$ of the special unitary group $\mathbb{SU}(N)$.

The Lie algebra $\mathbf{su}(N)$ for the special unitary group is the same as the Lie algebra of all $N \times N$ traceless skew Hermitian matrices, i.e., of all $N \times N$ skew Hermitian matrices M such that $trace(M) = 0$. A basis of the Lie algebra $\mathbf{su}(2^n)$ is
$$\{\xi_{j_1 j_2 \cdots j_n} \mid j_1, j_2, \ldots, j_n = 0, 1, 2, 3\} - \{\xi_{00 \cdots 0}\} .$$

20.4.3. Example: The Lie algebra $\mathbf{so}(3)$ of the special unitary group $\mathbb{SO}(3)$.

Finally, we should mention that the Lie algebra $\mathbf{so}(3)$ of the special orthogonal group $\mathbb{SO}(3)$ is the Lie algebra of all 3×3 skew symmetric matrices over the reals $\mathbb{R}$. The following three matrices form a basis for $so(3)$
$$L_1 = \begin{pmatrix} 0 & 0 & 0 \\ 0 & 0 & -1 \\ 0 & 1 & 0 \end{pmatrix}, L_2 = \begin{pmatrix} 0 & 0 & 1 \\ 0 & 0 & 0 \\ -1 & 0 & 0 \end{pmatrix}, L_3 = \begin{pmatrix} 0 & -1 & 0 \\ 1 & 0 & 0 \\ 0 & 0 & 0 \end{pmatrix}.$$

20.5. Lie groups as transformation groups on manifolds.

DEFINITION 26. *Let M be a smooth manifold. Then a **group of transformations acting on** M is a Lie group $\mathbb{G}$ together with a smooth map*

$$\mathbb{G} \times M \longrightarrow M$$
$$(g, x) \longmapsto g \cdot x$$

such that

1) *For all $x \in M$, and for all $g_1, g_2 \in \mathbb{G}$*

$$g_1 \cdot (g_2 \cdot x) = (g_1 g_2) \cdot x$$

2) *For all $x \in M$,*

$$e \cdot x = x \; ,$$

where e denotes the identity of $\mathbb{G}$.
$\mathbb{G}$ *is called a **transformation group** of M.*

DEFINITION 27. *Let M be a smooth manifold, and let $\mathbb{G}$ be a Lie group acting on M. Then the action*

$$\mathbb{G} \times M \longrightarrow M$$

*induces an **infinitesimal action***

$$\mathbf{Vec}_R(\mathbb{G}) \xrightarrow{\Psi_\mathbb{G}} \mathbf{Vec}(M) \; ,$$

where $\Psi_\mathbb{G}(v)|_x$ is the tangent vector to the curve

$$\gamma_v(t, x) = e^{tv} x$$

in M at x, i.e.,

$$\Psi_\mathbb{G}(v)|_x = \left. \frac{d}{dt} \left(e^{tv} x \right) \right|_{t=0} \; .$$

20.6. The big and little adjoint representations.

Let $\mathbb{G}$ be a Lie group, and let $\mathfrak{g}$ denote the corresponding Lie algebra.

For each element $h \in \mathbb{G}$, consider the inner automorphism:

$$\mathbb{G} \xrightarrow{\mathcal{I}_h} \mathbb{G}$$
$$g \longmapsto hgh^{-1}$$

and let

$$T\mathbb{G} \xrightarrow{d\mathcal{I}_h} T\mathbb{G}$$

denote the corresponding differential. We can now define the **big adjoint representation**

$$Ad : \mathbb{G} \longrightarrow Aut(\mathfrak{g})$$

by

$$Ad_h = (d\mathcal{I}_h)|_I$$

where I denotes the identity of G, and where $Aut(\mathfrak{g})$ denotes the group of automorphisms of the Lie algebra $\mathfrak{g}$.

We also can in turn define the **little adjoint representation**

$$ad : \mathfrak{g} \longrightarrow End(\mathfrak{g})$$

of the Lie algebra $\mathfrak{g}$ by

$$ad_v(u) = [u, v] \ ,$$

where $[-,-]$ denotes the Lie bracket, and where $End(\mathfrak{g})$ denotes the ring of endomorphisms of the Lie algebra $\mathfrak{g}$.

As the story goes, $End(\mathfrak{g})$ is actually the Lie algebra of the Lie group $Aut(\mathfrak{g})$, and we have the following commutative diagram

$$\begin{array}{ccc} \mathfrak{g} & \xrightarrow{ad} & End(\mathfrak{g}) \\ \exp \downarrow & & \downarrow \exp \\ G & \xrightarrow{Ad} & Aut(\mathfrak{g}) \end{array}$$

which relates the big and little adjoints. Little adjoint ad is actually the differential $d(Ad)$ restricted to the identity I of the big adjoint Ad.

Perhaps the following example would be of help:

EXAMPLE 1. *Let $\mathbb{G}$ be the special unitary group $\mathbb{SU}(2)$. Let $su(2)$ denote its Lie algebra. Then $Aut(\mathbb{G})$ is the special orthogonal group $\mathbb{SO}(3)$ and $End(\mathfrak{g})$ is the Lie algebra $so(3)$ of $\mathbb{SO}(3)$. Thus, we have the familiar commutative diagram*

$$\begin{array}{ccc} su(2) & \xrightarrow{ad} & so(3) \\ \exp \downarrow & & \downarrow \exp \\ \mathbb{SU}(2) & \xrightarrow{Ad} & \mathbb{SO}(3) \end{array}$$

used in quantum mechanics and in quantum computation[18].

REMARK 6. *The reader should verify that*

$$ad_{\xi_j} = L_j$$

20.7. The orbits of transformation Lie group actions.

Finally, we should remark that the entanglement classes defined previously in this paper are nothing more than the orbits of a group action. For completeness, we give the definition below:

DEFINITION 28. *A subset $\mathcal{O}$ of the smooth manifold M is an **orbit** of the action of the group $\mathbb{G}$ on M provided*
  1) *$x \in \mathcal{O} \Longrightarrow g \cdot x \in \mathcal{O}$ for all $g \in \mathbb{G}$, and*
  2) *If S is a non-empty subset of $\mathcal{O}$ which satisfies condition 1) above, then $S = \mathcal{O}$.*

In other words, an orbit is a minimal nonempty invariant subset of M.

---

[18]The well known Dirac belt trick is a consequence of the above commutative diagram.

References

[1] Belinfante, Johan G.F., and Berbard Kolman, **"A Survey of Lie Groups and Lie Algebras with Applications and Computational Methods,"** S.I.A.M., (1989).

[2] Bell, J.S., **"Speakable and Unspeakable in Quantum Mechanics,"** Cambridge University Press (1987).

[3] Bell, J.S., Physics, 1, (1964), pp. 3475 - 3467.

[4] Bennett, Charles H., David P. DiVincenzo, Tal Mor, Peter W. Shor, John A. Smolin, and Barbara M. Terhal, **Unextendible product bases and bound entanglement**, http://xxx.lanl.gov/abs/quant-ph/9808030.

[5] Bennett, Charles H., David P. DiVincenzo, Christopher A. Fuchs, Tal Mor, Eric Rains, Peter W. Shor, John A. Smolin, and William K. Wootters, **Quantum nonlocality without entanglement**, http://xxx.lanl.gov/abs/quant-ph/9804053.

[6] Bennett, Charles H., Herbert J. Bernstein, Sandu Popescu, and Benjamin Schumacher, **Concentrating partial entanglement by local operations**, Phys. Rev. A, Vol. 53, No. 4, April 1996, pp 2046 - 2052.

[7] Boerner, H., **"Representation of Groups with Special Considerations of Modern Physics,"** North-Holland, (1970).

[8] Borel, A., **"Linear Algebraic Groups,"** Springer-Verlag, (1991).

[9] Bourbaki, N., **"Elements of Mathematics: Lie Groups and Lie Algebras: Chapters 1-3,"** Springer-Verlag, (1989).

[10] Brassard, Gilles, Richard Cleve, and Alain Tapp, **The cost of exactly simulationg quantum entanglement with classical communication**, http://xxx.lanl.gov/abs/quant-ph/9901035.

[11] Carteret, H.A., and A. Sudbery, **Local symmetry properties of 3-qubit states**, http://xxx.lanl.gov/abs/quant-ph/0001091.

[12] Carteret, H.A., A. Higuchi, and A. Sudbery, **Multipartite generalisation of the Schmidt decomposition**, http://xxx.lanl.gov/abs/quant-ph/0006125.

[13] Cerf, Nicholas J. and Chris Adami, **"Quantum information theory of entanglement and measurement,"** in Proceedings of Physics and Computation, PhysComp'96, edited by J. Leao T. Toffoli, pp 65 - 71. See also http://xxx.lanl.gov/abs/quant-ph/9605039.

[14] Chevalley, C., **"Theory of Lie Groups I,"** Princeton University Press, (191946).

[15] Cohn, P.M., **"Lie Groups,"** Cambridge University Press, (1961).

[16] Cox, Davis, John Little, and Donal O'Shea, **"Ideals, Varieties, and Algorithms,"** Springer-Verlag (second edition) (1992).

[17] Deutsch, David, and Patrick Hayden, **Information flow in entangled quantum systems**, http://xxx.lanl.gov/abs/quant-ph/9906007.

[18] Einstein, A., B. Podosky, and N. Rosen, **Can quantum mechanical description of physical reality be considered complete?**, Phys. Rev. **47**, 777 (1935); D. Bohm, "Quantum Theory," Prentice-Hall, Englewood Cliffs, NJ (1951).

[19] Eisert, Jens, and Martin Wilkens, **Catalysis of entanglement manipulation for mixed states**, http://xxx.lanl.gov/abs/quant-ph/9912080.

[20] Englert, Berthold-Georg, and Nassaer Metwally, **Separability of entangled q-bit pairs**, http://xxx.lanl.gov/abs/quant-ph/9912089.

[21] Fulton, W., and J. Harris, **"Representation Theory, a First Course,"** Springer-Verlag, (1991).

[22] Goodman, Roe, and Nolan R. Wallach, **"Representations and Invariants of the Classical Groups,"** Cambridge University Press (1998).

[23] Gruska, Jozef, **"Quantum Computing,"** McGraw-Hill, (1999).

[24] Helgason, S., **"Differential Geometry, Lie Groups, and Symmetric Spaces,"** Academic Press, (1978).

[25] Helgason, S., **"Groups and Geometric Analysis,"** Academic Press, (1984).

[26] Horodecki,Michal, Pawel Horodecki, Ryszard Horodecki, **Limits for entanglement measures**, http://xxx.lanl.gov/abs/quant-ph/9908065.

[27] Horodecki, Pawel, Michal Horodecki, and Ryszard Horodecki, **Binding entanglement channels**, http://xxx.lanl.gov/abs/quant-ph/9905058.

[28] Horodecki, Michal, Pawel Horodecki, and Ryszard Horodecki, **Separability of n-particle mixed states: necessary and sufficient conditions in terms of linear maps**, http://xxx.lanl.gov/abs/quant-ph/0006071.

[29] Horodecki,Michal, Pawel Horodecki, Ryszard Horodecki, **Mixed-state entanglement and quantum communication**, in "Quantum Information: An Introduction to Basic Theoretical Concepts and Experiments," by G. Alber, T. Beth, M. Horodecki, P. Horodecki, R. Horodecki, M. Rotteler, H. Weinfurter, R. Werner, amd A. Zeilinger, Springer-Verlag, (2001), pp 151 - 196.

[30] Howe, R., **"The First fundamental Theory of Invariant Theory and spherical subgroups,"** in Proceedings of the Symposia on Applied Mathematics, volume 48, American Mathematical Society, Providence, Rhode Island, (1988), pp 133 - 166.

[31] Humphreys, J.E., **"Linear Algebraic Groups,"** Springer-Verlag, (1975).

[32] Humphreys, J.E., **"Introductions to Lie Algebras and Representation Theory,"** Springer-Verlag, (1980).

[33] Jacobson, N., **"Lie Algebras,"** Wiley-Interscience, (1962). (Reprinted Dover 1979.)

[34] Jonathan, Daniel, and M. Plenio, **Entanglement-assisted local manipulation of pure quantum states**, Phys. Rev. Lett. 83, 3566 (1999). (http://xxx.lanl.gov/abs/quant-ph/9905071)

[35] Jonathan, Daniel, and Martin B. Plenio, **Entanglement-assisted local manipulation of pure quantum states,** http://xxx.lanl.gov/abs/quant-ph/9905071.

[36] Kus, Marek, and Karol Zyczkowski, **Geometry of entangled states**, http://xxx.lanl.gov/abs/quant-ph/0006068.

[37] Lewenstein, M., D. Bruss, J.I. Cirac, B. Kraus, M. Kus, J. Samsonowicz, A. Sanpera, and R. Tarrach, **Separability and distillability in composite quantum systems - a primer**, (2000), http://xxx.lanl.gov/abs/quant-ph/0006064.

[38] Linden, N., and S. Popescu, **On multi-particle entanglement**, http://xxx.lanl.gov/abs/quant-ph/9711016.

[39] Linden, N., S. Popescu, and A. Sudbery, **Non-local properties of multi-particle density matrices**, http://xxx.lanl.gov/abs/quant-ph/9801076.

[40] Linden, N. and Sandu Popescu, **Good dynamics versus bad kinematics. Is entanglement needed for quantum computation?**, http://xxx.lanl.gov/abs/quant-ph/9906008.

[41] Lo, Hoi-Kwong, Sandu Popescu, and Tim Spiller, **"Introduction to Quantum Computation and Information,"** World Scientific (1998).

[42] Lomonaco, Samuel J., Jr., **A Rosetta stone for quantum mechanics with an Introduction to Quantum Computation**, in "Quantum Computation," in this AMS Proceedings of the Symposia of Applied Mathematics (PSAPM). (http://xxx.lanl.gov/abs/quant-ph/0007045)

[43] Lomonaco, Samuel J., Jr., **Quantum hidden subgroup algorithms: A mathematical perspective**, in "Quantum Computation and Information," edited by S. J. Lomonaco, AMS Contemporary Mathematics Series, American Mathematical Society, Providence, Rhode Island, (2002). (http://xxx.lanl.gov/abs/quant-ph/0201095).

[44] Makhlin, Yuriy, **Nonlocal properties of two-qubit gates and mixed states and optimization of quantum computations**, http://xxx.lanl.gov/abs/quant-ph/0002045.

[45] Nielsen, Michael A., **Conditions for a class of entanglement transformations**, Physical Review Letters, Vol 83 (2), pp 436–439 (1999). (http://xxx.lanl.gov/abs/quant-ph/981105)

[46] Nielsen, Michael A., **Majorization and its applications to quantum information theory**, preprint.

[47] Nielsen, M.A., **Continuity bounds for entanglement**, Phys. Rev. A, Vol. 61, (2000)

[48] Nielsen, M.A., **Characterizing mixing and measurement in quantum mechanics**, http://xxx.lanl.gov/abs/quant-ph/0008073.

[49] O'Connor, Kevin M., anf William K. Wootters, **Entangled rings**, http://xxx.lanl.gov/abs/quant-ph/0009041.

[50] Nielsen, Michael A., and Isaac L. Chuang, **"Quantum Computation and Quantum Information,"** Cambridge University Press (2000).

[51] Olver, Peter J., **"Applications of Lie Groups to Differential Equantions,"** Springer-Verlag, (1993).

[52] Pontrjagin, Leon, **"Topological Groups,"** Princeton University Press, (1958).

[53] Popov, V.L., **"Groups, Generators, Syzygies, and Orbits in Invariant Theory,"** American Mathematical Society, Providence, Rhode Island, (1992).
[54] Rains, Eric M., **Polynomial invariants of quantum codes**, http://xxx.lanl.gov/abs/quant-ph/9704042.
[55] Samelson, Hans, **"Notes on Lie Algebras,"** Springer-Verlag, (1990).
[56] Sattinger, D.H., and O.L. Weaver, **"Lie Groups and Algebras with Applications to Physics, Geometry, and Mechanics,"** Springer-Verlag, (1993).
[57] Schlienz, J., and G. Mahler, Physics Letters A 39 (1996).
[58] Shor, Peter W., **Polynomial time algorithms for prime factorization and discrete logarithms on a quantum computer**, SIAM J. Computing, 26(5) (1997). pp 1484 - 1509.
[59] Shor, Peter W., John A. Smolin, and Ashish V. Thapliyal, **Superactivation of bound entanglement**, http://xxx.lanl.gov/abs/quant-ph/0005117.
[60] Spivak, Michael, **"A Comprehensive Introduction to Differential Geometry,"** Volumes 1-5, Publish or Perish, Inc. (1979).
[61] Sternberg, S., **"Group Theory and Physics,"** Cambridge University Press, (1994).
[62] Sudbery, Anthony, **On local invariants of three-qubit states**, quant-ph/0001116.
[63] Sudbery, Anthony, **The space of local equivalence classes of mixed two-qubit states**, http://xxx.lanl.gov/abs/quant-ph/0001115.
[64] Terhal, Barbara M., and Pawel Horodecki, **A Schmidt number for density matrices**, http://xxx.lanl.gov/abs/quant-ph/9911117.
[65] Virmani, S., and M.B. Plenio, **Ordering states with entanglement measures**, http://xxx.lanl.gov/abs/quant-ph/9911119.
[66] Wallach, N.R., and J. Willenbring, **On some q-analogs of a theorem of Kostant-Rallis**, Canad. J. Math., Vol. **52** (2), 2000, pp. 438-448.
[67] Warner, Frank W., **"Foundations of Differential Manifolds and Lie Groups,"** Scott, Foresman and Company, Glenview, Illinois, (1971).
[68] Weyl, H., **"The Theory of Groups and Quantum Mechanics,"** Dover, (1931).
[69] Weyl, H., **"The Classical Grouops, their Invariants and Representations,"** Princeton University Press, (1946).

UNIVERSITY OF MARYLAND BALTIMORE COUNTY, BALTIMORE, MD 21250
E-mail address: Lomonaco@UMBC.EDU
URL: http://www.csee.umbc.edu/~lomonaco

Index

$3 - SAT$, 195
BPP^{NP}, 195
G-connections, 271
$NP \subseteq BQP$, 208
$P = PSPACE$, 204
$P^{\#P}$, 194, 204
$\mathcal{P}$-tensor product decomposition, 45
$c - local$, 199
n-qubit register, 35
1-chain, 270
1-cochain, 270
3-SAT, 208, 213
4-dimensional amplitudes, 294

abelian anyons, 268, 270
abelian Chern-Simons theories, 268
abstract tensor expression, 285
accuracy threshold, 110
action, 280
adjoint, 12, 13, 247
advanced encryption standard, 241
AES, 241
Aharonov and Nave, 212
Aharonov-Bohm interaction, 268
Alexander polynomial, 288
algebra of polynomial entanglement
    invariants, 319
algebra of real valued functions, 340
Alice, 41, 43, 239
all-optical quantum information processors,
    101
amplitude amplification, 188
ancilla, 45, 109, 224
ancilla qubit, 228
ancilla qubits, 224
annihilation, 283
annihilation bra, 283
anyonic model, 268
anyons, 267, 268
Artin braid group, 290, 292, 301
atlas, 335
authentication, 242
automorphism group, 51
average value, 23
averaged observed value, 248

B92, 258

B92 quantum cryptographic protocol, 258
B92 quantum protocol, 259
basic set of entanglement invariants, 319
BB84 protocol, 255
BB84 protocol with noise, 256
BB84 protocol without Eve present, 254
BB84 protocol without noise, 253
beamsplitter, 68, 69, 73, 93
beamsplitters, 101
Bell, 308
Bell basis, 88
Bell basis states, 332
Bell-state analyzer, 90, 91
Bell-state analyzers, 125
Bell-state measurement, 94, 97, 98
Bell-state projective measurement, 96
Bell-state synthesizer, 87
Bell-state synthesizers, 124
Bell-states, 88
Benioff, 145
Bennett, 40, 50, 201
Bernstein, 203, 205
big adjoint action, 307, 316
big adjoint representation, 32, 345
bit & phase, 223
bit flip, 223
bit flip error, 223
black box, 151
black box model, 209
blackbox, 33
Bloch sphere, 328
Blum, 210
Bob, 41, 43, 239
Boolean states, 106
Bose condensate quantum computer, 119
Bose condensate quantum computers, 118
Bose condensates, 119
bose condensates, 124
Bose-Einstein condensation, 118
Bose-Einstein statistics, 91
bounded-error probabilistic polynomial
    time, 203
bounded-error quantum polynomial time,
    203
BPP, 152, 194, 203, 204, 207
BQNP, 195, 212, 214

BQNP-complete problem, 212
BQNP-completeness, 214
BQP, 149, 152, 194, 203, 204, 207
bra, 9
bra vectors, 9
bra's, 246
bracket, 9, 246
bracket model, 290
bracket polynomial, 292
braid generator, 292
braid group, 272, 292, 293
braiding, 268, 272
braiding cancellation, 287
bras, 246, 279
BSM, 94

Caesar cipher, 240
Calderbank, 231
cancellation of maxima and minima, 284, 285
cap, 283, 286
caps, 297
catalysis, 45
Catch 22, 240, 242, 260
categorical physics, 297
category, 297
cavity QED, 111, 112, 125
cavity quantum electrodynamics, 111
cavity-QED quantum computers, 112
certificate complexity, 209
chart, 335
Chern-Simons Lagrangian, 296
Chern-Simons three-form, 295
Church's thesis, 145
Church-Turing principle, 103
Church-Turing thesis, 144, 193, 194
ciphertext, 240
circle in spacetime, 282
circuit family, 149
circularly polarized, 7
class BQP, 150
classical computing device, 50
classical crypto systems, 242
classical entropy, 45, 46, 48
classical randomized class BPP, 149
classical reversible computation, 51
classical Turing machine, 196
CNOT, 148
CNOT, see also controlled-NOT, 51
coding space, 223
coherence, 112
coherences, 76
coherent superposition, 105
collapse of the wave function, 246
collision-intractible hashing, 208
commutator, 24, 32, 248
compatible, 23, 248
compatible operators, 23
complete linear operators, 15

complete set of entanglement invariants, 318, 328
complex general linear group, 342
complex projective $(n-1)$ space, 11
complex projective space, 146
complex projective space $\mathbb{C}P^{n-1}$, 11
complex special linear group, 342
complexity class BQP, 149, 150
complexity class P, 144, 145
computable, 144
computational basis, 202
computational degrees of freedom, 103, 110
computational security, 242
computational step, 50
computationally secure, 242
constituent part, 31
CONT', 53
CONT'', 54, 56
continued fraction, 169
continued fraction expansion, 155
continued fractions, 169
control bit, 52
controlled NOT, 83, 148, 201
controlled-Z operations, 228
controlled-NOT, 51, 53, 54, 84, 101, 102
convergent, 169, 172
Cook's theorem, 214
Cook-Levin theorem, 212
Cooley-Tukey algorithm, 153
Cooley-Tukey fast Fourier transform, 156
Cooley-Tukey FFT, 156
Cooper-pair charge, 116
counter measures, 260
CRCD, see also Classical Reversible Computing Device, 51
creation, 283
creation ket, 283
crossings, 297
CSS code, 231
CSS construction, 231
cup, 283, 286
cups, 297

Dan Simon, 145
dark counts, 257
Data Encryption Standard, 183
David Deutsch, 145
DeBroglie, 274
decoherence, 77, 87, 112, 125, 150, 222, 224
decoherence-free subspaces, 110
degenerate code, 227
degenerate eigenvalues, 15
degenerate operator, 15
degenerate quantum codes, 228
density (2^1), 328
density operator, 26–32, 315
derivation, 339
DES, 183, 240
deterministic finite state control, 196

deterministic query complexity, 195
Deutsch, 35, 194, 196, 205
Deutsch-Jozsa problem, 205
deviation density matrix, 113
diagonalization, 15, 27, 48
diagonalized, 15
diagonalizes, 27
Dieks, 57
diffeomorphism, 342
different perspectives, 314
differential manifolds, 335
differential operator, 324
Diffie, 242
digital encryption standard, 240
Dirac amplitude, 294
Dirac brackets, 278
Dirac notation, 9–13, 23, 245, 246
Dirac's bra-ket notation, 146
directional derivatives, 327, 339
discrete Fourier transform, 155
discrete logarithms, 152
distance of a code, 269
dynamical invariant, 47

eavesdropper, 239
eavesdropping strategies, 260
Edward Witten, 295
efficient universal QTM, 197
efficiently computable, 144
eigenket, 14
eigenspace, 14
eigenvalue, 14
Einstein, 275
Einstein, Podolsky, and Rosen, 308
Einstein, Podolsky, Rosen, 40
Einstein-Podolsky-Rosen (EPR), 87
embedding, 50, 52, 296
energy spectrum, 278
entangled, 39, 166, 312
entangled multiparticle systems, 96
entanglement, 74, 104
entanglement class, 316, 321
entanglement classes, 43, 328
entanglement invariant, 323
entanglement invariants, 327
entanglement of quantum states, 301
entanglement swapper, 95
entanglement swappers, 94
entanglement swapping, 87, 94, 96
entanglement type, 43, 311, 316
environment, 87, 222
EPR, 40, 41, 97, 125
EPR pair, 38, 97, 310
EPR pairs, 259
EPR-pair, 87, 94
EPR-pair source, 87
EPR-pair sources, 124
erasure errors, 228
error estimation, 255

error syndrome, 230
Euclidean algorithm, 163, 178
Euler's constant, 172
Euler's totient function, 172
Eve, 239
excited state, 270
EXP, 195
expected value, 23
exponential map, 317, 343
exponentiation of vector fields, 339
extended local moves, 313

factoring algorithm, 145
factoring numbers, 123
fault-tolerant quantum computation, 234
Fermi-Dirac statistics, 91
ferromagnetic dot, 115
Feynman, 145, 194, 280
Feynman integral, 281
filtration, 21
final key, 255
fineness of a partition $\mathcal{P}$, 45
finite memoryless stochastic source, 45
flat connections, 296
flow, 339
Fourier sampling, 205
Fourier sampling tree, 206
Fourier transform, 152, 165, 166, 175, 202, 205
FPQE, 312
fractional quantum Hall effect, 267
framing cancellation, 287
free Boolean ring, 51
functional integrals, 296
fundamental problem of quantum entanglement, 312

G-qubit, 271
Gödel, 144
generalized amplitudes, 288
generalized Feynman integral, 295
generator matrix, 231
global quantum system, 31
global unitary transformation, 42
global unitary transformations, 316
Gottesman, 35
graph category, 298
group of local unitary transformations, 307, 315
group of transformations, 345
Grover, 157, 183, 208
Grover's algorithm, 114, 187, 189, 190
Grover's search algorithm, 156

Hadamard, 55
Hadamard gate, 82, 151
Hadamard gates, 84
Hadamard operator, 106
Hadamard transform, 82, 166, 186, 200, 202

Hadamard transformation, 151
Hadamard transforms, 184
Hamiltonian, 24
Hamming code, 231
Heisenberg, 275
Heisenberg model, 33
Heisenberg picture, 33, 35
Heisenberg uncertainty principle, 23, 24, 248
Heisenberg's uncertainty principle, 248, 249
Hellman, 242
Hermitian operator, 13
Hermitian operators, 247, 278
hidden variables, 40
high-temperature-superconductor Josephson junctions, 117
Hilbert space, 8, 245
homology classes, 270
Hopf algebra, 271, 288
hybrid argument, 208, 209

implementation weaknesses, 261
implementations, 262
implicit frame wiring diagrams, 56
incompatible, 24, 248
incompatible operators, 24
induced infinitesimal action, 308, 321
inefficiently computable, 144
infinitesimal action, 319–324, 345
infinitesimal generator, 339
information capacity of the transmission channel, 89
insecure channel, 240
integral curve, 339
interaction-free detector, 68
intrusion detection, 242
invariant of regular isotopy, 290
invariants of three-manifolds, 296
inversion, 184
ion-trap quantum computer, 110
ion-trap quantum computers, 125
IP, 195
irreversible, 313
irreversible computation, 52

J gate, 114
Jacobian matrix, 38
Jones polynomial, 288, 290, 296
Jordan curve theorem, 284
Josephson junction quantum computer, 116
Josephson junction SQUID quantum computers, 126
Josephson junctions, 124
Jozsa, 194, 205
Julian Schwinger's algebra of measurement operators, 20
juxtaposition, 35
juxtaposition of two quantum systems, 35

ket, 9
ket vectors, 9
kets, 245, 279
Key Problem 1, 242
Key Problem 2, 242
Key Problem 3, 242
Kitaev, 195, 203, 212
Knill-Laflamme bound, 228
knot, 296
knot amplitudes, 282
knot invariants, 268, 288
Kronecker sum, 43, 319, 329
Kuperberg, 268

left- and right-circularly polarized photons, 11
lg, 46
Lie algebra, 307, 315, 318, 339, 342–344
Lie algebra $\mathbf{Der}(C^\infty \mathbf{u}(2))$, 327
Lie bracket, 318, 339
Lie group, 42
Lie group $\mathbb{G}$, 341
Lie groups, 341
Lie subgroup, 341
Linden, 333
linearly polarized, 7
link invariants, 296
little adjoint representation, 32, 346
local equivalence, 43
local interaction, 41
local Lie algebra, 43
local move, 313, 315
local moves, 312
local subgroup, 42
local unitary group, 307
local unitary transformation, 42
local unitary transformations, 315
locally equivalent, 43, 316

MA, 195, 205, 207
Mach-Zehnder interferometer, 82, 84
macroscopic superconducting quantum states, 116
Max Born, 275
Max Planck, 275
maximal entanglement, 88
mean squared standard deviation, 248
measurement, 18, 22, 27, 28
Merlin-Arthur, 207
method of quantum adversaries, 195, 208, 210
Michelson interferometer, 68
Miller-Rabin, 162
mirrors, 101
mixed ensemble, 26, 28, 29
mixed ensembles, 329
modern Church-Turing thesis, 194
morphisms, 297
Morse diagrams, 289

Morse knot decomposition, 286
Mott transitions, 119
multi-tape quantum Turing machines, 199
multipartite quantum system, 31, 49

nanotechnology, 117
neutral atom quantum computers, 118
Niels Bohr, 274
Nielsen, 40
nine-qubit code, 223
NMR, 78, 113, 124
NMR quantum computer, 113, 125
no-cloning theorem, 57, 100, 222, 228
non-computable functions, 144
non-deterministic polynomial time, 212
non-local computation, 300
non-locality, 40, 41
non-standard form, 199
nonabelian anyons, 268, 271
nonabelian Chern-Simons theories, 268
nondegenerate code, 227
nondegenerate eigenvalues, 15
nondegenerate operator, 15
nondegenerate quantum codes, 228
nonunitary, 110
normal, 14
normal operator, 14
NOT, 55
NOT operator, 81, 107
NP, 212
NP-complete, 212
NP-complete problems, 195, 208
nuclear magnetic resonance, 78, 113
number field sieve, 162

observable, 13
observables, 247
one parameter subgroup, 341
one-time-pad, 241, 252
opaque eavesdropping, 260
operator, 13
optical lattices, 117, 119
oracle problem, 151, 181
orbit, 346
orbital degrees of freedom, 116
orbits, 43, 316, 346
output symbols, 46, 48

P, 194, 204
parity check matrix, 231
partial differential equations, 323, 327
partial trace, 29, 30
partially secret, 256
partition, 45
path qubit, 69
path-ordered integral, 25
Pauli errors, 224
Pauli exclusion principle, 115
Pauli group, 223, 230

Pauli matrices, 18
Pauli operators, 227, 230, 268
Pauli spin matrices, 17, 107
perfect secrecy, 241
perfectly secure, 241
period, 164, 175
permutation group, 51
PGS, 243
phase flip, 223
phase flip error, 223
photonic and atomic qubits, 112
PKS, 244
plaintext, 240
plaintext/ciphertext attack, 183
plaintext/ciphertext attack on DES, 183
Planck's constant, 24
polarized light, 6, 7, 11, 21
polarizers, 101
polynomial-time, 144
Popescu, 333
Popescu and Rohrlich, 45
populations, 76
positive operator valued measure, 21
positive operator valued measure (POVM, 70
POVM, 21, 70, 71, 124
POVM receiver, 259
practical secrecy, 240
practically secure, 240
preparations, 28, 46
pretty good security, 243
prime factorization problem, 163
principle of non-locality, 40, 41
principle of reality, 40
Prisoner's Dilemma, 79
privacy amplification, 256, 258
probabilistic polynomial time, 194
probabilistic Turing machine, 193, 194
probability amplitudes, 147
probability distribution, 46
probability operator valued measure, 21
projection operator, 13, 48
projection operator for the eigenspace, 14
projector, 13
pseudo-spin states, 116
PSPACE, 194, 204
public directory, 243
public key cryptographic communication
    system, 243
public key cryptographic system, 243
pure ensemble, 26, 48
pure ensembles, 329

Q5SAT, 213
QED, 111
QED cavity, 102
QIP, 195
QMA, 212
QSAT, 212, 214

QSPACE, 195
QTM, 197, 202
quantum, 124
quantum adversary, 208
quantum adversary argument, 210
quantum algorithm, 162
quantum alphabet, 254
quantum amplitude functor, 300
quantum amplitudes, 297
quantum cellular automata model, 146
quantum chaos, 103
quantum circuit model, 146, 150
quantum circuits, 196, 199
quantum code, 268, 269, 271
quantum complexity classes, 203
quantum computer, 110, 162
quantum computer communication networks, 112
quantum computers, 110
quantum computing device, 50
quantum controlled-NOT (or XOR), 84
quantum controlled-NOT gate, 82
quantum copier, 100, 101
quantum cryptographic communication system, 253
quantum cryptographic protocols, 259
quantum cryptography, 70, 71, 124, 239, 251
quantum decoherence, 75, 124
quantum dense coder, 90
quantum dense coding, 88
quantum double, 271
quantum entangled, 35, 39, 45
quantum entanglement, 39, 42, 312
quantum entanglement classes, 316
quantum entanglement invariant, 318
quantum entanglement invariants, 317, 318, 327
quantum entropy, 46, 48
quantum error correction, 103, 107, 222
quantum error correctors, 107
quantum error-correcting code, 225, 226
quantum factorizer, 103, 104
quantum field theory, 103, 295, 296
quantum Fourier transform, 153, 155
quantum games, 77, 78
quantum gate, 148
quantum gates, 80, 148
quantum Gilbert-Varshamov bound, 228
quantum group, 288
quantum Hamming bound, 227
quantum key receiver, 70, 124
quantum link invariants, 301
quantum lower bounds, 208
quantum measurement, 27
quantum measurement rubric, 18
quantum measurement theory, 18
quantum mechanics, 274

quantum memory, 112
quantum NOT gate, 81
quantum NP, 212
quantum register, 35, 36, 105, 106, 116
quantum repeater, 58
quantum replicator, 57
quantum robot, 119, 123
quantum robots, 119, 126
quantum secrecy, 252, 253
quantum simulators, 103, 125
quantum Singleton bound, 228
quantum state space, 146
quantum telephone exchanges, 125
quantum teleportation, 58, 59, 62, 97, 100, 124
quantum teleporter, 97
quantum Turing machine model, 146
quantum Turing machines, 196, 197
quantum witness, 212
quantum-dot quantum computer, 115
qubit, 7, 8, 68, 146, 196, 244, 246, 268
qubit device, 68
qubit entangler, 85, 87
qubit entanglers, 85
qubit errors, 224, 225
qubits, 124

raw key extraction, 255, 256
raw keys, 255
real general linear group, 341
real orthogonal group, 341
real special linear group, 342
receiver, 239
reconciled key, 257
recursive Fourier sampling, 206
recursive Fourier sampling problem, 207
reduced density matrix, 77
reflection, 185, 187
regular isotopy, 288, 289
Reidemeister move, 286, 287
repeated squaring, 154
Reshetikhin, 268
Reshetikhin-Turaev, 296
restricted fundamental problem of quantum entanglement, 307, 315, 317
reversible, 313
reversible classical transformation, 152
reversible computation, 50
reversible computing devices, 53
RFPQE, 307, 315, 317
right (left) elliptically polarized, 7
right invariant, 342
right invariant derivations, 318
right invariant directional derivatives, 318
right invariant smooth vector fields, 318
right multiplication map, 342
rotation, 55, 200
RSA, 243

same entanglement type, 43
Schrödinger equation, 24
Schrodinger, 274
Schrodinger equation, 277
Schrodinger picture, 33, 35
search problem, 182
searching a city phone book, 182
searching a database, 123
secret key, 240
secure channel, 240
security parameter, 258
selective measurement, 21
selective measurement operator, 21
self-adjoint operator, 13
self-adjoint operators, 247
semiconductor technology, 125
sender, 239
Shannon bit, 244
Shannon entropy, 45
Shor, 194, 231, 234
Shor's algorithm, 163, 165, 166
Shor's quantum factoring algorithm, 106, 163
silicon-based nuclear spin quantum computer, 114
Simon, 194
Simon's algorithm, 151
Simon's problem, 151
single-tape QTM, 203
smooth, 335, 336
smooth manifold, 335
smooth vector field, 337
Solovay, 203
SPACE, 195
spacelike distance, 41
special orthogonal group, 342, 346
special unitary group, 315, 346
spectral decomposition, 14, 17
spectral decomposition theorem, 14
square root of NOT, 55
square root of SWAP, 55
square-root-of-not gate, 80
SQUID, 117
SQUID quantum computer, 117
SQUIDs, 116, 124
stabilizer, 229, 231, 232
stabilizer code, 230
stabilizer codes, 228, 233
stabilizer operators, 269, 270
standard deviation, 23
standard form, 198
standard local moves, 313
standard unitary representation, 53
state bras, 9
state kets, 9
state of a quantum system, 10
Steane, 231
stochastic source, 46

sufficiently local, 45
sufficiently local operator, 45
sufficiently local unitary transformation, see also unitary, 45
superconducting cavities, 112
superconducting loops, 117
superconducting quantum interference devices, 116
superluminal communication, 41
superposition, 7, 36, 147, 245, 246
superselection rules, 270
SWAP, 55
symmetric group, 53

tangent bundle, 337
tangent bundles, 335
tangent space, 318, 323
tangent vector, 336
tangentially equivalent, 336
target bit, 52
teleportation, 58
Temperley Lieb algebra, 290, 291
tensor product, 10, 146
tentative final keys, 255, 257
the probabilistic generalization of NP, 205
three-strand braid group, 294
Toffoli gate, 52, 54, 148
topological amplitude, 294
topological amplitudes, 295
topological cancellation, 284
topological defects, 267
topological invariants, 296
topological number, 267
topological quantum field theory, 268, 294
toys for aging children store, 309
TQFT, 268
trace, 29
transformation, 13
transformation group, 345
translucent eavesdropping with entanglement, 261
translucent eavesdropping without entanglement, 261
transmission coefficient, 69
trap-door function, 242
Turaev, 268
Turing machine, 144
Turing machine model, 146
two-qubit entangler, 85

uncertainty, 23
uniform family of circuits, 149
uniform model, 146
uniformity conditions, 149
unitary group, 307, 342
unitary operator, 24
unitary operators, 247
unitary ribbon category, 268, 272
unitary transformation, 24

unitary transformations, 247
universal QTM, 202
universal quantum computer, 102, 103
universal quantum computers, 125
universal quantum copier, 101
universal quantum gate, 102
universal quantum Turing machine, 194, 198
universal set of gates, 147

vacuum to vacuum amplitude, 279
vacuum-vacuum computation, 300
vacuum-vacuum expectation, 295
Vazirani, 205
vector field, 337
vector fields, 326, 331, 335
Vernam cipher, 241
vertically and horizontally linearly polarized, 11
Viro, 268
von Neumann entropy, 46, 48
von Neumann-type projective measurement, 70
vortex, 117

wave-packet, 275
weight of an operator, 224
Wilson loop, 296
winding number, 267
wiring diagram, 51–56
Witten, 268, 296
Witten's integral, 295, 296
Witten's invariant, 268
Wollaston prism, 72, 73
Wootters, 40, 57

Yang-Baxter equation, 288, 301
Yao, 194, 199, 203
Yao's construction, 202
Yao's lemma, 207
Young's two slit experiment, 249–251
Yuri Manin, 145

Zalka, 155
Zurek, 57

Selected Titles in This Series

(Continued from the front of this publication)

29 **R. A. DeMillo, G. I. Davida, D. P. Dobkin, M. A. Harrison, and R. J. Lipton,** Applied cryptology, cryptographic protocols, and computer security models (San Francisco, California, January 1981)

28 **R. Gnanadesikan, Editor,** Statistical data analysis (Toronto, Ontario, August 1982)

27 **L. A. Shepp, Editor,** Computed tomography (Cincinnati, Ohio, January 1982)

26 **S. A. Burr, Editor,** The mathematics of networks (Pittsburgh, Pennsylvania, August 1981)

25 **S. I. Gass, Editor,** Operations research: mathematics and models (Duluth, Minnesota, August 1979)

24 **W. F. Lucas, Editor,** Game theory and its applications (Biloxi, Mississippi, January 1979)

23 **R. V. Hogg, Editor,** Modern statistics: Methods and applications (San Antonio, Texas, January 1980)

22 **G. H. Golub and J. Oliger, Editors,** Numerical analysis (Atlanta, Georgia, January 1978)

21 **P. D. Lax, Editor,** Mathematical aspects of production and distribution of energy (San Antonio, Texas, January 1976)

20 **J. P. LaSalle, Editor,** The influence of computing on mathematical research and education (University of Montana, August 1973)

19 **J. T. Schwartz, Editor,** Mathematical aspects of computer science (New York City, April 1966)

18 **H. Grad, Editor,** Magneto-fluid and plasma dynamics (New York City, April 1965)

17 **R. Finn, Editor,** Applications of nonlinear partial differential equations in mathematical physics (New York City, April 1964)

16 **R. Bellman, Editor,** Stochastic processes in mathematical physics and engineering (New York City, April 1963)

15 **N. C. Metropolis, A. H. Taub, J. Todd, and C. B. Tompkins, Editors,** Experimental arithmetic, high speed computing, and mathematics (Atlantic City and Chicago, April 1962)

14 **R. Bellman, Editor,** Mathematical problems in the biological sciences (New York City, April 1961)

13 **R. Bellman, G. Birkhoff, and C. C. Lin, Editors,** Hydrodynamic instability (New York City, April 1960)

12 **R. Jakobson, Editor,** Structure of language and its mathematical aspects (New York City, April 1960)

11 **G. Birkhoff and E. P. Wigner, Editors,** Nuclear reactor theory (New York City, April 1959)

10 **R. Bellman and M. Hall, Jr., Editors,** Combinatorial analysis (New York University, April 1957)

9 **G. Birkhoff and R. E. Langer, Editors,** Orbit theory (Columbia University, April 1958)

8 **L. M. Graves, Editor,** Calculus of variations and its applications (University of Chicago, April 1956)

7 **L. A. MacColl, Editor,** Applied probability (Polytechnic Institute of Brooklyn, April 1955)

6 **J. H. Curtiss, Editor,** Numerical analysis (Santa Monica City College, August 1953)

For a complete list of titles in this series, visit the
AMS Bookstore at **www.ams.org/bookstore/**.

PSAPM/58